SUSTAINABLE UTILISATION OF ENERGY AND BIODIVERSITY

Resources for Wealth Creation and Development

SUSTAINABLE UTILISATION OF ENERGY AND BIODIVERSITY

Resources for Wealth Creation and Development

Editor

B.O. SOLOMON

CENTRE FOR SCIENCE & TECHNOLOGY OF THE
NON-ALIGNED AND OTHER DEVELOPING COUNTRIES
(NAM S&T CENTRE)

SCITECH PUBLICATIONS (INDIA) PVT. LTD.

Chennai • Hyderabad • Kolkata • Pune • Bhubeneswar • New Delhi

Centre for Science & Technology of the Non-Aligned and other Developing Countries (NAM S&T Centre)
Core-6A, 2nd Floor, India Habitat Centre, Lodhi Road,
New Delhi-110 003 (India)
Phone: +91-11-24644974, 24645134, Fax: + 92-11-24644973
E-mail: namstct@gmail.com • Website: www.namstct.org

SCITECH PUBLICATIONS (INDIA) PVT. LTD.

Corp. Office: No. 4, Anna Street, Chitlapakkam, Chennai-600 064
E-mail: scitechcorp@scitechbooks.in
Regd. Office: 7/3c, Madley Road, T. Nagar, Chennai-600 017
E-mail: scitech@scitechbooks.in
Visit us at: **www.scitechbooks.in**

Branches:

BHUBANESHWAR:	scitech_bbsr@scitechbooks.in
CHENNAI:	scitech@scitechbooks.in
HYDERABAD:	scitech_hyd@scitechbooks.in
KOLKATA:	scitech_kol@scitechbooks.in
PUNE:	scitech_pune@scitechbooks.in
NEW DELHI:	scitech_del@scitechbooks.in

ISBN: 978 81 8371 312 2

Production Co-ordinator: K. Madhu Sudhan

Published by: M.R. Purushothaman for Scitech Publications (India) Pvt. Ltd.
3-4-249/1, Street No. 2, Lingampally, Kachiguda, Hyderabad - 500 027.
Printed at: Nice Printing Press, Sahibabad, Ghaziabad, U.P.

Foreword

The energy crisis in the developing Nations of the World is currently at an alarming rate and the need to proffer solutions is not only expedient but must be of high priority. It is based on this that the Nigerian Federal Ministry of Science and Technology has some agencies such as National Biotechnology Development Agency, (NABDA), Energy Commission of Nigeria (ECN) and National Research Institute for Chemical Technology, (NARICT) who recognize the importance of energy production and utilization for the benefit of Mankind. The increased world population and the quest to improve living conditions have resulted in an increased demand for energy. This has placed an increased demand in the exploration and depletion of energy resources (mainly fossil fuel). Fossil fuels have a huge negative impact on the environment. Combustion results in the emission of green house gases which are responsible for global warming. This consequently results in extreme change in climatic conditions, uncontrollable rise in pollution of air, water, and soil which poses a serious threat to human survival on earth. The forecast that Africa would suffer the greater impact of climatic change coupled with the inadequate energy supply, has placed Africa and other developing countries on the quest for alternative/ sustainable energy.

This book is an outcome of a special workshop organised by the Centre for Science & Technology of the Non-Aligned and Other Developing Countries (NAM S&T Centre), India, which jointly with the Ugandan National Academy of Sciences (UNAS) organized an International Roundtable on 'Sustainable Utilisation of Energy and Biodiversity Resources for Wealth Creation and Development' at Kampala, Uganda during 10–13 March 2009. This scientific meeting brought together different stakeholders in the developing countries in the fields of Energy Efficiency, Renewable Energy, Bio-energy and Maintenance of Ecology and Biodiversity to deliberate on various aspects for better utilization of energy and biodiversity resources.

The book, which is a compendium of case study presentations and research papers, represents a snapshot of our understanding of the concept of efficient energy production and utilization of these bioenergy resources in replacing the existing fossil sources in a sustainable and eco-friendly manner thus providing developing countries the template to acquire wealth through conservation of their energy and its resources. It also offers a clear view of the challenges posed and which they face. For energy companies, the challenge is to find a way to meet the public demand for energy, while at the same time addressing society's expectations for corporate social and environmental responsibility. For environmental organizations, the challenge is to be a strong voice for biodiversity conservation while working with industry to find the balance between the potential threat that energy development represents and the opportunities for harnessing the influence, expertise and resources of energy companies for conservation efforts.

I am pleased and honoured to have been asked to write the foreword for this book. The authors (the Centre for Science & Technology of the Non-Aligned and Other Developing Countries (NAM S&T Centre), India), Students, Library, all active researchers in the area of Energy and Biodiversity conservation should be congratulated for providing valuable references for the research Community.

Dr. ALHASSAN BAKO ZAKU
Honourable Minister
Federal Ministry of Science & Technology
Abuja-Nigeria

Preface

Energy is a key issue in the present day scenario because it ensures the survival of life on planet earth. Renewable (green) and non renewable sources of energy both have been used to meet these needs. Renewable energy has found favour with the environmentalists and the scientists. Non-renewable energy has been implicated for causing a rise in pollution (air, water, soil, noise, electro-pollution and electronic waste) levels. These pollutants which include among others, SO_2, NH_3 are increasing rapidly as a result of industrialization. Their transformations have long atmospheric lifetimes and can be carried weather systems to distances of up to thousands of kilometre from their point of emission. These pollutants cause not only health problems but also leads to acidification, eutrophication, green house gas effect, global warming and subsequently the climate change, which is a pointer toward the possibility of a major catastrophe. If steps are not taken immediately it may threaten the very survival of the planet earth. It is therefore pertinent for countries to develop technologies, policies and strategies to optimize energy generation, distribution and efficient use for sustainable development. This book presents the efforts made by some developing countries in this respect.

The opportunities and the potential biogas energy sources available, cooperation between the private sector and government to promote its use and barriers to its production in Botswana are discussed. The different technologies for tapping biogas are also discussed, highlighting the one adapted for Botswana conditions. Biogas systems that are no longer operational, currently operational and upcoming systems are also discussed. The report shows that biogas research and practices started in Botswana in the 1980s, funded by the government, but the projects eventually collapsed. After 25 years, the technology is coming to life again, this time around initiated by the private sector.

In Egypt, like other countries, the foremost driver of energy is fossil fuel which is likely to be exhausted within few decades. Within a population estimated at 85 million in 2007, a demand for green energy (bio-fuel) has been considered, but the quantum and form of its exploitation has not advanced. The Supreme Egyptian Council commissioned a body to develop a strategy to increase energy generation from renewable sources to 20% of total generated energies by 2020; i.e. around 13,500 MW. Egyptian rural areas are expected to profit from biofuel production provided this does not affect food production and the environment. However, production of biofuel from non-food crops such as forestry residues should be encouraged. Actions that reduce overall energy consumption, such as increasing the fuel economy of automobiles and expanding mass transit options in cities, among others, are discussed.

The increasing population and fast-growing economy of India necessitated a greater demand for energy which is mostly met by fossil fuels. The excessive use of fossil fuel is causing considerable environmental and resource degradation. India heavily depend on imported energy particularly petroleum products. The country had a negative energy balance for decades, and this has forced the purchase of energy from outside the country. The present energy scenario in India, status of renewable energy programmes, research and technology development, government policies and issues related to renewable energy financing, institutional strengthening and infrastructure development are also discussed. Various projected scenarios and Indian initiatives to reduce the CO_2 emissions, the challenges in renewable energy development and approach in dealing with these challenges are presented.

The paper from Iran proposed that application of certain nuclear techniques to measure long-term variation in deposition rate of atmospheric aerosols earlier mentioned as particles and/or droplets. The application of these techniques is easier to carry out than the direct measurement by artificial collectors over extended period of time.

Mauritius, a small island and developing state is working on a number of energy policies and strategies to be able to achieve sustainable development. The Maurice Ile Durable Fund (MIDF) was set up on the 19 of June 2008 under the Finance and Audit (Maurice Ile Durable Fund) Regulations 2008, in line with programme-based budgeting to achieve Government vision and goals for sustainable development. Accordingly, the overarching objective of the MIDF is to pave the way for Mauritius to move towards a 'green' island in the next twenty years by enlisting the collaboration of all stakeholders. The policies and strategies are directed towards the preservation of local natural resources with a view to adapting to climate change. Projects are financed to explore and harness the potential for local sources of renewable energy, reduce dependency on imported fossil fuels and improve the energy mix of the country. Actions are taken to allow households to generate their own electricity from either solar or wind and sell any excess to the national grid.

In Myanmar, the balance utilization of different energy types has been emphasized and increased production of commercial energy prioritized in order to replace traditional energy types and at the same time meeting the requirement of industrial development programmes. Biomass fuels are produced from biomass resources such as animal waste, agricultural waste, industrial waste, municipal waste and by-product of the forest. The produced biomass fuels, which reduce air pollution, greenhouse gas emissions, global warming are producer gas, biogas, biodiesel and bioethanol. Use of bio-fuels prevents deforestation and at the same time controls pollution and improves health and sanitation standards of the area.

Nepal is rich in biodiversity and energy resources and these resources could be harnessed in sustainable manner. The most potential renewable energy technologies are hydropower, biomass energy, wind energy, bio-briquette, solar water heating, improved cooking stoves, solar dryer, solar cooker and solar PV. These technologies are at the stage of commercial applications or developmental stage in the country. Policy measures for exploiting these technologies on sustainable basis are discussed.

Prior to the advent of colonialists, Africans depended entirely on African traditional medicines for the management of all their prevailing diseases. Currently, about two thirds of Western medicines are derived from natural sources attesting to their therapeutic potentials. Although Africa has over 50,000 plant species and a rich cultural heritage of indigenous medical knowledge (IMK), less than 2% of African medicinal plants have been subjected to scientific/clinical validation. Policy measures and programs for utilizing this IMK in healthcare delivery in Nigeria are recommended.

Nigeria is endowed with enormous biodiversity resources, many of which have been successfully domesticated. Bioresources such as municipal and industrial wastes and high energy plants are important and inimitable biological resources. The strength of most of the developing countries lies in the richness of their bioresources. It is therefore crucial that these bioresources be sustainably managed and utilized to improve the country energy mix without ignoring the conservation of the same for future generations. Effort of the National Biotechnology Development Agency at setting up various Bioresources Development Centres (BioDeC) across the ecological zones and the diffusive potential toward empowerment are highlighted. Institutional arrangements and efforts designed to develop different technologies, regulation, economic and market incentives to drive a sustainable strategy in the utilization of energy for wealth creation and development in Nigeria are also presented.

The utilization of renewable resources such as solar, wind, dendro, and mini-hydro are relatively small in Sri Lanka. The transport sector is entirely depending on petroleum based fuels such as gasoline and the diesel. The escalating prices of petroleum-based fuels in the recent past have resulted into a great burden to the economy. This has created a trend to explore alternative avenues to fulfill energy requirement with a view to tackling the energy challenges. The available options are the utilization of renewable energy sources and conservation of energy. Policies and programs for doing these are recommended.

The wild biodiversity resources in South Africa fulfill a number of functions in the rural livelihoods of citizens in developing countries and this has helped in alleviating or mitigating poverty. These include daily consumption value, the safety-net value in adverse conditions and the value of income generating activities based on sale of biodiversity resources in local or distant markets. The contribution of biodiversity resources constitutes between one-fifth and one-quarter of total livelihoods income. A number of government interventions and policies steps are identified to optimise the benefits for each of these different values.

The government of Tanzania, like other governments in developing countries, has opened doors for investors to invest in biofuel production. Areas for investment have been identified and more than nine companies have already shown their interest to invest in this sector. The issue of biofuel production, growth in the 'new industry' and sustainability of biodiversity in the country are highlighted.

Biofuels research and development situation in Vietnam, as well as opportunities and challenges are also reported in this book. Biofuels project goals, major measures for the development and utilization of bio fuels in Vietnam up to 2020 are also discussed.

Zambia has decided to explore alternatives to oil and one such alternative considered is biofuels, which include ethanol and biodiesel. In 2007, the Zambian Government adopted a new National Energy Policy. The policy adequately covers issues to deal with biofuels. The government through its appropriate institutions has developed both standards for ethanol and biodiesel production as well as draft of National Biofuels Strategy in Zambia. The feedstocks which can be used to produce biofuels have been assessed. The cultivation of Jatropha, one of the preferred feedstocks has been demonstrated both at small scale and commercial level.

Uganda's population is currently 30.2million people. Over 90% of Uganda's population relies heavily on traditional bio-fuels for energy. These mainly consist of firewood, charcoal and crop residues. This dependency on traditional bio-fuels is exerting enormous pressure on natural resources resulting in the depletion and degradation of natural forest cover. The country has lost over 1.3million hectares of forest cover in the last 15 years as a result of human encroachment for settlements, farmland, commercial purposes and the search of fuel-wood. Currently, there is an increasing aspiration of investing in biogas and energy crops for biodiesel production. However, these aspirations and trends are limited, because of low knowledge and technical skills among the populace to harness these resources. Policy interventions and measures for harnessing these bioresources for improved energy supply in the country are highlighted.

Finally, the efforts by GTZ in research, development and production of improved stove in both Kenya and Uganda were also presented in this book.

PROF. B.O. SOLOMON
Director-General/CEO
National Biotechnology Development Agency
Federal Ministry of Science & Technology
Abuja-Nigeria

Introduction

The increasing population pressure and demand for a better life style have led to a greater use of energy. As a consequence, the rate of energy consumption each day is becoming much higher than it was ever before resulting in escalating pressures on the use of available resources, which are mostly fossil fuels in the form of coal, oil, gas and minerals for getting nuclear power. A negative balance in the equilibrium of process source depletion and replenishment by natural processes, which has also led to uncontrollable rise in air, water, soil, noise, electrical and other types of pollution, is posing a challenge for healthy survival of life on Planet Earth.

The energy resources are categorized as non-renewable (limited) and renewable (cyclic). The latter comprise solar, hydro, wind, geothermal, biomass and tidal power. Bio-mass resources used as fuel include organic materials such as wood, roots and other plant and animal material, and the energy derived from these is termed as bio-energy. Crops grown for bio-energy include not only the traditional crops like wheat and oilseed rape, but also dedicated energy crops including short-rotation willow-coppice, unusual grasses and forestry products. Crops can be converted to energy either by processing these into liquid fuel for the transport sector (bio-fuels) or by directly burning in power plants as biomass by gasification. With combustion, carbon dioxide (CO_2) is released which is absorbed by plant material through the process of photosynthesis. Effectively, producing energy from bio-fuels or biomass could be seen as 'recycling' CO_2. However, Bio-energy production is never a neutral process when it comes to 'greenhouse gases'. Thus, for example, one of the most potent greenhouse gases N_2O is released from fields that are intensively fertilized. In fact, life-cycle analyses of bio-energy production have shown that with bad agricultural and management practices, production can actually result in a net increase in the emission of greenhouse gases. The production of bio-energy is gathering more and more attention as a feasible way of reducing dependence on imported oil and gas and is even being hailed as one of the potential key weapons in the battle against global warming.

Various life forms of plants and animals within a given ecosystem, or for the entire Earth, are integrally known as Biodiversity. The biodiversity found on Earth today consists of many millions of distinct flora and fauna, which is the product of nearly 3.5 billion years of evolution. Diminishing the ability of Earth's biodiversity support systems to absorb within limits the impacts of diseases, shortages and other enhanced assaults is a very serious issue. Any further consequences of large scale exploration, development and use of fossil carbon could drastically promote conditions beyond our ability to control them. We are already in imminent danger of major climatic and environmental changes with irreparable shocks to Earth's biodiversity. Humankind is clearly mismanaging its economic, energy and environmental resources with potentially disastrous results. Attempts to find methods of generating large amounts of cheap energy are having inevitable unintended but not unexpected consequences.

Constant economic growth is both a financial benefit to our future and civilization's major environmental challenge. It may destroy our civilization unless we act fast with preventive and proactive measures. One possible way of averting shortages of fossil fuels is simply to use less of them. This may be achieved by increasing the utilisation efficiency of fossil fuels and minimizing the energy wastage. Many such fuels, such as coal and oil, are used in machinery with very low conversion efficiencies. Raising the efficiency of some common forms of machinery like automobile engines may contribute to a significant reduction in use of fossil fuel. Popularisation of mass transport over individual use of cars would also contribute to such savings. Efficient design of buildings would save energy and money on heating and cooling bills and also reduce depletion of fossil fuels supplies. One possible solution is cogeneration, where the excess heat generated from machinery is harnessed to provide another source of energy.

The juxtaposition of energy needs and biodiversity values has led to challenges for both the energy industry and the conservation community. For energy companies, the challenge is to find a way to meet the public demand for energy, while at the same time addressing society's expectations for corporate social and environmental responsibility. For environmental organizations, the challenge is to be a strong voice for biodiversity conservation while working with industry to find the balance between the potential threat that energy development represents and the opportunities for harnessing the influence, expertise and resources of energy companies for conservation efforts.

In order to deliberate on the above issues, the Centre for Science & Technology of the Non-Aligned and Other Developing Countries (NAM S&T Centre) jointly with the Uganda National Academy of Sciences (UNAS) organized an International Roundtable on 'Sustainable Utilisation of Energy and Biodiversity Resources for Wealth Creation and Development' at Kampala, Uganda during 10–13 March 2009, which was attended by 66 experts and senior professionals from 15 countries. The participants were welcomed by Prof. Paul Edward Mugambi, President, Uganda National Academy of Sciences (UNAS) and the event was inaugurated by Hon. Eng. Simon D'Ujanga, MP and Minister of State for Energy, Republic of Uganda.

The Roundtable brought together different stakeholders in the developing countries in the fields of energy efficiency, renewable energy, bio-energy and maintenance of ecology and biodiversity to deliberate on various aspects for better utilization of energy and biodiversity resources. The case study presentations and research papers led to an understanding of the concept of efficient energy production and utilization and how the bio-energy resources can replace the existing conventional sources of energy in a sustainable and cleaner way for the protection of ecology and biodiversity. The present publication is based on the deliberations of this Roundtable and also includes papers contributed by eminent scientists in this field. It comprises of 21 scientific articles and review papers.

I would like to acknowledge the dynamic involvement and untiring efforts of Prof. B.O. Solomon, Director-General/CEO, National Biotechnology Development Agency, Federal Ministry of Science & Technology, Nigeria for the technical editing of this publication, supplying the Preface and for suggesting the future course of action while reviewing the contents. I am also indebted to H.E. Dr. Alhassan Bako Zaku, Federal Minister of Science & Technology, Nigeria for sparing his valuable time in writing the Forward. Last but not the least the valuable service provided by the entire team of the NAMS&T Centre, particularly by Mr. M. Bandyopadhyay, Dr. V. P. Kharbanda, Mr. Pankaj Bhuttan and Ms. Debanjana Dey in compiling the presented papers and giving a shape to this volume are also deeply appreciated.

I hope that this publication will find use with the entire scientific community and provide a valuable reference material on utilisation of energy and biodiversity resources to meet the needs of the present without compromising the ability of future generations to meet their own needs.

ARUN P. KULSHRESHTHA

Director, NAMS&T Centre

Contents

Foreword — *v*

Alhassan Bako Zaku

From the Editors' Desk — *vii*

B.O. Solomon

Introduction — *xi*

Arun P. Kulshreshtha

BIOENERGY: SOURCES & MANAGEMENT

1. Opportunities for Tapping Biogas Potential in Developing Countries—The Case of Botswana — 3

   Aaron Aupa Somolekae, Bostwana

2. Poverty Mitigation and Local Economic Development through Use of Natural Resources: Examples from South Africa — 25

   Charlie Michael Shackleton and Sheona Shackleton, South Africa

3. Bio-energy Conservation Policy Options in Uganda: A Bio-Economic Perspective — 44

   Buyinza Mukadasi and Masaba Sowedi, Uganda

ENERGY & SUSTAINABLE DEVELOPMENT

4. Optimizing Green Energy: Recipe for Sustainable Development 79
Jitendra Behari, India

5. "Maurice Ile Durable"-Green Mauritius- Towards sustainable development 95
Doonunjoy Dassaye, Mauritius

6. Sustainable Utilization of Energy for Wealth Creation and Development: The Nigerian Scenario 107
B.O. Solomon, R.S.M. Gidado, A.A. Akinrelere, N.C. Ezebuiro, G.A. Etuk-Udo and O.A. Adetunji, Nigeria

7. Trends in the sustainable utilization of renewable energy resources of Sri Lanka for facing the energy challenge 126
J.G. Shantha Siri, Sri Lanka

8. Energy Efficiency as a Source of Income and Improved Livelihoods: The Case of GTZ Improved Stove Projects in Kenya and Uganda 136
Ms. Myra Mukulu, Kenya

ALTERNATIVE FUELS

9. Sustainable Energy 145
Lat Lat Tun, Myanmar

10. Bio-fuels and energy security in Vietnam 153
Ms. Ngo Thi Loan, Vietnam

11. Uganda's Energy Crisis and Its Quest for Biofuel Solutions 166
M. Bazira Henry, Uganda

12. Brief Report on the Current Status of Biofuels Industry in Zambia 178
Hakachite Christopher, Zambia

RENEWABLE ENERGY

13. Adoption of Energy Efficiency Measures by Households in Shared-Meters Apartments 201
Michael Kiza, Uganda

14. Need and Challenges in Developing Renewable Energy Resources: A Networked Approach of CSIR 209

Sandeep Kumar Tiwari, India

15. Status and Potential of Renewable Energy in India 234

Arvind Kumar and Rajiv Sharma, India

16. The Renewable Energy Policy for Uganda 251

Albert Rugumayo, Uganda

BIODIVERSITY: EMERGING SCENARIO & APPLICATIONS

17. Status and Initiatives of Biodiversity and Energy Resources Utilization in Nepal 265

Dilip Subba, Nepal

18. Sustainable Utilization of Biodiversity Resources for Wealth Creation and Development 282

B.O. Solomon, R.S.M. Gidado, A.A. Akinrelere, N.C. Ezebuiro and O.A. Adetunji, Nigeria

19. Acidification of Natural Resources: Concern for the Future of West Asia 296

Masoud Vahabi Moghaddam, Iran

20. Exploring the therapeutic potentials of African medicinal plants 310

Charles Wambebe, Nigeria

21. Biofuel and Biodiversity Prospects in Tanzania 324

Hulda Gideon, Tanzania

Index 337

Bionergey: Sources & Management

1

Opportunities for Tapping Biogas Potentials in Developing Countries—The Case of Botswana

Aaron Aupa Somolekae
Botswana Technology Centre, Gaborone, Botswana
E-mail: Somolekae@botec.bw, soms.001@gmail.com

In Botswana and other developing countries, most communities use wood as a source of fuel for domestic and commercial purposes. Eighty five percent of the wood fuel is consumed by household sector, 75% of which is accounted for by the rural household sub-sector. High dependency on fuel wood has led to unsustainable supply and consequent scarcity of this resource, hence the degradation of the forests. It is therefore necessary to have affordable alternative energy sources. This paper discussed the performance of the biogas technologies in terms of cost-benefit analysis, environmental and social impacts assessments. It also discussed the opportunities and the potential biogas energy sources available, cooperation between the private sector and government to promote biogas use. Barriers that affected the biogas production were further discussed with possible methods addressing them. The different technologies for tapping biogas were discussed highlighting the one adapted for Botswana conditions. Biogas systems that were no longer operational were discussed together with the currently operational and upcoming systems. The report showed that biogas research and practices started in Botswana in the 1980s, funded by the government, but the projects eventually collapsed. Over 25 years later, the technology is coming to life again, this time around initiated by the private sector. With sources such as abattoir animal waste, hotel wastes, cow dung from cattle ranches, municipal waste water treatment plants and landfills, the stock feed for the biogas production is abundant. Environmental, social and economical impacts of biogas production were discussed. Unfortunately, the cost of establishing a biogas system in rural households is high, from USD 1,000 for a 1000 litre capacity digester and may not be affordable to many households without government's assistance. In conclusion, private-public partnership proved to be the way forward in increasing the use of biogas technology in developing countries.

KEY WORDS: *Renewable energy, biogas technology, biodigester, anaerobic digestion*

1.0 INTRODUCTION

The paper's objectives are to identify potential opportunities in developing countries to utilize the resources that are abandoned to produce biogas, and to look into ways in which various governments can help in dissemination of biogas technology.

Botswana's energy sector constitutes both traditional biomass energy sources and the commercial energy sources. Traditional biomass energy sources contribute significantly to Botswana's energy balance of which it constitutes 30% of primary energy supply as in Figure 1 (EAD, 2003). Ninety seven percent of junior secondary schools use fuelwood for their daily cooking (Batlhapi and Wright/EAD, 2008). This is by far too much dependence on this energy source and it therefore tends to deplete the woodlands because there is no aggressive woodland management which would otherwise counter this harvesting of fuel wood. The dry wood which would be ideal for harvesting is diminishing and people now tend to harvest live trees which are becoming scarce in the surrounding villages. It therefore forces the villagers to travel long distances in search of fire wood. This is uneconomical to the households because women and children (who in the traditional set-up are the ones who normally are available to collect fire wood) end up using most of their time collecting this scarce resource.

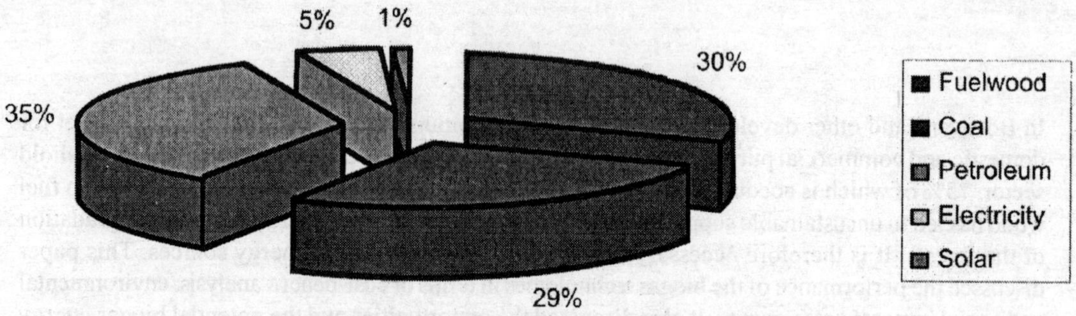

FIGURE 1: Total Primary Energy Supply in Botswana

At the 1992 Earth Summit, in Rio de Janeiro, Brazil, the United Nations (Botswana is a member country), adopted Agenda 21 which is a comprehensive plan of action to be taken globally, nationally and locally by organizations of the United Nations System, Governments, and Major Groups in every area in which human impacts on the environment (Agenda 21, Brazil). It was agreed that to sacrifice the earth's environment at the cost of developing economies is unsustainable. By developing ecological agriculture, we should exploit and utilise natural resources reasonably and reduce environmental pollution to reach sustainable development as well as integrated benefits of society, economy and environment (EAD/EECG, 2008).

From the wood energy supply and demand (Figure 2), it can be seen that the difference of supply versus demand is little and with continued use of the wood as a source of energy without good woodland management, the demand will surpass the supply, and this will increase the desertification of Botswana, already in Gaborone and surrounding villages, wood demand is higher than the supply (EAD, 2003).

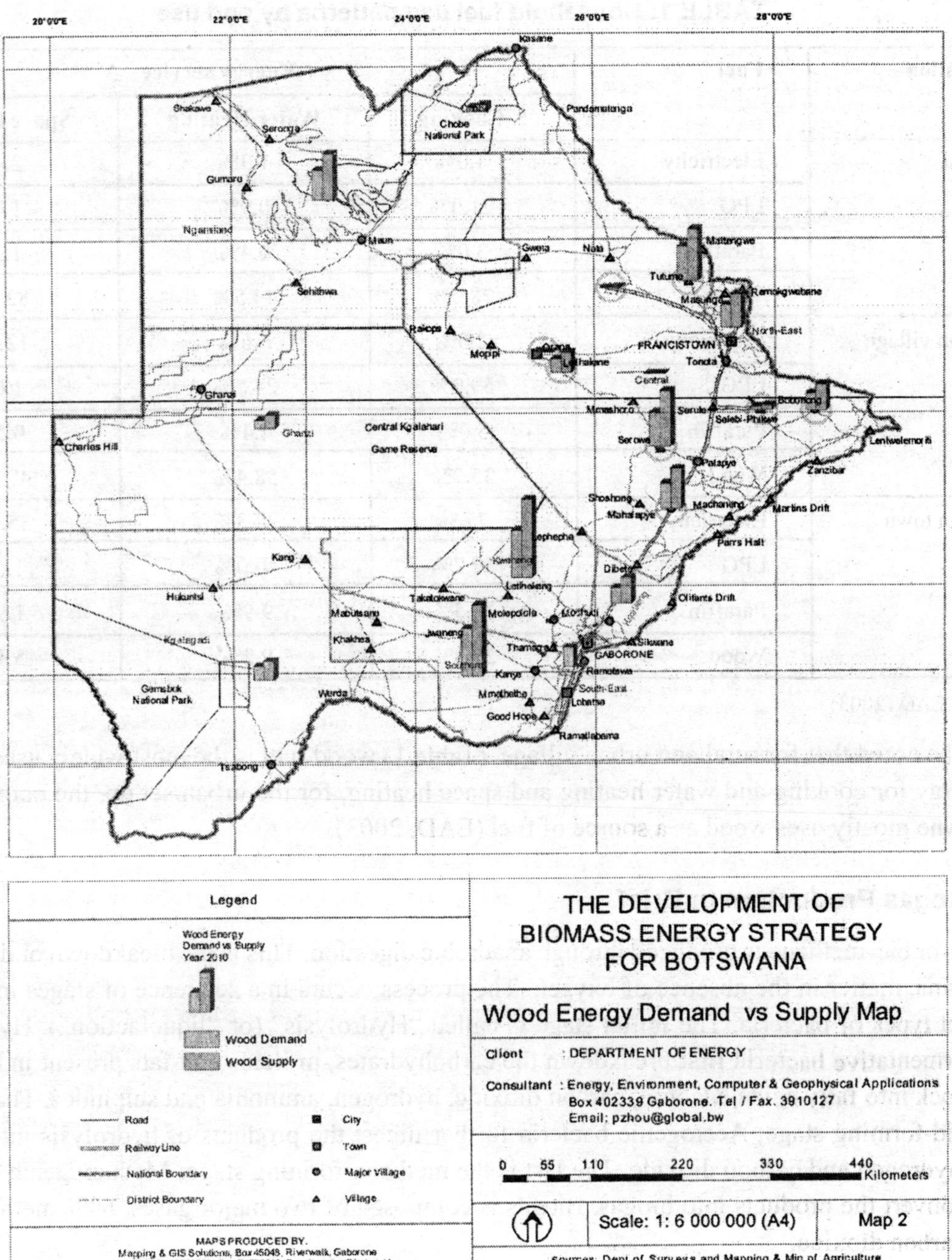

Courtesy of EECG

FIGURE 2: Wood Energy Demand versus Supply

TABLE 1: Household fuel use patterns by end use

Location	Fuel	Energy service		
		Cooking	Water Heating	Space Heating
Rural	Electricity	1.0%	1.3%	2.4%
	LPG	21.0%	0.9%	1.0%
	Paraffin	3.0%	0.4%	1.3%
	Wood	78.0%	44.5%	82.7%
Urban village	Electricity	2.0%	6.6%	12.6%
	LPG	59.0%	24.5%	1.6%
	Paraffin	5.0%	4.4%	0.5%
	Wood	35.0%	58.4%	48.5%
Urban town	Electricity	7.6%	30.3%	35.3%
	LPG	57.7%	40.0%	1.6%
	Paraffin	10.5%	9.9%	1.9%
	Wood	22.8%	9.2%	5.1%

Source: EAD (2003)

It can be noted that for rural and urban village, (Table 1) wood fuel is the most widely used source of energy for cooking and water heating and space heating, for the urban set up, the poor are the ones who mostly uses wood as a source of fuel (EAD, 2003).

1.1 Biogas Production in Brief

Biogas or bio-methane is produced through anaerobic digestion. This is the breakdown or digestion of organic matter in the absence of oxygen. The process occurs in a sequence of stages involving distinct types of bacteria. The initial stage is called "Hydrolysis" (or "liquefaction"). Hydrolytic and fermentative bacteria first breakdown the carbohydrates, proteins and fats present in biomass feedstock into fatty acids, alcohol, carbon dioxide, hydrogen, ammonia and sulphides. The next is the acid-forming stage; Acetogenic bacteria further digest the products of hydrolysis into acetic acid, hydrogen and carbon dioxide. The last is the methane-forming stage; Methanogenic bacteria then convert the products into biogas. Biogas is composed of two major gases, 60% methane and 40% carbon dioxide.

The combustion of biogas produces useful energy in the form of heat. After filtering and drying, biogas is also suitable for use in an internal combustion engine, which, when combined with a generator produces electricity. Other applications of biogas may include electric power production from gas turbines or fuel cells. Biogas can be used as a substitute for natural gas or propane in space heaters, refrigeration equipment, cooking stoves or other equipment. Compressed biogas can be used as an alternative transportation fuel.

1.2 Biogas Technologies

The biogas plant is, commonly known as the biogas digester. Since various biochemical reactions take place in the biogas digester, it is also called biodigester, or bio-reactor or anaerobic reactor. The main function of this structure is to provide anaerobic condition within it. It can be made of various construction materials and in different shapes and size. Construction of this structure forms a major part of the investment cost. There are three basic digester designs but they differ in cost, climate suitability and the concentration of manure solids they can digest.

1.2.1 Floating drum digester

Experiment on biogas technology in India began in 1937, but it was not until 1962 when, Jashu Bhai J. Patel's design was approved by the Khadi and Village Industries Commission (KVIC) of India. In this design, the digester chamber is made of brick masonry in cement mortar. A mild steel drum is placed on top of the digester to collect the biogas produced from the digester. There are two separate structures for gas production and collection. The system has a high investment costs.

1.2.2 Fixed dome digester

Fixed dome Chinese model biogas plant was built in China as early as 1936. It consists of an underground brick masonry compartment (fermentation chamber) with a dome on the top for gas storage. In this design, the fermentation chamber and gas holder are combined as one unit. This design eliminated the use of the mild steel gas holder and the life of the fixed dome type plant is longer (from 20 to 50 years) compared to KVIC system.

1.2.3 Plug flow digester

A typical design for a plug-flow system includes a manure collection system, a mixing pit and the digester. In the mixing pit, the addition of water adjusts the proportion of solids in the manure slurry to the optimal consistency. The digester is a long, rectangular container, usually built below-grade, with an airtight, expandable cover. New material added to the tank at one end pushes older material to the opposite end. Coarse solids form a viscous material as they are digested, limiting solids separation in the digester tank. As a result, the material flows through the tank in a "plug." Average retention time (the time a manure "plug" remains in the digester) is 20-30 days. Anaerobic digestion of the manure slurry releases biogas as the material flows through the digester. A flexible, cover on the digester traps the gas. Pipes then carry the biogas from the digester to the point of use. The digester requires minimal maintenance. The first documented use of this type of design was in South Africa in 1957.

Each household can produce their own biogas by installing a toilet-attached biogas plant. The technology is simple: the manure and toilet waste are mixed with water and dumped in an airtight underground pit of about 6 cubic metres. In these anaerobic conditions, methane starts forming and it is led via a narrow pipe into the gas stove in the kitchen (Bajgain, The Nepal biogas support program).

1.3 The main factors that influence the selection of a design or model

1.3.1 Economic factors

An ideal plant should be as low-cost as possible (production cost per unit volume of biogas). The current prices are shown in Table 2, derived from Botswana currency and costs (1 US$ = 8 BWP) (Renewable Energy Pty (Ltd) t/a Biogas, 2008).

TABLE 2: Digester capacities and cost of construction

Digester (L)	Cost (USD)	Gas holder capacity(L)	Gas (litres)	Running time (hours)
1000	1 000	750	700	1.5
1500	1 250	1000	875	2.0
2000	1300	1500	1400	2.5
2500	1375	2000	1750	3.0
3000	1563	2500	2000	3.5
6000	1875	5000	4000	4.0
12000	2500	10000	9000	8.0
16000	3375	14500	13000	12.0

1.3.2 Simple design

The design should be simple not only for construction but also for operation and maintenance. This is important in a community where the rate of literacy is low and the availability of skilled human resource is scarce.

1.3.3 Utilization of local materials

Use of easily available local materials cannot be over emphasized in the construction of a biogas plant, this helps in lowering the costs of the plant.

1.3.4 Durability

Construction of a biogas plant requires certain degree of specialized skill which may not be easily available. A plant of short life could also be cost effective but such a plant may not be reconstructed once its useful life ends. Especially in situation where people are yet to be motivated for the adoption of this technology as is the case in Botswana and the necessary skill and materials are not readily available, it is necessary to construct plants that are more durable although this may require a higher initial investment, it should be noted that, in the case of Botswana, most materials for manufacturing are imported from South Africa.

2.0 BIOGAS PLANTS IN BOTSWANA IN THE 1980S

In 1979, research on three types of biogas plants was conducted in Botswana. These were; i) Plug-flow type, ii) Chinese fixed dome and iii) Indian floating dome were conducted at Rural Industries Innovation Centre (RIIC), The Indian floating dome performed better than the two types under Botswana conditions. RIIC adopted the floating dome type of biogas plant, which was later erected in different parts of Botswana. The projects were funded by the Botswana government. Table 3 shows the distribution of biogas digesters in the 1980s.

TABLE 3: Distribution of Biogas Plants in Botswana in the 1980s

Location	Capacity m^3	First Used	Purpose	Ownership	Hrs of use	Hrs to Pump	Estimated output m^3
Diphawana	75	1982	Water pumping	Syndicate	-	4	23
Mogwalale 1	110	1987	Water pumping	Syndicate	-	6 – 8	27.5
Mogwalale 2	15	1987	Cooking	Individual	6	-	3.75
Kang Clinic	40	1986	Cooking	Local authority	6	-	10
BH 1500 Kang	45 x 2	1988	Water pumping	Syndicate	-	10	22.5
Kang Brigade	45	1986	Bakery	Institution	none	-	11.25
Lehututu	15	1986	Cooking	Individual	6	-	3.75
BAC	40	1987	Demo.	Institutional	?	?	10
Kgotswana	45	1987	Water Pumping	Syndicate	-	6 – 8	11.25
Serowe South	15	1987	Cooking	Individual	6	-	3.75

Source: EAD (2003)

Out of the ten systems, five were sampled by Energy Affairs Department (EAD), and the digesters are currently not working and have been abandoned. Below is a summary of the plants and their current status as sampled by EAD.

2.1 Diphawana Biogas Plant

The system is a 75 m³ Indian floating dome. The digester is still intact with the gas collector collapsed to the bed (Figure 3). The biogas generated by the digester was used to power a diesel engine for the purpose of water pumping. Water would then be collected in the reservoir for watering domestic livestock. It has been said that the engine which was used to pump water was stolen. The engine was never replaced leading to the failure of the project. The water reservoir appears to have been abandoned for a long time as well (Figure 4).

FIGURE 3: Biogas digester and the engine room in Diphawana

FIGURE 4: Diphawana water reservoir

2.1 Mogwalale Biogas Plant

The system was a 110 m³ digester. It is also an Indian floating dome. The digetser is still intact with the gas collector collapsed to the bed. The biogas generated by the digester was used to power a diesel engine for the sole purpose of water pumping. Water would then be collected in the reservoir (Figure 5) for watering domestic livestock. The Southern District council (local authority) then decided to take over the operations of the plant on behalf of the villagers. The council used diesel to operate the generator for water pumping, hence the biogas digester was no longer used.

2.3 Kang Clinic Biogas Plant

The 40 m³ plant was constructed at what used to be an old clinic in the village of Kang. It started operating in 1986 under the Kgalagadi District council (local authority). It was used solely for

FIGURE 5: Mogwalale biogas digester and the water reservoir

cooking. One villager was trained to operate the digester. He was responsible for the collection of cow dung. Unfortunately many farmers started selling cow dung as they saw some value in it. It became difficult to get the cow dung as the council did not budget for the procurement of cow dung. The same villager was also employed as a cook in the new clinic, therefore, when it was his turn to cook for the entire week, no one was assigned to take up his duty as a biodigester operator. The project also ended up failing as there was no support forthcoming from the clinic and council authorities.

2.4 Kang brigade Biogas Plant

Kang Brigade digester was 45 m^3, installed for the purpose of baking bread. The floating dome has been dismantled from the digester body. The oven is rusty and warned out (Figure 6). The facilities were never used.

FIGURE 6: Kang Brigade biogas operated oven

2.5 Kang BH 1500 Biogas Plant

The Kang biogas digesters were two 45 m³ plants (Figure 7). The systems were used for water pumping purposes by a group of farmers. Just like in the case of Mogwalale, the plants were later transferred to the local authority which used diesel instead of biogas. Cow-dung collection is labour intensive and requires that the plant be manned at all times. The local authority did not cater for this. The digesters were ultimately abandoned.

FIGURE 7: Kang BH 1500 Biogas Digesters

RIIC and Botswana College of Agriculture (BCA) operates two demonstration units on site without budget. There is no information on how they perform, but they are still functioning.

3.0 BARRIERS THAT AFFECTED BIOGAS TECHNOLOGY DESSEMINATION IN BOTSWANA

a) Compared to other technologies developed and supplied by RIIC, biogas is not amongst the priority areas. Perhaps this is due to the nature of the technology. The majority of the technologies at RIIC are based on the development of a product that can be sold, mainly in agriculture, but with biogas, the development of the technology remained with the supplier.

b) Biogas production has been more of a "social project" than a "financially viable project".

c) The takeover of some of the plants by the local authority contributed to the failure of the biogas generating plants because these local authorities would then use other sources of energy (diesel) instead of continuing with the current source and put in place measures to sustain the operations.

d) In some instances the operators of the biogas digesters worked part time on them, therefore when it comes to the time when they will be at work, there would be no one to operate plants, the government did not put in place measures to make sure the plants continued to be operated.

e) In one area, the engine which was used for water pumping was stolen and never replaced; therefore the biogas digester had no functionality.

f) There was no provision for the maintenance of the plants or it was inadequate, as such when the plants needed servicing, no resources were available.

g) There were no policies in place that encouraged the dissemination of biogas technology.

h) The technology adopted was expensive compared to other technologies.

i) Water in Botswana is a scarce resource which makes it difficult to sustain the technology.

4.0 BIOGAS PLANTS CURRENTLY IN OPERATION

Figures 8-9 are biogas plants still in operation while Figure 10 is under construction.

4.1 Pitsane Biogas Plant

The biogas below (Figure 8) is owned by an individual and is used mainly for cooking and at night it is also used for lighting, using a generator; this can provide light for to 2 h. The gas holder has a capacity of 5 m^3.

FIGURE 8: Pitsane Household Biogas plant for cooking and generator for lighting

- The generator is fitted with other gas purifying equipment in order to obtain approximately 100% methane gas that is passed to the generator.
- Requires between 500 g – 1 kg of Kitchen waste and cow dug per day

4.2 Cumberland Hotel Biogas Plant

The Cumberland Hotel biogas plant is a floating drum type which uses kitchen waste from the hotel as its stock feed. It replaces LPG gas for cooking and saves the hotel close to 40% costs of LPG. Figure 9 is the biodigester plant. The gas holder is made of steel with a capacity of 20 m³.

FIGURE 9: Cumberland hotel biogas plant

The system has the following attributes:-

- Processes 60 kg/day of food waste
- Waste used is generated from the Hotel Kitchens
- Waste is processed daily.
- Provides gas for cooking up to 8 h daily.
- Provides about to 6 (48 kg) cylinders of Gas/Month.
- Cost saving of around 35 -40%.

4.3 Richmark Poultry Plant – under construction

The project came about as a solution to cut the costs of diesel which was mainly used in an incinerator to dispose waste from the poultry plant and also as a way of cutting down the cost of water used in order to recycle the water. It will also replace LPG for cooking.

The primary stockfeed is the poultry droppings and below is a diagram (Figure 9) which shows how the plant will work (Renewable Energy Pty (Ltd) t/a Biogas, 2008).

The following is the business case for the abattoir:-

- 4000 Kg of abattoir waste destroyed in a diesel incinerator (cost US$ 120/day)
- 200 Abattoir staff fed daily.

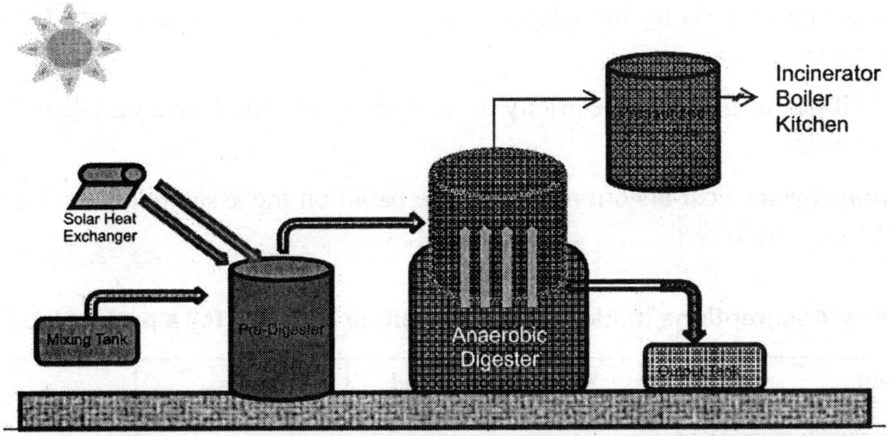

FIGURE 10: Richmark poultry plant in Moletamane –Tuli block

- Coal used to heat Chicken coops
- Waste water cannot be recycled due to contamination from blood

The following is the suggested solutions:-

- Wastes to be sent to a biogas system
- Gas generated used to replace Diesel in the incinerator
- Gas generated to replace the diesel fueled water heater
- Replace LPG use in the abattoir kitchens
- Replace coal.

Anticipated savings with the installation of the biogas system

- US$3,700 on diesel fuel alone per month
- Water can be recycled into environment for farming purposes
- Organic fertilizer to be sold to nearby farmers for use

5.0 RESULTS

5.1 Cost-benefit Analysis

The choice of assessment was based on the data available:-

i. Small, medium and large biogas digesters using poultry manure and municipal wastewater based on Inter-governmental Panel on Climate Change (IPCC) and Renewable Energy (RE) (Pty)Ltd

ii. Biogas digesters using municipal liquid waste from Gaborone City Council (GCC) experience

iii. Landfill gas to generate electricity at municipal landfills based on Clean Development Mechanism.

The calculations for the cost-benefit analysis were based on the assumptions in Table 4 adopted from (EAD/EECG, 2008).

TABLE 4: Key Assumptions made in the cost benefit analysis for a project life of 20 years

Biogas Yield	Volatile Suspended Solids (kg/hr/day)	Biogas (m^3CH_4/kg)	
Poultry	0.01	0.36	IPCC/ RE(Pty)Ltd
Cattle	1.5	0.1	IPCC/ RE(Pty)Ltd
Dairy	1.9	0.13	IPCC 2006
Sheep	0.32	0.13	IPCC 2006
Goat	0.35	0.13	IPCC 2006
Municipal Liquid Waste	11.9		GCC 2008
Municipal Solid Waste (Landfill gas)			
Volume of LPG generated (m^3 CH_4/ton waste)		50	EU/Synergy
Amount of Waste (kg/person day)		2.38	EU/Synergy
LFG recovery rate and CH4 Content (%)		50	EU/Synergy
System Operating Hours(hr/yr)		8664	EU/Synergy
System failure Rate (% yr operating hrs)		10	EU/Synergy
Lower Heating Value of CH_4 (kWh/Nm^3CH_4)		9.05	EU/Synergy
CH4 density (ton/m^3)		0.00067	EU/Synergy
Financial Assumption			
Discount Rate		15.5%	Bank of Botswana
Fuel Properties			
Coal Calorific Value (kJ/kg)		22,700	Morupule colliery
Diesel Calorific Value (kJ/kg)		46,000	
Diesel Density (kg/m3)		0.880	
Waste Generation			
Biodegradable Waste generation per beast (kg)		20	Botswana Meat Commission.

Waste per bird for 6wks (kg)	3.4	Star Poultry
Wastewater generation per bird (l)	17	Star Poultry
Fuel Prices		
Diesel (BWP/l)	8	
LPG (BWP/kg)	16	
Coal (BWP/kg)	0.04	

The following Tables 5–8, shows the results of calculations for the cost benefit analysis using the assumptions in table 4 adopted from (EAD/EECG, 2008).

TABLE 5: Poultry biogas digesters

Establishment size (Broilers/year)	Fuel substituted	Net present value (NPV)	Internal rate of return (IRR)	Pay-back Period	Cost-Benefit Ratio
14,000	Coal	-12,508	Negative	85	13.98
	Diesel	-7197	2%	16	2.61
	LPG	-7346	4%	13	2.16
240,000	Coal	-46897	Negative	80	13.09
	Diesel	34598	27%	4	0.59
	LPG	66207	37%	3	0.44
3,500,000	Coal	-586192	Negative	59	9.73
	Diesel	596713	31%	3	0.52
	LPG	1086410	44%	2	0.38

The net present value is negative when the substitution energy source is coal for all the sizes of poultry establishments, and the payback period is high. For substituting diesel and LPG, there is some profit on the medium and large scale establishment.

TABLE 6: Municipal Waste water biogas project

Digester	Fuel substituted	NPV	IRR	Pay-back Period	Cost-Benefit Ratio
1,000	Coal	-83,121	Negative	72	11.85
	Diesel	-56387	Negative	23	3.78
	LPG	-48433	2%	16	2.60

20,000	Coal	-973399	Negative	77	12.59
	Diesel	-618483	0%	19	3.18
	LPG	-492487	4%	13	2.20
874,000	Coal	-34289528	Negative	401	65.86
	Diesel	-4128928	13%	7	1.14
	LPG	7936951	20%	5	0.81

The NPV is negative when substituting all the energy sources for the municipal wastewater projects, and the payback period is high for coal and diesel for small and medium sized projects. For substituting, LPG, there is a considerable profit on the large scale project.

TABLE 7: Economic assessment of landfill biogas

Variable	Value	Unit
Electricity tariff	0.4	kWh
NPV(excl. Certified Emission Reductions)	-34,122,160	US$
NPV(incl Certified Emission Reductions)	-19,012,976	US$
IRR(excl Certified Emission Reductions)	Negative	%
IRR(incl Certified Emission Reductions)	Negative	%
SPP(excl Certified Emission Reductions)	29	Years
SPP(incl Certified Emission Reductions)	13	Years

For the landfills, it is uneconomical to venture into the projects. The pay bay period is long and by the time the landfill gas is depleted, the project will not have yet paid for its self.

TABLE 8: Abattoir waste

Digester	Fuel Substituted	NPV	IRR	Pay-Back Period	Cost-Benefit Ratio
1,000	Coal	-14,656	negative	105	17.27
	Diesel	8,075	25%	4	0.66
	LPGs	15,327	31%	3	0.53
20,000	Coal	-181,428	negative	43	7.02
	Diesel	266,815	37%	3	0.44
	LPG	452,056	51%	2	0.32

87,400	Coal	-670,821	Negative	34	5.63
	Diesel	1,281,046	42%	2	0.39
	LPG	2,097,680	59%	2	0.28

For abattoir waste, the pay back period is too large for substituting coal, but is a good business venture when substituting diesel and LPG.

5.2 Environmental and social impacts

Outlines the effects of energy technologies based on the levels of pollution, emissions and other environmental factors and the economical development, gender and health impacts.

a) Production of energy (heat, light, electricity)

The calorific value of biogas is 6 kWh/m^3 - this corresponds to about half a litre of diesel oil. The net calorific value depends on the efficiency of the burners or appliances (Green power, India).

Biogas is able to substitute most of the consumption of firewood in rural households.

1 m^3 Biogas (approximately 6 kWh/m^3) is equivalent to (Green power, India):

- Diesel (approx. 12 kWh/kg) 0.5 kg
- Wood (approx. 4.5 kWh/kg) 1.3 kg
- Cow dung (approx. 5 kWh/kg dry matter) 1.2 kg
- Plant residues (approx. 4.5 kWh/kg dry matter) 1.3 kg
- Hard coal (approx. 8.5 kWh/kg) 0.7 kg
- LPG gas (approx. 5.3 kWh/m^3) 1.1 m^3
- Propane (approx. 25 kWh/m^3) 0.24 m^3

The biogas generated from small and medium sized units (up to 6m^3) is generally used for cooking and lighting purposes. Large units can be used to power engines and generators for power generation or mechanical work.

b) Transformation of organic wastes into high quality organic fertilizer

i. The biogas digester is fed with stock feed, which may be cow dung, poultry products or abattoir waste. The output from the digester is a high quality organic fertilizer or bio-fertilizer. Bio-fertilizer is a 100 % natural organic fertilizer.

ii. This fertilizer is very important, especially in a country where the farmers do not have the resources to buy chemical fertilizers frequently. It has been calculated through university laboratory tests in India that the fertilizer which comes from a bio-gas plant contains three times more nitrogen than the best compost made through open air digestion. If you compost chicken manure, for example, the finished compost will have in it only 1.58 to 2% nitrogen. The same manure digested in a bio-gas plant will generate 6% nitrogen (Green power, India).

c) Health benefits of biogas

i. The availability of biogas can have effects on nutritional patterns. With easy access to energy, the number of warm meals increases.

ii. Biogas burns clean therefore homes do not fill with smoke and ash.

iii. Women and children experience less bronchial problems.

d) Reduction of workload

i. To gather wood, women and children can spend up to 2-4 h per day searching and carrying the firewood. Once a biogas plant is installed, they will have that much extra time for themselves. This will help in improving the quality of women's and children's lives. They will now have more time for education and other interesting activities outside the home, i.e. they will have more time for social and economic activities.

ii. Cooking with gas takes less time than with wood or charcoal or any other commonly used fuel and it is easier to cook with a gas stove.

iii. The annual time saving for firewood collection and cooking averages to almost 1000 hours in each household provided with a biogas plant per year.

e) Environmental advantages: through protection of forests, soil, water and air

i. Estimating an average per capita consumption of 3 kg of wood per day for energy (cooking, heating and boiling water) in rural areas, the daily per capita demand of energy equals about 13 kWh which could be covered by about 2 m^3 of biogas. A biogas plant therefore directly saves forest.

ii. Annually, each biogas plant can save more than four tons of firewood and 32 litres of paraffin/kerosene.

iii. A single biogas system with a volume of 2.8 m^3 can save as much as 0.12 ha of forest (woodland) each year (Green power, India).

iv. A recent study found that each biogas plant can mitigate about five tons of carbon dioxide equivalents per year (Green power, India).

v. The credits thus earned could provide alternative financing for the sustainability of biogas program in that particular region.

vi. Global Environmental Impact

Methane is not only the second most important greenhouse gas (it contributes 20% to the effect while carbon dioxide causes 62%), it has also a 25 times higher global warming potential compared with carbon dioxide in a time horizon of 100 years (Green power, India).

i. The Biogas plant effectively reduces the amount of methane directly released into the atmosphere, by trapping it and facilitating its use as a green fuel. After burning, methane only releases harmless gases in air.

1m³ cattle manure = 22.5 m³ biogas = 146 kWh gross = 36 kg CO_2- emissions (Green power, India)

ii. Smaller agricultural units can additionally reduce the use of forest resources for household energy purposes and thus slow down deforestation, soil degradation and resulting natural catastrophes like flooding or desertification.

1 m³ biogas (up to 65% CH4) = 0.5 L fuel oil = 1.6 kg CO_2 = 5.5 kg fire wood (Green power, India)

iii. The reduction of **1 kg methane is equivalent to the reduction of 25 kg CO_2**. The reduction of greenhouse gases with a high global warming potential can be more efficient compared with the reduction of CO_2.

iv. CO_2 generation by burned biogas only amounts to 80% of the CO_2 generation of fired fuel oil and 50% of the CO_2 generation from coal (per kWh electrical energy).

6.0 DISCUSSION

6.1 Cost-Benefit analysis

a) small household biodigesters are not cost effective

b) Medium-scale biodigesters and those using municipal liquid waste are marginally profitable

c) large scale biodigesters are pofitable for substituting LPG or Diesel

d) When displacing fuelwood or coal, biogas is not profitable even for large biodigesters.

e) It is therefore noted that for small or household biogas plants to be significantly used, governments should subsidise or give incentives to those who venture into the technology in order for them to sustain the use of the technology.

f) Producing electricity using landfill gas would not be profitable even with carbon financing, although the latter makes a significant contribution.

6.2 Environmental and social impact

a) With anaerobic digestion, a renewable source of energy is captured, which has an important **climatic double effect**:-

   i) The use of renewable energy reduces the CO_2-emissions through a reduction of the demand for fossil fuels.

   ii) At the same time, by capturing uncontrolled methane emissions, the second most important greenhouse gas is reduced:

b) The widespread production and utilization of biogas makes a substantial contribution to soil protection.

   i) Biogas could increasingly replace firewood as a source of energy, conserving forests.

   ii) Biogas systems yield more and better fertilizer. As a result, more fodder becomes available for domestic animals. This, in turn, can lessen the danger of soil erosion attributable to overgrazing.

c) Whole grain and beans may be cooked longer, increasing their digestibility, especially for children, and water may be boiled more regularly, thus reducing waterborne diseases

d) Homes become more hygienic since there is no smoke and no need to store wood. Thus, using biogas has a direct effect on global atmosphere by considerably reducing the greenhouse gasses, and also helps in alleviation of standard of living for households.

7.0 CONCLUSIONS

As seen above, there is no doubt that the use of biogas technologies has a positive impact on rural and poor households, it reduces deforestation, and therefore reduces the effects of global warming, which is a concern to every nation. Biogas technology dissemination in the developing countries can be most effective if the respective governments put in place support mechanisms in order for the private sector and communities to invest in the biogas technologies and for the financiers to confidently lend money to prospective investors for the projects.

Private-Public Partnership is the only way that can help in sustainable dissemination of biogas technologies in developing countries, like in the case of Nepal biogas support program (Bargain, 2009), whereby the government and financial banks have got subsidies specifically for the installations of biogas systems.

8.0 RECOMMENDATIONS

In order for the biogas technology to be sustainably implemented, government support is crucial and this can be achieved by having the following regulations and systems in place:-

a) To encourage biogas system dissemination, there should be strict regulations that govern the fabrication, installation and after sales services of Biogas plants from design to commissioning of the installation through international and local standards. These will minimize breakdowns and maintenance costs therefore giving a potential investor a foreseeable good return on investment.

b) Price is the key barrier to the commercialization of generation of electricity from municipal waste, as the cost of producing power from municipal waste is higher than that generated from coal in the case of Botswana. A price support mechanism for electricity generated from municipal solid waste should be developed.

c) There should also be price support for communities and individuals who venture into the biogas resource activities, in order for them to overcome the initial price of installing the systems.

d) The government should then put in place regulations and laws that ensure that the local power utility companies purchases power from those who produce the power through renewable energy means, thereby making sure that the power produced has a market and hence promotes the use of renewable energy sources. The power produced should of cause be of high quality as that of the utility company.

e) Tax incentives could be introduced to encourage the use of biogas plants, there could be tax rebates for those facilities which are built for the purpose.

i) Loans with subsidized interest rates should be specially formulated for these projects. This will make it easier and cheaper for project developers to obtain finance, and will help to overcome the difficulties caused by financiers' reluctance to invest in the perceived higher-risk area of renewable energy projects. The loans could include quality standards and inspections as a condition of the loan.

REFERENCES

Energy Affairs Department (EAD), 2003, the statistical bulletin.

Mabowe, B. R., 2008, Biogas utilization in Botswana, A review of experiences and way forward, pp 2, pp. 5-11.

Renewable Energy Pty (Ltd) t/a Biogas, 2008, phamphlet.

EAD/EECG, 2008, Draft Biomass Energy Strategy (BEST) for Botswana, pp. 109-119, 155.

Pre-feasibility study of a methane landfill gas recovery project,Gaborone landfill, EU- Synergy.

Green power, India, Biogas benefits. www.greenpowerindia.org/**Biogas**.htm (accessed 26th Jan. 2009)

Batlhapi, T. and Wright G./ EAD, 2008, Utilization of fuel wood by government primary and junior secondary schools in Botswana, pp 8.

Agenda 21 declaration on environment and development, Reo Dejanaero, Brazil. www.un.org/esa/sustdev/**agenda21**.htm (accessed 14th Jan. 2009).

Bajgain, S. The Nepal biogas support program. www.snvworld.org/en/Documents/Biogas%20Nepal%20-%20Successful_model_of_public_private_partnership.pdf (accessed 15 Jan. 2009).

Trishna Gurung, 2007, Biogas — saving nature naturally in Nepal. http://www.panda.org/news_facts/newsroom/index.cfm?uNewsID=95320 (accessed 15 Jan. 2009).

2

Poverty Mitigation and Local Economic Development through use of Natural Resources: Examples from South Africa

Charlie Michael Shackleton* and Sheona Shackleton
Department of Environmental Science,
Rhodes University, Grahamstown 6140, South Africa
*E-mail: c.shackleton@ru.ac.za

Wild biodiversity resources fulfil a number of functions in rural livelihoods of citizens in developing countries which serve to alleviate or mitigate poverty. These include daily consumption value, the safety-net value in adverse conditions and the value of income generating activities based on sale of biodiversity resources on local or distant markets. The contribution of biodiversity resources constitutes between one-fifth and one-quarter of total livelihoods income. Degradation of these resources or restriction of access would consequently result in a marked deepening of poverty for a significant proportion of rural households. This is especially so for the poorer sectors of rural society who make greater use of biodiversity resources in the first place. A number of government interventions and policies steps are identified to optimise the benefits for each of these different values. The most important being the need for greater recognition of and investment in the biodiversity resource sector.

KEY WORDS: *Biodiversity resources; Policies; Poverty; Safety-net; South Africa; Trade; Value*

1. INTRODUCTION

Economic development requires a large range of building blocks to be in place. The precise magnitude and composition of those blocks varies in time and space, but includes intellectual resources, financial resources, infrastructural resources, human resources and an accommodating institutional environment. Underpinning all of these requirements are the actual environmental goods and services, the raw materials and processes that are the very basis of all human economic activities and general well-being. Raw material resources such as timber, fibre, water, minerals, soils and animal and fish proteins (wild or cultivated) are the ultimately the foundation of all economic activity.

Consequently, the use of ecosystem goods and services offers both opportunities for local economic development as well as constraints. The opportunities are offered by the multitude of resources available at any place, which can be used for local consumption, trade, value addition and processing. The overarching constraint is that extraction of such resources and harnessing of beneficial ecosystem services must not exceed the capacity of the ecosystem to maintain its own health and to continue to provide these goods and services into the future. Damage to local environmental goods and services will ultimately undermine human well-being. The challenge therefore, is how to use natural resources to drive and support economic growth without undermining the ecosystem services that they deliver as a benefit to society as a whole.

South Africa boasts an immense wealth of biodiversity resources, having hundreds of thousands, if not millions, of species, making it the third most biodiverse country in the World (Wynberg, 2002). At the ecosystem level, this biodiversity supports the production of goods and services required by all humankind; water, air, soil fertility, wood, food and so on (Scholes and Biggs, 2005). At a more local level the harvesting of numerous natural resources provides consumptive products for millions of poor South Africans, as well as income for equally significant numbers (Twine *et al.*, 2003; Lawes *et al.*, 2004). Consumption of and trade in these resources is the very mainstay of their well-being, and crucial in preventing deeper poverty levels (Shackleton *et al.*, 2007; Shackleton *et al.*, 2008). For example, despite massive improvements in the provision of electricity, most rural (and a significant proportion of urban) South Africans continue to use fuelwood as a key energy source for cooking; approximately 75 % of the population use medicinal plants for medicinal or cultural reasons; millions of urban and rural households make use of wild edible herbs. With such high demand for these resources, it is not unsurprising that there are large and established trade networks, spanning local, regional, national and, for some resources (e.g. specific medicinal plants, mopane worms), international boundaries. The total value of this trade is unknown, and unrecorded in local or national economic statistics. It certainly equates to billions of Rands per year. The direct-use and trade values are substantial and provide a cost saving to the State. Where biodiversity resources are overused or exhausted, people have to purchase alternatives, which reduce their scarce cash resources thereby increasing their likely dependency on State welfare grants.

However, despite this wealth of biodiversity resources, a significant proportion of the South Africans population lives in poverty, which is concentrated in, although not exclusive to, the rural areas, especially amongst the black African population as a direct result of the apartheid policies of the previous non-democratic government (Aliber, 2003). Approximately 70% of South Africa's poor live in rural areas, and likewise, approximately 70% of rural dwellers are poor and 18-24% are categorized as chronically poor (Government of South Africa, 2000). Approximately 40 % of rural households rely on cash handouts from either the State or urban-based kin. Poverty levels are being exacerbated by the HIV/AIDS pandemic, which is eroding many of the traditional coping mechanisms and safety nets (Aliber, 2003).

However, the extensive government focus on creating jobs in the formal sector, capital intensive local economic development (the big fix), and cash welfare grants in the interim totally ignores the importance of environmental goods and services to rural livelihoods and in preventing deepening poverty. It has been argued by Shackleton *et al.* (2001) that income-based measures fail to acknowledge the diversity of livelihood strategies in which rural households engage as well as the relative magnitude of returns from land-based livelihood strategies. Because of this, policy responses to poverty, especially in rural areas, fail to strengthen current strategies to reduce vulnerability and consequent poverty.

This paper provides a brief overview of a range of examples from South Africa, demonstrating the importance of wild biodiversity resources for contributing to poverty reduction in rural areas. Consequently, government policies, programmes and interventions aimed at securing biodiversity resource access and promoting sustainable use represent an investment in poverty alleviation. Clearly this is specific to the South African situation but some issues and interventions may resonate with potential programmes and policies that are in place or being tested elsewhere in the developing world.

2. TERMINOLOGY

2.1. Poverty and Poverty Alleviation

Understanding what is meant by poverty, and the different definitions of poverty alleviation and related concepts, is crucial to interpreting whether natural resources can play a positive role in addressing poverty (Arnold, 2001; Shackleton *et al.*, 2007). This section provides these definitions, drawing primarily on the thorough discussions of this subject in the context of forests and natural resources provided by Angelsen and Wunder (2003), Sunderlin *et al.* (2005) and the State of the Forests Report (FAO, 2003).

2.1.1. Poverty and well-being

Poverty, in its broadest sense, is seen as the pronounced deprivation of well-being related to a lack of material income or consumption (the conventional measures of poverty), low levels of education and health, poor nutrition and low food security, high levels of vulnerability and exposure to risk, and a profound lack of opportunity to be heard (Chambers, 1988; World Bank, 2000; Sunderlin *et al.*, 2005). The inability of the poor to give voice to their needs can be related to their powerlessness within existing social, economic, political and cultural structures (Chambers, 1988). May (1999) defined it simply as "the inability of individuals, households or entire communities to command sufficient resources to satisfy socially acceptable minimum standards of living". Both these definitions recognise that poor people's concerns go beyond just adequate income to include aspects of security, capability, independence, choice, health and well-being, and the ability to devise appropriate coping strategies when faced with shocks and crises. Yet, these dimensions of poverty are often neglected in poverty assessments or studies dealing with poverty, including those in the biodiversity resource sector, possibly because they are difficult to measure quantitatively (Kepe, 2002; Kingdon and Knight, 2003). Vulnerability, in particular, needs more attention, as it is believed that poor people are being exposed to increasing risk due to, amongst others, climate change, HIV/AIDS, land degradation, changes in trade, globalisation, and increasing violence and crime (Aliber, 2003; World Bank 2000).

2.1.2. Poverty alleviation, mitigation and reduction

Poverty alleviation can be defined as the lessening of the deprivation of well-being (Sunderlin *et al.* 2005). However, this is more complicated than it first appears. Oksanen and Mersmann (2003) write about poverty alleviation being "a multi-dimensional problem requiring holistic solutions that are sustainable over time. Economic growth and increased income are important but not the only routes to poverty reduction. These need to be complemented by more equitable distribution of income and assets, reduction of insecurity and vulnerability, better access to basic services, reduction of social exclusion and empowerment of the poor and good governance".

Several commentators (FAO, 2003; Angelsen and Wunder, 2003; Sunderlin *et al.*, 2005) have suggested that two types of biodiversity resource based poverty alleviation apply at the household level: (a) poverty avoidance, prevention or mitigation and (b) poverty elimination or reduction. Poverty prevention or mitigation refers mainly to the situation where biodiversity resources serve a safety-net function or as a gap filler, including as a source of petty cash, and contribute to the reduction of risk and vulnerability. Under this scenario, the use of biodiversity resources would help people to maintain a minimum standard of living (even if below the poverty line) preventing them from slipping further into poverty, or ensuring that they are less poor than they might otherwise have been. Poverty elimination or reduction, on the other hand, refers to the situation where biodiversity resources help lift the household out of poverty by functioning as a source of savings, accumulation and asset building, resulting in a lasting improvement in income and welfare. Poverty alleviation, as used in this paper, is thus presented as a broad term that includes addressing all the dimensions of poverty outlined above, and that encapsulates both poverty prevention and poverty reduction or elimination.

2.2. Biodiversity Resources

We use the term biodiversity resources to refer to any raw or processed product that is produced from a wild biological resource that is harvested for either domestic consumption or small-scale trade, excluding marine and freshwater resources. In some instances, the resource may be cultivated or sourced from modified agro-ecosystems but cannot be regarded as a typical agricultural crop. As such it is largely synonymous with the term 'non-timber forest product' widely used in the literature (Belcher, 2003).

2.3. Safety-Net

A safety-net is generally regarded as a source of emergency sustenance in times of hardship, or a fallback option in the absence of alternatives. Safety-nets are mechanisms that mitigate the effects of poverty and other risks on vulnerable households (World Bank, 2004). In the poverty literature the term is often used to refer to public social security transfers such as unemployment benefits, child grants, pensions, subsidies, etc., as well as the informal transfer of goods between households (Devereux, 1999; World Bank, 2004). In the biodiversity resource literature, it is the existence of biodiversity resources that can be drawn on for sustenance and income in times of need that constitutes the safety net. Biodiversity resources can provide a form of natural insurance or cushion against risk when little else is available (Angelsen and Wunder, 2003; McSweeny, 2004; Paumgarten, 2005), often helping to prevent households, particularly the poorest (Takasaki *et al.*, 2004), from sinking deeper into poverty during difficult times.

3. THE CONTRIBUTION OF BIODIVERSITY RESOURCES TO RURAL LIVELIHOODS IN SOUTH AFRICA

3.1. Daily Consumptive Use

A strong body of literature indicates that rural communities, and to some extent urban ones, make extensive use of biological resources as part of the daily or seasonal consumptive uses. Most reports focus on one or two specific products in a project area, for example use of firewood (Liengme, 1983; Banks *et al.* 1996; Ham and Theron, 2001), edible fruits (van Eck *et al.*, 1997; Shackleton *et al.*, 2000; Cunningham and Shackleton, 2004), medicinal plants (Mander, 1998; Williams *et al.*, 2000; Botha *et al.*, 2004), or weaving fibres (Cunningham, 1987; Cawe and Ntloko, 1997; Gyan and Shackleton, 2005; Shackleton, 2005), to name a few. Fewer studies deal with use of all resources at a project site (Dovie *et al.*, 2002; Shackleton *et al.*, 2004a,b, Cocks and Wiersum, 2003). Random household surveys indicate that the resources used by the majority of households include fuelwood, wooden utensils, wild fruits, wild vegetables and twig and grass brushes (Table 1)

TABLE 1: Prevalence of use (mean ± SE) of wild resources from South African savannas (n = 14 villages; 30 – 60 hh per village) (from Shackleton & Shackleton 2004b)

Resource	Percent (%)	Resource	Percent (%)
Wild spinaches	95.6 ± 1.3	Bushmeat	51.6 ± 8.4
Fuelwood	95.9 ± 1.9	Wild honey	50.5 ± 10.6
Wooden utensils	95.1 ± 1.9	Medicinal plants	49.4 ± 7.5
Grass brushes	90.7 ± 4.6	Wood for ousing poles	49.0 ± 8.1
Wild fruits	88.2 ± 4.0	Thatching material	48.8 ± 9.0
Twig brushes	87.1 ± 5.1	Wild mushrooms	25.2 ± 9.2
Wood for fencing	62.0 ± 5.5	Reeds for construction	14.6 ± 6.5
Weaving materials	54.4 ± 9.6	Wood for furniture	6.7 ± 1.7
Edible insects	53.5 ± 9.5	Seeds for decoration	3.2 ± 1.8

Not only is consumption of biodiversity resources widespread, but the direct-use values are significant, representing a considerable costs saving to the State. Averaged across all households (i.e. user and non-users) it equates to approximately US$60 per household per month (Table 2). The annual, gross, direct-use value of these resources from the savanna biome alone has been conservatively estimated to be over R8 billion, most of which is unaccounted in the formal national and provincial economic statistics (Shackleton *et al.*, 2001). Nationally, R8 billion represents over 2 % of the annual government budget, at the time of calculation (Currently the exchange rate is US$1 = R9.00).

When quantified relative to all other incomes sources (cash and non-cash) the use of biological resources generally contributes between one-sixth and one-quarter of total livelihood income streams (Table 3), either for direct-use or for income generation. Frequently it may not be the primary livelihood activity. Yet it is often greater than the value of livestock husbandry and arable production combined (Shackleton *et al.*, 2001). At approximately 20% of total livelihood, the contribution of biological resources is keenly appreciated by rural communities, and the loss of these resources requires significant changes to livelihoods to cope and adapt.

Although relatively little work in South Africa has specifically disaggregated resource use and income data based on socio-economic characteristics, there are clear indicators that poorer and more isolated communities, as well as households that are less well off or headed by women, are often more dependent on wild biodiversity resources (Shackleton *et al.*, 2002a; Twine *et al.*, 2003; Shackleton and Shackleton, 2006). For example, poor households in the Kat River valley use approximately 3.3 ± 0.8 kg of fuelwood per day, whereas wealthier households use only 1.6 ± 0.3 (Shackleton and Shackleton, 2006). Corresponding figures for wild fruits are 57 ± 10 litre per year for the poorest households and 36 ± 7 for the wealthier households. This finding is not unusual and a significant number of studies across the tropics have demonstrated that it is generally the poorest households who are most directly reliant on wild natural resources for both subsistence and cash income (Clarke *et al.*, 1996; Campbell *et al.*, 1997; Qureshi and Kumar, 1998; Cavendish, 2000; Neumann and Hirsch, 2000; Cavendish, 2002).

TABLE 2: Gross annual direct use values (adjusted to 2008 values) to randomly selected rural households at different sites (averaged across users and non-users)

Province	Site	2008 Value (R)	Reference
Limpopo	Hagondo	7,810	Shackleton *et al.* (2002a)
	Mogano	15,619	
	Bushbuckridge	4,457	Shackleton and Shackleton (2000)
	Throndale	5,989	Dovie *et al.* (2002)
	Mametja	9,035	Twine *et al.* (2003)
KwaZulu-Natal	Scattered hhs*	7,910	Beukman *et al.* (1998)
	Kwajobe	6,083	Shackleton *et al.* (2002a)
	Mtubatuba	1,692	Magasela *et al.* (2001)
Eastern Cape	Scattered hhs*	7,075	Hassan & Havemann (1997)
	Fairbairn	4,404	Shackleton *et al.* (2002b)

	Ntilini	2,868	
	Tidbury	2,802	
	Cwebe	6,215	Shackleton *et al.* (2007)
	Ntubeni	16,251	
Mean		7,015 ± 1,118	

*Original data did not include all resources. Therefore, conservatively adjusted gross direct use value as 50 % of the mean value for the missing NTFPs from the other studies.

TABLE 3: Contribution of direct use of wild natural resources to household accruals (from Shackleton *et al.*, 2007)

% contribution to total hh accruals	Gross/nett	Country/Province	Site	Reference
22.0	Net	Limpopo	Manganeng	Crookes (2003)
28.2	Net	Limpopo	Makua	
19.3	Net	Limpopo	Thorndale	Dovie (2001)
15.2	Net	Zimbabwe	Romwe & Mutangi	Campbell *et al.* (2002)
17.2	Net	Zimbabwe		Cavendish (2002) in ibid.
22.0 (1993/94) *	Gross	Zimbabwe	Shindi	Cavendish (2000)
23.0 (1996/97) *				
4.0 – 20.0	Net	Cameroon	Mt Cameroon	Ambrose-Oji (2003)

(* The original values were 35 % and 37 % respectively, but we have deducted the contribution of livestock fodder and gold panning)

Similarly, a number of studies have indicated that among households headed by women wild biodiversity resources often contribute significantly more to total household income than is the case for households headed by men (Clarke *et al.*, 1996; Cavendish, 2000). Thus, women with few other sources of income on which to rely and no means to leave their family to seek employment, have transformed previously subsistence activities into a means of income generation on either an *ad hoc* or full-time basis (Shackleton and Shackleton, 1997; Shackleton, 2005).

The benefits of this use are not just home consumption but are also important in terms of cash saving. Being able to collect and use wild biodiversity resources to meet daily needs for energy, shelter, food and medicine, allows scarce cash resources to be used to secure other household needs and the accumulation of the necessary assets for a more secure livelihood. This includes education of children, investment in agricultural tools, capital for income generation activities, and the like. For example, the direct-use value of fuelwood is approximately R2,000 per household per year. Summing such a cost saving across the wide number of resources used represents several thousands of Rands per household per year (Shackleton and Shackleton, 2004a). The magnitude of the cost saving is greater to poorer households than for wealthier households simply by virtue of the reduced total income sources and sizes for poor households (Cavendish, 2000; Shackleton and Shackleton, 2006). Moreover, the cost saving has benefits not only at the household level, but also the national level (Shackleton, 2001). The role of daily use of wild natural resources in the provision of energy, food, medicine and shelter to the rural poor alleviates some of the costs that the government would incur had it to provide these services in rural areas (although at a higher social cost). Thus, the government has a vested interest in ensuring the sustainable supply and use of these resources until it is capable of providing such services.

3.2. Safety-Net Functions of Wild Biodiversity Resources

One of the most important roles of wild biodiversity resources in poverty mitigation is assisting households cope during times of stress or shock, thereby preventing them sliding deeper into poverty. This may take three forms (Barany *et al.*, 2001; Shackleton and Shackleton, 2004b). The first is the use of wild resources not usually used by that household, either species or types; e.g. use of wood poles collected from the environment for building purposes rather than the purchase of commercial poles or cement blocks. The second is increased consumption (either on a relative or absolute basis) of wild biodiversity resources already a component of their livelihood. Typically this involves substitution of purchased commodities with harvested ones; e.g. increased use of wild spinaches, or a decline in use of paraffin in favour of fuelwood (Paumgarten, 2006). Lastly, it may be the temporary sale of wild biodiversity on local and regional markets; e.g. roadside wood-carvers, and fuelwood vendors. Such trade often develops into a full-time activity in the absence of alternatives (Shackleton, 2005).

Because the safety-net function of biodiversity resources is temporally variable, little information exists regarding the prevalence of this activity throughout rural communities, and hence its contribution to poverty mitigation. Recently, Paumgarten (2006) reported that all households (100) in her sample at two sites had experienced some crisis or shock within the last two years that had affected their usual livelihood activities, and that 70% of these households had turned to wild resources to help them cope during that period. Overall, the percentage experiencing hardships and the proportion turning to biodiversity resources was not influenced by wealth of the household. However, for poor households, use of biodiversity resources as a safety-net was their second most used coping strategy, whereas for wealthy households it was the fifth most used coping strategy. Additionally, whilst some wealthy households were engaged in trade of wild biodiversity resources, none did so as a response to a short-term crisis. In comparison, several poor households opted to sell wild biodiversity resources as a safety-net during such times. Indeed, many other studies have shown that most small-scale vendors of such resources recount that the initial impetuous for them taking to vending was some household hardship, especially for those with limited or no education (Ndabeni, 2001; Rogerson and Sithole, 2001; Shackleton and Shackleton, 2004b; Shackleton *et al.*, 2008). Nearly all detailed case studies of specific biodiversity product industries or descriptions of rural household dynamics reveal the use of a coping strategy based on one or more of the three safety-net forms mentioned above, at some time or other. It is a ubiquitous phenomenon, the contribution of which to poverty alleviation or mitigation requires further examination and appreciation. Of particular concern in South Africa and neighbouring countries is the high prevalence of HIV/AIDS. The harrowing effects of this pandemic have resulted in large-scale changes in rural and urban livelihoods, and there is growing evidence that wild biodiversity resources are a key coping mechanism for affected households. For example, the incidence of consumption of wild meats (small mammals and birds) by children in vulnerable and affected households is 2 – 4 higher than in non-affected households (McGarry and Shackleton in press).

3.3. Local Trade in Wild Biodiversity Resources

There is widespread trade in biodiversity resources and products within rural communities and through external markets at local, regional, national and international scales. There are no national or regional estimates of the numbers of people or households involved in South Africa, on either an *ad hoc* or semi-permanent basis. But, by all accounts, across the numerous types of products it is probably millions (Shackleton *et al.*, 2008).

The high participation in small-scale trading in wild biodiversity resources belies the frequently poor returns to labour. Many commentators have noted that participation in such industries does not lift most participants out of poverty, but that it does play a significant role in poverty mitigation, i.e. preventing the worsening of poverty (Shackleton *et al.*, 2008). Yet, in making such a conclusion, many commentators fail to consider the alternatives available to the rural poor, locally or further afield, as they are already marginalised by poor skills, education, infrastructure, remoteness and limited access to real employment opportunities. So, whilst returns to labour are low in absolute monetary terms for some products, they frequently compare favourably to immediate alternatives of local wage labour, *assuming that it is available*, as well as providing social benefits not captured in direct economic analyses.

Although generally the cash returns to small-scale trading in wild biodiversity resources are low, it is true to say that they are also highly variable, being influenced by a host of local and external factors. A key one, not much discussed within the literature, is simply the actual amount of time the entrepreneurs devote to trading in wild biodiversity resources (Gyan and Shackleton, 2005, Shackleton 2005). The range in weekly or monthly incomes across any trading sector shows a high degree of variation, typically 3 – 5 times the mean (Krhger and Verster, 2001). For example, according to Marcus (2000) 63 % of crafters in KwaZulu-Natal earned less than R250 per month (gross) from craft, yet almost 10% earned over R1 000 per month. That is because many producers or vendors, engage in biodiversity resource product trade as a part-time activity, either a few hours a week, or at a particular time of the year. Consequently, it is not surprising that their absolute incomes are low. However, despite the low returns, individuals or households able and willing to devote energy and time to wild product trading can make their livelihoods more secure than through other land-based livelihood options, or the demoralising false security of insecure casual wage labour. For example, Botha (2001) reported some medicinal plant vendors with net incomes of over R40 000 per year, and Shackleton (2005) described the case of a single mother who had paid for her son to attend medical school from the income she earned from weaving reed mats. Typically returns per hour from biodiversity trade are 2 – 5 times greater than local wage labour rates (Table 4). Thus, for any single wild biodiversity enterprise, be it woodcarvers, medicinal plant collectors, fuelwood vendors, and the like, there are a large number that earn relatively small amounts, but there are also those that earn several or tens of thousands of Rands per year (Shackleton *et al.*, 2008).

TABLE 4: Returns per hours worked in local trade in biodiversity resources relative to the local wage labour rate

Product	Income earned per hour worked (R)	Local wage rate at same period (R)	Year	Source
Reed mats	2.50	1.50	1997	Shackleton & Shackleton (1997)
Thatch grass	7.50	1.50	1997	Shackleton & Shackleton (1997)
Marula beer	4.00	1.50	2001	Mander *et al.* (2002)
Palm brooms	3.00 – 26.00	3.50	2002	Gyan & Shackleton (2005)
Wild spinaches	17.50	7.00	2004	Mavimbela (2004)
Reed craft	6.00	3.50	2005	Pereira *et al.* (2006)
Grass brooms	3.00	2.00	2005	Shackleton (2005)
Reed mats	4.00	2.00	2005	Shackleton (2005)

3.4. Regional and International Trade

Biodiversity resources are not marketed solely at the local level, but also at wider scales, i.e. regionally, nationally and internationally. Typically these are targeted at higher value markets. The potential returns from marketing products in these distant, but higher value markets, are greater than local ones, but at times the primary benefits are more often than not captured elsewhere in the market chain so that the rural poor are seldom better off unless particular care is taken such as the pro-poor and ethical approaches of Phytotrade Africa (www.phytotrade.com). Nonetheless, they do offer opportunities to market goods and earn an income from resources and value-added products that may have no, or only small, markets locally. Many examples exist in South Africa, of which perhaps the most well known are marula (*Sclerocarya birrea*, subsp. *caffra*) products, mopane worms (*Imbasia belina*), rooibos tea (*Aspalathus linearis*), sap from several *Aloe* species (mainly *Aloe ferox*) and several medicinal plants products such as *Harpagophytum procumbens* and *Pelargonium reniforme*.

The actual cash flows to the rural harvester or producer vary enormously depending upon the nature of the resource or product, the size of the market, how the market chain is organised, as well as its length and number of stakeholders. For example, Wynberg *et al.* (2003) contrasted the financial returns and other benefits to actors along different market chains for a variety of marula products, ranging from fruit harvested and sold locally with no value addition, through to large corporates buying fruit locally and processing and exporting it. At the one end of the spectrum, cash returns to rural households were low (albeit still better than local agricultural wage labour rates), but there were no barriers to entry so large numbers were involved, especially women. At the other end, returns were 2-3 times greater but only for far fewer households and those with a vehicle were at an advantage.

In another example Newton and Vaughan (1996) provided the first examination of the stakeholders involved in the *Aloe ferox* industries in South Africa and the distribution of benefits between them. They estimated that in 1996 the value of the *Aloe ferox* industry to rural harvesters alone was R4 million per year from producing 400 – 700 tonnes of bitters, equivalent to at least R 10 million in today's terms assuming no growth of the industry. Given that there has been growth, its worth to small-scale rural harvesters is now in the region of R 12 million per year. Mark up through the retail chain suggests that the value of the entire industry in South Africa would be close to R50 million per year. The incomes to harvesters are distributed between a small number of full-time *Aloe* harvesters, or tappers, and a larger number of casual tappers who collect irregularly to supplement other income sources. Net annual incomes were in the order of R9 700 for more-or-less full-time tappers in 1996, or approximately R1 700 per month in today's terms. But with growth of the industry there is scope for a greater number of beneficiaries and better equity in income distribution if appropriate market chain models are developed.

More recently the role of payment for ecosystem services as a means of poverty mitigation has been mooted. In South Africa this is still in its infancy. However, there are some promising pilot models in the water sector, as well as for restoration of degraded lands (Mills and Cowling, 2006), which do suggest that economic viability can be attained. Under such a programme, rural land owners will be compensated by wider regional stakeholders for maintaining functional ecosystem services on their land, or rural people employed to restore those services.

These examples alone indicate that it is difficult to draw generic conclusions regarding the benefits of more distant markets for poverty alleviation. Suffice to say, it is not a silver bullet, but nor should it be ignored. In each case it has the potential to reduce poverty for a number of participating households. The participation in distant markets has some advantages over local marketing, although these do not come without potential complications. The ideal scenario is one where harvesters and/ or producers can sell their goods on both local and more distant markets, provided that there are sufficient raw materials to supply both.

3.5. Importance for the Poor

Most work to date suggests that it is the poorest members of society who are most dependent on wild biodiversity resources (Fereday *et al.*, 1997; Cavendish, 2000; Neumann and Hirsch, 2000; Beck and Nesmith, 2001; Fisher, 2004; Shackleton *et al.*, 2007). The collection of biodiversity resources is an activity that is generally available to all households, but that is more likely to be exploited by poorer groups with limited land resources and other assets, minimal education and skills, and few other income sources, contributing a greater proportion of total income to these households (Arnold, 2002; Fisher 2004; McSweeny, 2004). The low barriers to entry to the trade in many natural products means that this activity provides an important option for poor and marginalised households who would have difficulty accessing other employment opportunities, or who are less able to cope with or insure against risk than better-off households (Fisher, 2004; Takasaki *et al.*, 2004). Women in particular benefit widely from the use and sale of biodiversity resource products, as do older and less educated people who cannot compete as effectively in the job market (Falconer, 1992, 1996; Terry, 1999; Kaimowitz, 2003). In South Africa, these resources are particularly important for female-headed households, the elderly and poor urban and peri-urban women as well as rural women (Shackleton and Shackleton, 2004b; Cocks and Dold, 2004b; Shackleton, 2005; Pereira *et al.*, 2006; Mander and le Breton, 2006; Shackleton *et al.*, 2007). Recently, there is evidence that HIV/AIDS affected households, which are amongst the most vulnerable of all, are increasingly relying on wild biodiversity resources for a range of 'free' direct household needs as well as a source of cash income. Small enterprises based on forest products provide an ideal opportunity for home-based income generation; a necessity for many households given the increasing burden of care for sick household members and orphans (Barany *et al.*, 2001; Shackleton, 2005; Shackleton, 2006; Wiegers *et al.*, 2006).

4. CONCLUSIONS

The above review provides ample evidence regarding the importance of wild biodiversity resources for direct provisioning of rural households, as well as local and wider-scale income generation activities. Key conclusions are that:

- Wild biodiversity resources provide goods for daily lives and trade.
- Local markets offer only small value transactions but they are widespread, with many many participants.
- Returns per hour worked are typically greater than alternative options available to low skilled workers.
- Investing in poverty prevention is the same as wealth creation – therefore it must not be ignored.

With the world focus on poverty alleviation in developing countries, many government and external agencies overlook the importance of wild biodiversity resources and the role they play in poverty mitigation. They frequently perceive poverty alleviation initiatives and programmes as large and capital intensive, financed by external donors or entrepreneurs. Yet the data provided above show that wild biodiversity resources offer considerable value to local livelihoods, such that if they were diminished or depleted, already vulnerable and poor households would be significantly worse off. Consequently, development planners, NGOs and government departments should focus on positive means to reinforce and sustain the existing benefits. Precisely how that should be achieved varies in time and place, but a few overarching ones to start with would include:

- Recognise the multiplicity of local livelihoods, especially of the poor – they engage in many different livelihood strategies
- Provide adequate support (finances, personnel, regulations) for land management, not just fines for transgressions, nor a blind focus on the formal economy
- Consider resource supply to poor households when making zoning decisions (a new golf course, dam or residential suburb can add hours to a collection trip)
- Allow access for collection on municipal and State land, with necessary controls as required
- Strengthen local organisational and institutional capacity to manage resources, remove constraints and market products
- Make trade in natural resource an integral component of Local Economic Development Plans; don't dismiss is because it is small and informal
- Build on what is already happening; support existing and proven entrepreneurs rather than starting new groups from scratch or welfare approaches

REFERENCES

Aliber, M., 2003. Chronic poverty in South Africa: incidence, causes and policies. *World Development* 31, 473-490.

Ambrose-Oji, B. 2003. The contribution of NTFPs to the livelihoods of the 'forest poor': evidence from the tropical forest zone of south-west Cameroon. *International Forestry Review* 5, 106-117.

Angelsen, A. & Wunder, S. 2003. Exploring the forest-poverty link: key concepts, issues and research implications. CIFOR, Bogor. 58 pp.

Arnold, J.E.M. 2001. Forestry, poverty and aid. *CIFOR Occasional Paper* No. 33. Centre for International Forestry Research (CIFOR), Bogor.

Arnold, J.E.M. 2002. Clarifying the links between forests and poverty reduction. *International Forestry Review*, 4: 231-234.

Banks, D.I., Griffin, N., Shackleton, C.M., Shackleton, S.E. & Mavrandonis, J. 1996. Wood supply and demand around two rural settlements in a semi-arid savanna, South Africa. *Biomass and Bioenergy* 11, 319-331.

Barany, M., Hammett, A.L., Sene, A. & Beyhan, A. 2001. Nontimber forest benefits and HIV/AIDS in sub-Saharan Africa. *Journal of Forestry*, 99: 36-41.

Beck, T. & Nesmith, C. 2001. Building on poor people's capacities: The case of common property resources in India and West Africa. *World Development*, 29: 119-133.

Belcher, B.M. 2003. What isn't an NTFP? *International Forestry Review* 5, 161-168.

Beukman, R., Rivers-Moore, N., Mander, M., de Wit, M & Hassan, R. 1998. Economic valuation of KwaZulu-Natal woodlands based on household use: an input-output (I-O) framework. Unpubl. Report, CSIR, Pretoria.

Botha, J., Witkowski, E.T. & Shackleton, C.M. 2004. Market profiles and trade in medicinal plants in the Lowveld, South Africa. *Environmental Conservation*, 31: 38-46.

Campbell, B.M., Jeffery, S., Kozanayi, W., Luckert, M., Mutamba, M. & Zindi, C. 2002. *Household livelihoods in semi-arid regions: options and constraints*. CIFOR, Bogor. 153 pp.

Campbell, B.M., Luckert, M. & Scoones, I. 1997. Local-level valuation of savanna resources: a case study from Zimbabwe. *Economic Botany* 51, 59-77.

Cavendish, W. 2000. Empirical irregularities in the poverty-environment relationship of rural households: evidence from Zimbabwe. *World Development* 28, 1979-2003.

Cavendish, W. 2002. Quantitative methods for estimating the economic value of resource use to rural households. In: Campbell, B.M. & Luckert, M.K. (eds). *Uncovering the hidden harvest: valuation methods for woodland and forest resources*. Earthscan, London. pp. 17- 65.

Cawe, S.G. & Ntloko, S.S. 1997. Distribution, uses and exploitation patterns of *Flagellaria guineensis* Schumach. with particular reference to Port St. Johns, South Africa. *South African Journal of Botany* 63, 233-238.

Chambers, R. 1988. Poverty in India: concepts, research and reality. *IDS Discussion Paper* 241. Institute for Development Studies (IDS), University of Sussex, Brighton.

Clarke, J., Cavendish, W. & Coote, C. 1996. Rural households and miombo woodlands: use, value and management. In: Campbell, B.M. (ed.). *The miombo in transition: woodlands and welfare in Africa*. CIFOR, Bogor. pp. 101-136.

Cocks, M.L. & Dold, A.P. 2004a. A new broom sweeps clean: the economic and cultural value of grass brooms in Eastern Cape Province, South Africa. *Forests, Trees and Livelihoods*, 13: 33-42.

Cocks, M.L. & Dold, A.P. 2004b. The informal trade of *Cassipourea flanaganii* as a cosmetic in South Africa. In: Sunderland, T. and Ndoye, O. (Eds). *Forest products, livelihoods and conservation. Case studies of non-timber forest product systems. Volume 2: Africa.* Centre for International Forestry Research (CIFOR), Bogor. Pp 73-90.

Cocks, M.L. & Wiersum, K.F. 2003. The significance of plant diversity to rural households in the Eastern Cape province of South Africa. *Forest, Trees & Livelihoods* 13, 39-58.

Crookes, D. 2003. The contribution of livelihood activities in the Limpopo Province: case study evidence from Makua and Manganeng. *Development southern Africa* 20, 143-159.

Cunningham, A.B. 1987. Commercial craftwork: balancing out human needs and resources. *South African Journal of Botany* 53, 259-266.

Cunningham, A.B. & Shackleton, C.M. 2004. Use of fruits and seeds from indigenous and naturalised plant species. In: Lawes, M.J., Eeley, H.C., Shackleton, C.M. & Geach, B.S. (eds). *Indigenous forests and woodland in South Africa: policy, people and practice.* University of KwaZulu-Natal Press, Pietermaritzburg. pp. 603-626.

Devereux, S. 1999. Making less last longer: Informal safety nets in Malawi. *IDS Discussion Paper* 373. Institute of Development Studies (IDS), University of Sussex, Brighton.

Dovie, D. 2001. *Woodland resource utilisation, valuation and rural livelihoods*. M.Sc. thesis, University of the Witwatersrand, Johannesburg. 149 pp.

Dovie, B.K., Shackleton, C.M. & Witkowski, E.T. 2002. Direct-use values of woodland resources consumed and traded in a South African village. *International Journal of Sustainable Development and World Ecology* 9, 269-283.

Falconer, J. 1992. *Non-timber forest products in southern Ghana*. Natural Resources Institute, Chatham Maritime.

Falconer, J. 1996. Developing research frames for non-timber forest products. Experience from Ghana. In: Ruiz Perez, M. and Arnold, J.E.M. *Current issues in non-timber forest product research. Centre for International Forestry Research* (CIFOR), Bogor. Pp 143-160.

FAO. 2003. *State of the World's Forests 2003*. Part II. Selected current issues in the forest sector. Food and Agricultural Organisation (FAO), Rome.

Fereday, N., Gordon, A. & Oji, G. 1997. Domestic market potential for tree products from farms and rural communities: Experience from Cameroon. *NRI Socio-economic Series* 13. Natural Resources Institute, The University of Greenwich, Greenwich.

Fisher, M. 2004. Household welfare and forest dependence in Southern Malawi. *Environment &Development Economics*, 9: 135-154.

Government of South Africa. 2000. *Integrated Rural Development Strategy*. Government Printer, Pretoria. 53 pp.

Gyan, C.A. & Shackleton, C.M. 2005. Abundance and commercialisation of *Phoenix reclinata* in the King Williamstown area, South Africa. *Journal of Tropical Forest Science* 17, 334-345.

Ham, C. & Theorn, F. 2001. Community forestry resources: a case study of selected woodlots in the Eastern Cape province. *Southern African Forestry Journal* 191, 65-74.

Hassan, R. & Havemann, J. 1997. The values and rates of harvesting natural forest and woodland products for direct use by communities in the Eastern Cape province. Unpubl. Report, Development Bank of Southern Africa, Midrand.

Kaimowitz, D. 2003. Not by bread alone ... Forests and rural livelihoods in sub-Saharan Africa. In: Oksanen, T., Pajari, B. & Tuomasjakka, T. 2003. (eds). *Forests in poverty reduction strategies: capturing the potential*. European Forest Institute, Torikatu. pp. 45-63.

Kepe, T.V. 2002. *Grassland vegetation and rural livelihoods: a case study of resource value and social dynamics on the wild coast South Africa*. PhD thesis, University of the Western Cape, Cape Town. 267 pp.

Kingdon, G.G. & Knight, J. 2003. Well-being poverty versus income and capabilities poverty. *CSAE WPS*/2003-16. Global Poverty Reduction Group, Centre for the Study of African Economics (CSAE), University of Oxford, Oxford.

Kotze, D.C. 2001. Promoting crafts woven from wetland plants: Guidelines for fieldworkers and other stakeholders. Unpublished report, Institute of Natural Resources (INR) and Centre for Environment and Development (CED), University of Natal, Pietermaritzburg.

Krhger, S. & Verster, R. 2001. An appraisal of the Vulamehlo handcraft project. *Development southern Africa* 18, 239-252.

Lawes, M., Eeley, H., Shackleton, C.M. & Geach, B. (eds). (2004). *Indigenous forests and woodlands in South Africa: policy, people and practice*. University of KwaZulu-Natal Press, Pietermaritzburg. 863 pp.

Liengme, C.A. 1983. A study of wood use for fuel and building in an area of Gazankulu. *Bothalia* 13, 501-518.

Magasela, S.B., Watson, H.K., Mhango, B.J., Bob, U. 2001. Direct use values of non-timber forest products from the rural areas of Hlabisa, KwaZulu-Natal. Unpubl. Report, geography department, University of Durban –Westville, Durban.

Mander, M. 1998. *Marketing of indigenous medicinal plants in South Africa: a case study of KwaZulu-Natal*. FAO, Rome. 151 pp.

Mander, M., Cribbins, J., Shackleton, S.E. & Lewis, F. 2002. The commercial marula industry in South Africa: a sub-sector analysis. Unpubl. Report, Institute for Natural Resources, Pietermaritzburg. 71 pp. (http://www.nerc-wallingford.ac.uk/research/winners/literature)

Mander, M. & le Breton, G.L. 2006. Overview of the medicinal plant industry in southern Africa. In: Diederichs, N. (ed). Commercialising Medicinal Plants: A southern African Guide, 1-9. Sun Press, Stellenbosch, South Africa.

Marcus, T. 2000. Crafting in the context of AIDS and rural poverty: a livelihood strategy with prospects. *Transformation* 44, 17-35.

May, J. (ed). 1999. *Poverty and inequality in South Africa: meeting the challenge*. David Phillip, Cape Town. 304 pp.

Mavimbela, S. 2004. *Local level trade in wild spinaches in Nelspruit town, Mpumalanga Province of South Africa: opportunities and constraints.* Honours dissertation, Rhodes University Grahamstown. 18 pp.

McGarry, D.K. & Shackleton, C.M. in press. Is HIV/AIDS jeopardising biodiversity? *Environmental Conservation.*

McSweeny, K. 2004. Tropical forests product sale as natural insurance: The effects of household characteristics and the nature of shock in Eastern Honduras. *Society and Natural Resources* 17, 39-56.

Mills, A.J. & Cowling, R.M. 2006. Rate of carbon sequestration at two thicket restoration sites in the Eastern Cape, South Africa. *Restoration Ecology*, 14: 38-49.

Ndabeni, L. 2001. Forest-product SMMEs in the Wild Coast Spatial Development Initiative. *South African Geographical Journal* 83, 159-166.

Neumann, R.P. & Hirsch, E. 2000. *Commercialisation of non-timber forest products: review and analysis of research.* CIFOR, Bogor.

Newton, D.J. & Vaughan, H. 1996. *South Africa's* Aloe ferrox *plant, parts and derivatives industry.* TRAFFIC, Johannesburg. 61 pp.

Oksanen, T. & Mersmann, C. 2003. Forestry in poverty reduction strategies – An assessment of PRSP processes in sub-Saharan Africa. In: Oksanen, T., Pajari, B. and Tuomasjukka, T. (Eds.). *Forestry in poverty reduction strategies: Capturing the potential.* EFI Proceedings No. 47. European Forest Institute, Joensuu. Pp 121-158.

Paumgarten, F. 2005. The role of non-timber forest products as safety-nets: a review of evidence with a focus on South Africa. *Geojournal*, 64: 189-197.

Paumgarten, F. 2006. *The significance of the safety-net of NTFPs in rural livelihoods, South Africa.* M.Sc. thesis, Rhodes University, Grahamstown. 25 pp.

Pereira, T., Shackleton, C.M. & Shackleton, S.E. 2006. Opportunities and constraints to trade in reed-based craft products in rural villages in the Eastern Cape, South Africa. *Development southern Africa*, 23: 477-496.

Qureshi, M.H. & Kumar, S. 1998. Contributions of common lands to household economies in Haryana, India. *Environmental Conservation*, 25: 342-353.

Rogerson, C.M. & Sithole, P.M. 2001. Rural handicraft production in Mpumalanga, South Africa: organization, problems and support needs. *South African Geographical Journal* 83, 149-158.

Scholes, R. & Biggs, R. (eds). (2004). *Ecosystem services in Southern Africa: A regional assessment.* CSIR, Pretoria.

Shackleton, C.M. 2001. Re-examining local and market orientated use of wild species for the conservation of biodiversity. *Environmental Conservation* 28, 270-278.

Shackleton, C.M., Dzerefos, C.M., Shackleton, S.E. & Mathabela, F.R. 2000b. The use of and trade in indigenous edible fruits in the Bushbuckridge savanna region, South Africa. *Ecology of Food and Nutrition* 39, 225-245.

Shackleton, C.M. & Shackleton, S.E. 1997. *The use and potential for commercialisation of veld products in the Bushbuckridge area.* Unpubl. rep., to DANCED, Pretoria. 125 pp.

Shackleton, C.M. & Shackleton, S.E. 2004a. The use of woodland resources for direct household provisioning. In: Lawes, M.J., Eeley, H.C., Shackleton, C.M. & Geach, B.S. (eds). *Indigenous forests and woodland in South Africa: policy, people and practice.* University of KwaZulu-Natal Press, Pietermaritzburg. pp. 195-226.

Shackleton, C.M. & Shackleton, S.E. 2004b. The importance of non-timber forest products in rural livelihood security and as safety nets: evidence from South Africa. *South African Journal of Science* 100, 658-664.

Shackleton, C.M. & Shackleton, S.E. 2006. Household wealth status and natural resource use in the Kat River Valley, South Africa. *Ecological Economics*, 57: 306-317.

Shackleton, C.M., Shackleton, S.E., Buiten, E. & Bird, N. 2007. The importance of dry forests and woodlands in rural livelihoods and poverty alleviation in South Africa. *Forest Policy & Economics* 9: 558-577.

Shackleton, C.M., Shackleton, S.E. & Cousins, B. 2001. The role of land-based strategies in rural livelihoods: the contribution of arable production, animal husbandry and natural resource harvesting. *Development southern Africa*, 18: 581-604.

Shackleton, C.M., Shackleton, S.E., Ntshudu, M. & Ntzebeza, J. 2002b. The role and value of savanna non-timber forest products to rural households in the Kat River valley, South Africa. *Journal of Tropical Forest Products* 8, 45-65.

Shackleton, S.E. 2005. The significance of local level trade in natural resource products for livelihoods and poverty alleviation in South Africa. PhD thesis, Rhodes University. 287 pp.

Shackleton, S.E. 2006. Forests as safety nets for mitigating the impacts of HIV/AIDS in southern Africa. Forest Livelihood Briefs, No. 4, October 2006. CIFOR, Bogor.

Shackleton, S.E., Campbell, B., Lotz-Sisitka, H. & Shackleton, C.M. (2008). Links between the local trade in natural products, livelihoods and poverty alleviation in a semi-arid region of South Africa. *World Dev.* 36, 505-526.

Shackleton, S.E., Shackleton, C.M., Netshulvhi, T.R., Geach, B.S., Ballance, A. & Fairbanks, D.H. 2002a. Use patterns and value of savanna resources in three rural villages in South Africa. *Economic Botany* 56, 130-146.

Sunderlin, W.D., Angelsen, A., Belcher, B., Burgers, P., Nasi, R., Santoso, L. & Wunder, S. 2005. Livelihoods, forests and conservation in developing countries: an overview. *World Development*, 33: 1383-1402.

Takasaki, Y., Barham, B.L. & Coomes, O.T. 2004. Risk coping strategies in tropical forests: floods, illnesses, and resource extraction. *Environment & Development Economics*, 9: 203-224.

Terry, E.M. 1999. The economic and social significance of the handicraft industry in Botswana. Unpublished PhD thesis, School for Oriental and African Studies, University of London, London.

Twine, W., Moshe, D., Netshiluvhi, T. & Siphungu, V. 2003. Consumption and direct-use values of bio-resources used by rural households in Mametja, a semi-arid area of Limpopo Province, South Africa. *South African Journal of Science*, 99: 467-473.

van Eck, H., Ham, C. & van Wyk, G. 1997. Survey of indigenous tree uses and preferences in the Eastern Cape region. *South African Forestry Journal* 180, 61-64.

Wiegers, E., Curry, J., Garbero, A. & Hourihan, J. 2006. Patterns of vulnerability to AIDS impacts in Zambian households. *Development and Change*, 37: 1073-1092.

Williams, V.L., Balkwill, K. & Witkowski, E.T. 2000. Unravelling the commercial market for medicinal plants and plant parts on the Witwatersrand, South Africa. *Economic Botany* 54, 310-327.

World Bank. 2000. *World Development Report (WDR) 2000/2001: Attacking poverty*. World Bank Group, Washington D.C. http://siteresources.worldbank.org/intpoverty/resources/WDR/.

World Bank. 2004. Social safety nets. World Bank social protection. http://www1.Worldbank.org/sp/safetynets/

Wynberg, R. (2002). A decade of biodiversity conservation and use in South Africa: tracking progress from the Rio Earth Summit to the Johannesburg World Summit on Sustainable Development. *S. Afr. J. Sci.* 98, 233-243.

Wynberg, R.P. & Laird, S.A. 2007. Less is often more: governance of a non-timber forest product, marula (*Sclerocarya birrea* subsp. *caffra*) in southern Africa. *International Forestry Review*, 9: 475-490.

Wynberg, R. P., Laird, S.A., Shackleton, S., Mander, M., Shackleton, C., Du Plessis, P., den Adel, S., Leakey, R.R.B., Botelle, A., Lombard, C., Sullivan, C., Cunningham, T. & O'Regan, D. 2003. Marula Policy Brief. Marula commercialisation for sustainable and equitable livelihoods. *Forests, Trees and Livelihoods*, 13: 203-215.

3

Bio-energy Conservation Policy Options in Uganda: A Bio-economic Perspective

Buyinza Mukadasi[1*] and Masaba Sowedi[2]
[1]Department of Community Forestry & Extension, Faculty of Forestry and Nature Conservation
Makerere University, P.O.Box 7062 Kampala, Uganda
[2]Institute of Environment and Nature Resources,
Makerere University, P.O.Box 7062 Kampala, Uganda
*E-mail: buyinza@forest.mak.ac.ug

This study examined the benefits and costs of selected policy options for increasing fuelwood supplies or decreasing fuelwood demand in Hoima district, Uganda. On the supply side, a benefit-cost analysis was carried out on a government sponsored tree farming project. In order to reduce the demand for fuelwood, two demand-side options were considered, namely, introduction of an improved energy-efficient woodstove, and the substitution of a kerosene stove for a traditional woodstove. Greater understanding of the linkages among these factors requires a systems approach. We have proposed such an approach using a non-linear dynamic programming model to explore the system behaviour of forest degradation. Results showed that tree-farming is one of the possible approaches to increase the supply of fuelwood (energy), while the woodstoves and kerosene substitution are policies that reduce the demand for fuelwood. This can help to alleviate the rural energy shortage and take some pressure off existing protected forest areas. The tree cover in the forest areas declined by 6% in the BASE scenario, 4.8 % in POPG scenario and 4.7% in TECH scenario, indicating an overall trend of forest degradation in the Hoima district under each of these scenarios. Reductions in the population growth rate, introduction of improved agricultural technology and increase in the prices of major agricultural crops can help slow down the rates of forest decline. This study did not analyse the wider energy planning program that would be needed to understand accurately the various alternatives available in Uganda.

KEY WORDS: *Fuelwood, Woodstoves, Tree-farming, Kerosene, Uganda*

1.0 INTRODUCTION

Fuelwood occupies an enviable place for providing many people, especially the poor and rural households, with a primary source of energy (Shackleton, 1998). Fuelwood is used for both domestic and industrial purposes in rural and urban areas of most economies of the developing world (Tietema, 1993). Wood consumed annually for fuel energy in sub-Saharan Africa increased from 1,500 million

m^3 to 3,500 million m^3 between 1950 and 2002 (Moyini and Muramira, 2002).

The energy sector is characterised by a heavy dependence on biomass resources, and studies have demonstrated that woodfuel (firewood and charcoal) accounts for more than 90% of energy used in Uganda (MoEMD, 2002). Traditionally, the majority of Ugandans depend on fuelwood for their domestic heating, cooking and lighting purposes. Total wood production (monetary and non-monetary) has registered a steady increase over the period between 1995 and 2000.

The importance of fuelwood becomes clearer when for example kerosene is substituted for charcoal in urban households, which would result in an increase in the national import bill by US$180 million (Ushs. 324 billion) annually (Falkenberg and Sepp, 1999). Such strategy will in fact lead to loss of jobs by poor people who are involved in fuelwood income generating activities.

It has been argued that in Southern Africa, there is a fuelwood 'crisis' in many areas (Dovie *et al.*, 2004). The 'crisis' refers to a shortage in fuelwood supplies to meet households' requirements and is manifested as an ever increasing distance covered by rural households to obtain and harvest wood suitable for fuel energy. With increasing deforestation, the distance travelled to collect firewood increases. This affects the productivity of households, when the time spent on collecting firewood would have been used for productive activities. The average distance travelled by households to collect firewood has increased dramatically between 1992 and 2000 from 0.06 km to 0.73 km (MoEMD, 2002). Fuelwood is one of the commodities that households sell to raise income especially in areas where free access to forestry resources is declining. Unfortunately, the income from sale of fuelwood not clearly recorded in the national accounting system as part of the forestry contribution to household income. Hence, forestry remains just recognised as extremely high economic importance but mainly in the informal sector (MWLE, 2002b).

1.1 Conceptual framework

Factors affecting the pressure on fuelwood use have been oversimplified as type and quality of wood, stove technology, population growth and availability of substitutes (Jacovelli and Caevalho, 1999; Kalumian and Kisakye, 2001). However, their link to supply-side approach to providing fuelwood through tree-farming projects and demand-side options aimed at reducing firewood demand by either introducing more efficient woodstoves and use of kerosene as a substitute fuel are still unclear and with little empirical justification. The objective of this paper is therefore to examine, using revealed preference technique, the various policy options for meeting the demand for fuelwood and rural energy supplies in Hoima district, Uganda.

Traditionally, the major land-use types in Hoima district are dichotomized into agriculture for food production and forestry for production of wood and fiber. Also included in the latter land-use type, is the protection forestry for stabilizing fragile upland systems and water catchment areas. However, the two-faced aims of forestry development also call for attention to the provision of the human needs which are: to increase rural people's self-sufficiency in producing food, fodder, fuel and timber on their cultivated lands; in addition to conserve the forest resource and land resources by reducing the current pressure on them to supply these commodities. The growing population demands a sustainable supply annual food, fodder, fuel and timber.

The gradual but steady population growth in Hoima district has inevitably caused a gap between the demand for fuelwood and sustainable supplies. The exponential growth in demand with population growth and growing burden placed on users of having to search ever further a field for fuelwood and of having to divert crop and animal residues needed for soil working or as livestock feed to fuel use have given rise to further rural problems.

Secondly, the existing wood stock has been widely mined to meet the food, fodder, fuelwood, and timber demand, that there remains no other feasible alternative to wood fuel other than other bio-mass such as crop residues and dung, and that the principal means of averting growing shortages, and the attendant deforestation and human suffering, was to initiate widespread planting of additional trees. As a consequence, a very large part of the initial investment in community forest management was in form of afforestation projects to increase fuelwood supplies. The causes of fuelwood shortages, consequences and possible responses are shown in Figure 1.

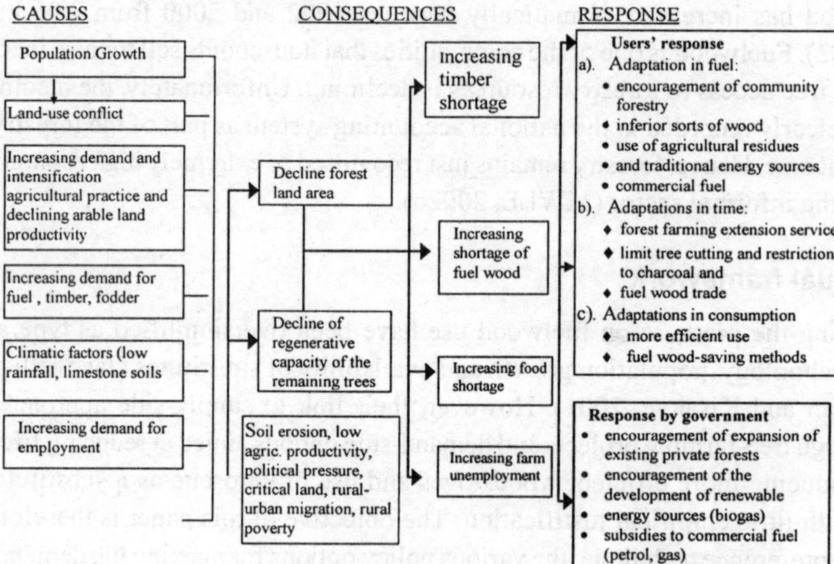

FIGURE 1: Diagnosis of the causes, consequences, and response to forest product shortages

2.0 METHODS AND MATERIALS

2.1 Description of study area

The study site was Kigorobya sub-county, Hoima district (1°00'- 2°00'N; 30°30'- 31°45'E). The district covers an area of 59,327 km^2 of which 22,686 km^2 is occupied by Lake Albert. Forests constitute a relatively smaller portion of about 12% (7123 sq. km) of the total area. The district generally has a pleasant climate with small variations in temperature, humidity and winds throughout the year. Rainfall ranges from 700-1000mm per annum and mean annual temperature is 28°C with a range of 15-32°C.

Vegetation is very varied ranging from medium altitude moist forests through forest/savannah mosaic, swamp to post cultivation communities. Grass savannah is derived from forests, wooded savannah and thicket by repeated cultivation, grazing and burning. They occur in sub-counties of Buseruka and Kigorobya on off-shore tips of Lake Albert. The soils are mainly weathered basement, yellowish-red clay loams on sedimentary beds. There are also complex formations of the pre-cambrian age, which consist mainly of metamorphic and igneous rocks, largely composed of gneisses and granites.

Hoima district has a total human population of 340,000, with a high population density (> 300 persons/km^2 in the west and approximately 200 in the east) (UBOS, 2002). Subsistence agriculture has remained the major economic activity contributing 48% on districts gross production. Approximately 83% of the population depend on agriculture for their livelihood with a small landholding size (0.48 ha per family). Given the limited opportunities for rural employment and low agricultural production, a few households have migrated to other adjoining districts.

According to the land-use pattern, the study area has permanent crops cover (24%), savannah and pasture cover (11%), woodland and forest (28%) and others (43%) (MWLE, 2001b). Susceptible to accelerated soil erosion and declining soil fertility, the sources of livelihood of the rural people have been threatened. Fuelwood is an important form of rural energy in Hoima district, nearly 97 percent of all household cook with fuelwood at least part of the time. The average household fuelwood consumption is 65 kg/week/family; in addition, rural households burn small amounts of charcoal and agricultural residues (Jacovelli and Caevalho, 1999; Kalumian and Kisakye, 2001).

2.2 Data collection and analysis

The field work was conducted during September – December, 2004, and three data collection methods were employed, namely surveys, open-ended interviews and review of project records and other published literature. The interviews were combined with simple Participatory Rural Appraisal (PRA) tools in order to ease the acquisition of information (Dove *et al.*, 2004). Data collection emphasised household approaches rather than community level enumeration because fuelwood use takes place at the household level. The survey included 120 rural and urban households (63% sampling intensity).

Another sample consisted of 20 cottage industry firms including bricks burners, tobacco curers, bakeries, restaurants, producers of sugarcane molasses, ceramics and pottery.

The study was based on data from a large survey of the demand for fuelwood and an analysis of options for meeting that demand. This use of large-scale survey was intended to determine the magnitude of fuelwood demand and the various factors that affect this demand. The demand survey investigated consumption and uses of wood and charcoal fuels. These questionnaires elicited data on time spent in collecting firewood; proportion and source of firewood collected; and proportion grown for own use. All the amounts of fuelwood were converted into actual weights because of lack of standardization of commonly used units.

Special emphasis was placed on preferences for the types of woodstoves and attitudes toward tree growing in order to develop policies directed at solving the wood fuels shortage. Some questions in the consumption survey also were directed toward consumption of other biomass fuels such as bamboo, rice husks, molasses, and animal dung, as well as conventional such as kerosene, gas, and electricity, to determine the extent to which these currently serve as substitutes for fuelwood.

Another survey dealt with staff in National Forestry Authority (NFA), fuelwood sellers, charcoal burners, and tree farmers. The questionnaires for the fuelwood tree farmers covered background information of the tree farmer; size and quality of the tree farm site; tree spacing, planting and harvesting period; labour requirements for clearing land, planting, tending, and harvesting trees; tree growth and mortality; economic returns; use of co-products such as leaves, fruits, and nuts; market for the output; extension of tree farming beyond the project area; tree farmers' evaluation of the program and participating institutions; prospects and challenges of the project.

Some fuelwood was collected by household members and some was bought in the market. Thus, market prices were available for valuing purchased fuelwood, although there was considerable variation in prices. However, a shadow price was estimated for collected, own-consumption fuelwood. The shadow price of fuelwood was derived using on *revealed preferences* method. We assumed that the value people place on wood fuel was at least equal their costs in collecting it as a free good, including travel time to collection sites and related transportation costs. Thus, transport time was valued at its opportunity cost in terms of foregone labour. Based on relative productivity and opportunity cost factors, a child's labour time was valued at one-half of the adult wage rate. In cases where there was extensive surplus labour due to underemployment or unemployment, the opportunity cost of labour time for collecting fuelwood was low. Transport costs were very low because family members travelled on foot to collect fuelwood unless there were relatively affluent and grow fuelwood on lands their own.

2.3 Conceptual framework

Fuelwood use statistics often greatly underestimate true consumption because most wood is gathered directly by users rather than being purchased. The shortage of fuelwood leads to economic welfare loses, as well as environmental degradation from over-cutting of trees.

Fuelwood was measured by weight (in kg). Villagers would normally harvest and collect fuelwood in head loads, pick-up vehicles and wheelbarrows. Households provided their wood consumption data their records and observations. The mass of a wheelbarrow load (standard measure of pricing) of fuelwood was approximately 45 kg.

Based on the survey results and other secondary sources, the following data and assumptions were used:

- Discount rate is set at 15 percent per annum.
- Pure stands yielded about 123 m^3 per hectare annually.
- Rural adult wage rate is Ushs. 3,700 per day (US$ 1.87) (8 work hours / day; child labour is rated half the adult wage; proxy market values were estimated in case of wages in-kind).
- Kerosene stoves cost Ushs. 17,000 (US$ 8.6); improved woodstoves cost Ushs. 20,000 (US$ 10); domestic kerosene costs Ushs. 1,200 (US$ 0.6) (these prices exclude periodical taxes and subsidies).

  Tree-farming inputs and costs based on survey data are presented in Table 1.

TABLE 1: Farm inputs of tree farming (*Pinus caribaea Var. hondurensis*) enterprise in Hoima district, Uganda (ha^{-1}).

Operation or activity	Year	Labour requirement (man-days equivalent)	Input cost (Ushs)
1. Land clearing and preparation	1	35	
2. Seedlings (5,000/ha, spacing: 1 x 2 m)	1	—	86/seedling
3. Lining, digging, and planting seedlings	1	35	
4. Seedling replacement in year 2 based on 20% mortality in year 1			86/seedling
5. Replanting seedlings	2	5	
6. Fertilizer purchase	1		18,500
	2		267,000
7. Fertilizer application	1	7	
	2	3	
8. Weeding and brush removal	1	10	
	2	13	

Operation or activity	Year	Labour requirement (man-days equivalent)	Input cost (Ushs)
9. Maintenance	3 to 8	4	
10. Weeding and single coppice tending	5	4	
11. Harvesting	5 to 9	1.6/solid m^3	
12. Cost of loan administration	1		48,000
13. Cost of nursery establishment	1		3,200,000
	2		65,000
14. Cost of collecting loan payment	5, 6, 7, 8		2,700

Labour requirements

Year	2	3	4	5	6	7	8	9
Man-days	21	4	4	156	4	4	4	148

Note: Exchange rate at survey time (2004) US$1 = Ushs 1980

Costs are incurred in form of purchase of purchase of factor inputs such as seedlings, servicing loans, wages for labour, rent for land, and other payments made to the different factors of production that are employed by the tree farming enterprise.

2.4 The tree farming practice

The structural model (Figure 2) shows that fuelwood consumption is related to food consumption, in that for each unit of consumed food, a certain amount of energy is needed for preparation. The income and availability of fuel determine the actual wood use. The abundance of fuel increases fuel consumption while at the other hand; a shortage will be compensated through a rational use or a decrease of use. The high incomes result in more consumption pattern of food, which needs more energy for food preparation, in cases where households have substantial incomes the wood energy is substituted by commercial fuel sources.

The private tree farming has helped to generate additional employment and income in the rural areas. The local wood production areas take many forms: roadside plantings, trees planted along the borders of agricultural fields, cultivated areas with a mixture of agricultural and tree crops, and tree crop plantations of commercial species. These wood production areas are often of prime importance for the local population, as they enable the private collection of fuel and construction wood, thus obtaining wood products without cash outlays. In many statistics this subsistence wood collection has often been unrecorded.

The objectives of intensification in forestry are to maximize the forest functions for meeting the people's demand and maintain the environmental quality. It should be developed based on both

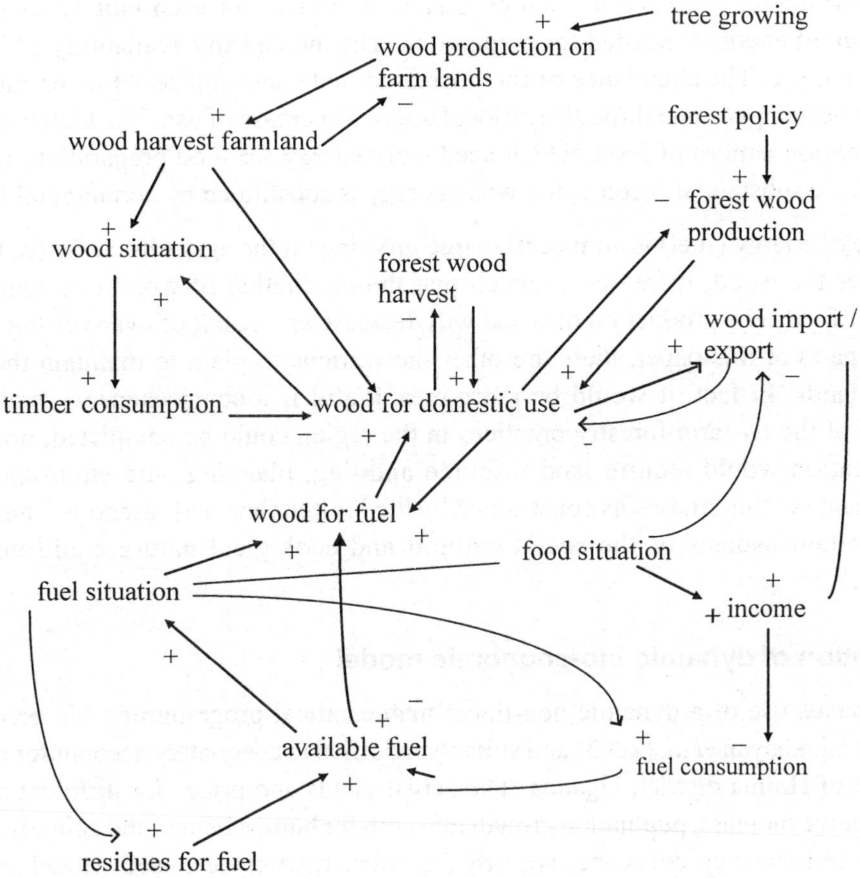

FIGURE 2: Structural model of the fuel production and consumption

forest ecosystem and human social ecosystem. Otherwise the created forest management regimes will not relevant to the existing conditions and requirements, and hence, it will fail to achieve its objectives. The components of forest ecological system are climate, soil, water, plants, and animals, meanwhile the components of the human social ecosystem are population, technology, socials structure, and ideology. Before the intensified forest management is developed, the roles of each of the components and their dynamics interactions should be understood.

The other interlinked problem is that apart from the wood for the fuel, wood is also used as timber construction and for several agricultural uses. The fuelwood is gathered from the own lands and the forest lands but it is assumed that the most important source of fuelwood are the agricultural. The production figures of wood on the agricultural lands vary but are substantial. For the regions where state forests are scarce, more fuelwood is produced on the own agricultural fields.

The fuel consumption is related to food consumption, in that for each unit of consumed food, a certain amount of energy is needed for preparation. The income and availability of fuel determine the actual wood use. The abundance of fuel increases fuel consumption while on the other hand a shortage will be compensated through a rational use or a decrease of use. The high incomes result in more consumption pattern of food, which need more energy for food preparation, in cases where households have substantial income; the wood energy is substituted by commercial fuel sources.

The shortage of energy (fuel) is an incentive tree growing on the agricultural fields, the prices and time to gather the wood, increases so much that the production of wood can compete with the production of food. The production of wood will decrease as a result of over-cutting. As explained in the other parts of this paper, there are other motivations to plant to maintain the trees on the agricultural lands. In fact, it would be more meaningful if a comprehensive evaluation of the performance of the on-farm forestry practices in the region could be completed, however, such a deeper evaluation would require land resource auditing, planning, and environmental impact assessment, and yet this study was constrained by the limited time and research funding, therefore several important aspects of the social-cultural and ecological nature could not be studied exhaustively.

2.5 Description of dynamic bio-economic model

This study makes use of a dynamic non-linear mathematical programming bio-economic model developed by Sankhayan *et al.* (2003) and suitably modified to adequately account for the conditions characteristic of Hoima district, Uganda. The actual yields and prices for different crops, growth rates of vegetative biomass, population growth rate both for human beings and animals, consumption rates of crops and forest products, etc., specific for Hoima district were used. Model calibration was performed to test the ability of the model to produce results in respect of area under different crops and density of vegetative biomass, reasonably close to those estimated on the basis of field surveys for the base year 2004. The model maximizes an aggregate utility objective function consisting of the following main components at village level:

(i) net cash income from various production activities;

(ii) leisure time for family labour; and

(iii) food grains, cash, fuel wood and fodder requirements for human and animal population.

These components, assumed to be exogenous and progressively increasing constraints with population growth enter the objective function indirectly through their hierarchical attainment. By using a positive wage rate as the opportunity cost of unused family labour, households have been assumed to have a reserve wage below which leisure is preferred over work. Due to lack of appropriate data to account for risk, it could not be incorporated in the model. The description of the model is as follows:

If the aggregate total net return (U) at time t is expressed as follows:

The objective function considered in the model is to:

$$Maximize \sum_{t=1}^{T} 1/(1+\tau))^{t} U_{t} \tag{1}$$

where

$$U_{t} = \sum_{i}^{u} \sum_{j=1}^{s} (a_{ijt} y_{ijt}^{ca}) + \sum_{i=u+1}^{v} (a_{it} y_{it}^{la}) + \sum_{i=v+1}^{w} (a_{it} y_{it}^{fa}) + \sum_{i=w+1}^{n} (a_{it} y_{it}^{lsa}) \quad t = 1, 2, \ldots, T \tag{2}$$

and;

$$a_{ijt} = (p_{it} q_{ijt}) - \sum_{r=1}^{m} (c_{rjit} x_{jit}),\ a_{it} = (p_{it} q_{it}) - \sum_{r=1}^{m} (c_{rit} x_{it}), \tag{3}$$

The abbreviations used in the above relations are as follows:

y_{ijt}^{ca} = level of ith crop activity grown in j[th] type of land unit during the tth year.

y_{it}^{la} = level of ith livestock and livestock product activity during t[th] year.

y_{it}^{fa} = level of ith forest activity.

y_{it}^{lsa} = level of ith leisure activity.

q_{ijt} = yield of the ith crop from jth land category and t[th] year.

q_{it} = yield of the ith non-crop activity during the tth year.

x_{ijt} is the quantity of rth resource per unit of the associated activity.

and are the per unit net returns from ith crop activity from j[th] land, and i[th] activity other than crop activity during the tth year.

p_{it} and c_{rjit} are the per unit prices of products and inputs.

τ= annual percent discount rate, taken as 5% per annum in this study.

2.6 Model assumptions and constraints

Households are assumed to make decisions under a number of constraints. Some of these constraints are:

- Annual land availability constraints were used according to relatively homogenous land units since annual crop rotations were considered as the major crop activities.

- Manpower and labor-force constraints (man equivalent days) are considered according to months, which increase with growth of population over the run of the model.
- It was assumed that the production function refers to "production + purchase – consumption - sale" and must be equal to zero during each production period.
- The tree growing enterprise is profit motivated whereby cost minimisation and profit maximization is the main goal.
- Enterprise operational capital during the current period is assumed to be a fixed percent of the cash proceeds of the previous year that can be supplemented through borrowing at the current rate of interest (taken to be 12%). Constraints are incorporated for important fertilizer types requiring that their use should not exceed the amount purchased.
- Some equations for variables such as crop area, land area cleared/abandoned, loss of vegetative biomass due to grazing, extraction of forest products and forest fires[1], and tree biomass in the uncultivated and cropland in each land category are also provided.
- Some of the other important constraints used in the model related to livestock carrying capacity, remaining stock of tree biomass in each land category at the end of the model period.

Some important relations, such as, stock of vegetative biomass and human and animal populations will, however, be described in more detail.

2.7 Activities used in the model

- **Production, consumption, sale and purchase activities:** A number of activities for production, consumption, sale and purchase are used for the crop, animal, forest and miscellaneous products. The level of consumption activity is assumed to be equal to annual average requirement per head times the population.
- **Land clearing/abandoning activities:** To enable addition of new land for cultivation from the uncultivated area, land clearing activities have been incorporated separately for each category of land. Abandoning of cultivated land activities are also allowed to get activated if needed. Of land clearing and abandoning activities, only one can enter into the model at a time.
- **Labour-hiring activities:** The model provides for labour hiring for all production activities for each of the 12 months when profitable. Depending upon the availability of employment opportunities, limited hiring out activities of labour during certain months is also incorporated.

1. Loss due to forest fires was treated as a fixed proportion of the stock of vegetative biomass during the previous year.

Population simulation model

The population projection model is a statistical extrapolation, it indicates what would be the expected trend under the prescribed assumptions. The more realistic the assumptions, the shorter the time of the projection. The longer the time period, the less likely is the predictive accuracy, but the more valuable are the projections towards the identification of the factors which may influence the future trend.

In order to achieve the study purpose, the devised models were expressed in an exchangeable medium based on definite assumptions, and the assumptions of the population sub-system model include:

a) The population increase is followed by an increase in the total labor-force, increase in the unemployment rate, declining incomes among rural dwellers (unemployed farmers), declining arable land-farmer ratio which leads to pressure on the forest and land resources, increasing the consumption of basic needs (food, fuelwood, and timber);

b) The total population is the state variable (level) which depends on the birth and death rates which are functions of the number of potential parents, population size and number of children born per woman;

c) Total population depends on the natural growth rate (birth and death) and the migration rate (in - out). The annual increase of population in Gunung Kidul is low (0.1%), initially and declines to - 0.08 (scenario value) after time (t+n);

d) The population comprises of children (0-15 years) with mortality rate 0.043%, adults (16 - 24) mortality 0.004%, parents (25 - 49) mortality rate 0.005%, and old (> 50 years) mortality rate 0.23%;

e) To predict the trend of rural laborforce, the proportion of rural dwellers to total population is assumed to be 80% at the beginning of the simulation time and will decline to 54%, this is because of the rural-urban migration of the young population in search of "white collar jobs" in the urban areas;

f) Rural unemployment increases but will to fall due to the success of the community forestry which is formulated to increase the job opportunities in the rural areas;

g) Rural off-farm employment will increase from 12% to 18% due declining land-man ratio, declining agricultural production and farmers incomes, and perhaps better wages and remuneration terms in the off-farming employment;

h) The population will increase until a time (t) when family planning and transmigration measures have been successful to control birth rate and transmigration program. In the year 2010, it is assumed that the total population, the total work force, the rural labor force, the rural unemployment will decline due to the transmigration and birth control programs, and the government instituted educational public awareness program;

i) The consequence of the increasing population is the emergence of decreasing farmer-land ratio, a rise in people's basic needs in the form of food and fodder, the wood for fuel and timber. The scarcity of these basic needs has encouraged increasing out-migration, massing laborforce that generates more disguised, under-employed and unemployed people, and increasing urbanization problems. The data required for the population simulation model is given in Table 3.

TABLE 3: Data needed for the population sub-system analysis

Attribute	Qty
1. Population groups :	
Children : persons	221806
mortality	0.043
Adult : persons	111964
mortality	0.004
Parents : persons	280885
mortality	0.005
Old : persons	115000
mortality	0.23
2. Growth rates (%)	
Base year	0.1
scenario (estimate)	- 0.08
3. Family size 5.1	
4. Population density	490
5. No. Households	142,061
6. Rural inhabitants (%)	
— stating year	80
— final year (scenario)	54
7. Ricefields (ha)	8067
8. dryland farm (ha)	86,193
9. Rural off-farm (%)	
— starting year	12
— final year (scenario)	18
10. Children / woman ratio	
— starting year	5
— final year (scenario)	2
11. Sex ratio	96%

Source: UBOS, 2002

The problems of the population sub-system include the rural unemployment and the consequent negative-feedback loop between the unemployment - out-migration, unemployment-urbanization, and the unemployment - income loop. The system will continue to produce the stable and goal seeking behavior over time.

The population sub-system model has also got the positive-loop cycles for example the relationship between the rural income and the agricultural development, population growth-land ratio-urbanization, and population density-land ratio-unemployment-out-migration cycle. In the positive-loop, the quantity continues to increase at a faster rate. However, if the quantities begin to decline, then the positive loop will also decline at an equivalent faster rate. The positive feedback loops in the model include the income and education loop, income and agricultural development cycle, the total population-farmland-out-migration loop, the population density-farmland ownership - urbanization loop, the population density - farmland ownership - unemployment - urbanization loops whereby in case one element increases then the other inter-related elements also increase giving a "snowball" effect, and likewise if one element declines then the inter-linked elements decline too at an accelerated rate.

There are negative feedback loops between the unemployment and out-migration loop, unemployment and urbanization loop, unemployment and income loop. This implies that these aspects tend to produce behavior over time that is stable and goal seeking whereby the sub-system return to an equilibrium following an increase in one of the interrelated elements of the sub-system.

This simple population model can only be used for the crude estimates of population development but may not produce accurate results where detailed demographic information is necessary. The significant errors follow from the fact that a homogeneous population is assumed, whose growth merely depends on the population size. Secondly, the human fertility rate does not only depend on the human potential but also on the social-cultural, the economic status, education standard, and the political developments. These factors indirectly do play an important role in the population system, however, based on the model purpose, they are regarded as exogenous variables.

For the modeling purposes, the population is classified into the four age groups:

1). Children (CHD) of ages 0 to 15 years;

2). Parents (PRT) of ages 16 to 24 years;

3). Adults (ADT) of ages 25 to 49 years;

4). Old people (OLD) of ages > 50 years.

The total population is the sum of the four age groups, each of which is a state variable (level). Each age group suffers annual losses by deaths; they are highest for old people (MOLD), however, new-born babies (INFT) are added each year to the children age class. The number of babies depends on the number of potential mothers in the parent (PRT) class, and their fertility (REPROD). The number of total deaths is calculated by using the age-specific mortalities. The total number of infants (INFT) is determined by the number of potential adult and parent women and their fertility. For modeling purposes, it is assumed that the male/female proportion is one and the number of mothers is half the adults plus half of the parents based on the assumption that all women will have baby.

Initially, the average woman-child ratio is five but and in course the simulation period (t+n) will declining to two due to improvement in education leading the change in the social cultural attitudes to a big family, better understanding of birth control program. The transitions between age classes follow directly from the number of years represented by the younger age group. Out of the first age group, 1/15 of the children population transfers (CCA) to the adult age group annually, and therefore, these transfers are deducted from the children age class, similar transition processes take place between the adult and parent class (ACP), which is 1/10 of the total adult and parent population, and between parent and old people age (PCO), which is 1/25 of the total parent and old population.

The mortality for children is assumed to be 4.3 percent (MCHD), 0.4 percent for adults (MADT), 0.5 percent for parents (MPRT), and 2.3 percent for old class (MOLD). The mortality of the old people age class follows from the fact that they "normally" leave this age group by death around the age of 66 years. The total population (POP) is the sum of the four age classes minus net migration, which is embedded in the growth rate figures. The total labour force is the number of men and women from age classes of adults and parents (72 percent), and a small percent from age class of old people (34 percent). The number of rural dwellers can be computed from the ratio between rural villages and total villages (RTP) multiplied by the total population.

The total rural laborforce (RWF) is therefore the total laborforce (TWF) multiplied by the percentage of the ration of the rural dwellers to the total population (RTOT). To predict the future trend of the rural laborforce, the proportion of rural inhabitants to the total population is assumed to decline from 80 percent to 54 percent by the time (t+n). In this study, the rural labourforce absorbed in agricultural activities is referred to as the farm laborforce (FWF). Therefore, the total unemployed labor-force refers to the total population minus the laborforce absorbed in the agricultural activities (farm laborforce) and those employed in the off-farm activities. The agricultural activities in this case are limited only to the cultivation of rice and other minor few common food-crops such as cassava and maize. The rest of rural laborforce unabsorbed in agriculture and not working in off-farm employment are categorized as rural unemployment (LOC-UN).

The number of the unemployment households (TRHH) is estimated by adding the initial farmers households (TRHHI) to the current population divided by the average family size of 5 members (POP1/5). The average household laborforce (ALRHH) can be got by dividing the total work force by the total peasant households (TWF/TRHH). The conversion of the agricultural laborforce absorptive capacity necessitates the conversion of the dry-land farming area to equivalent rice-field areas. The rice-field equivalent (REP) is equal to the area under rice-field plus the area occupied by dry-land farms (DLF) divided by 2 ((WR+DLF)/2).

Under normal circumstances the farming family is able cultivate 0.7 ha of rice-field and 0.3 ha of dry-fields and historically based on the statistics of landownership until the World War II, in the fertile regions of Java, a peasant family cultivated 0.3 ha of dry-land rice and 0.5 ha of dry-land. However, since Hoima district is a dry region, the required amount of land to sustain a peasant family is only calculated. The extent of rice equivalent needed by a rural peasant farmer to meet his subsistence needs is 0.2 ha of dry-land rice and 0.4 ha of dry-land cultivation divided by two (0.2 + 0.4/2 = 0.4). The total labor-force absorbed in the agriculture is derived from the areas of rice equivalent 0.4 multiplied by the ratio between total laborforce and the total households (TRHH). This ratio is assumed to represent the average number of work force per household (ALRHH).

The decreasing farmer-land ratio generates unemployment among the rural people because it is assumed agricultural sector provides employment to 80 percent of the total rural population. According to the study purpose, the number of rural unemployment is derived by subtracting the total rural laborforce with the number of laborforce absorbed in agriculture and the rural population working as traders, artisan, civil servants, and informal off-farm employment. Although, population pressure has not been fully tackled in this study, it is important to mention that population pressure on the land and forestry resources is a key factor in the underdevelopment of the Gunung Kidul region. The population pressure is obtained by the formulae:

$$PP = (1 - \alpha t) * zt * (ft * P_0(1+r)^t / \beta Lt$$

The population pressure (PP) is a function of the agricultural land (1 -at) multiplied by the minimal hectares (z) of the land required for a "decent standard of living", multiplied the total number of farmers expressed as a percentage of the total population in the area (f) multiplied by the total population of the base year [$P_{0=}(1+r)^t = pt$], exponential time lag is (t), all divided by the total arable land (L). The total land required for decent living (z) depends on the standard of living criteria, landuse type, technology, land productivity, cropping intensity, and economic value of agricultural activities. The agricultural development should therefore be integrated with efforts to create off-farm employment and family planning.

The total land required for decent living depends on the land-use type, technology, land productivity, cropping intensity, community economic status, and economic value of agricultural activities. In order to reduce population pressure to the minimal levels, the population growth rate should be reduced, and the total land required per household should be reduced, and available land quantity increased. However, the total land can not be increased without causing environmental hazards. Therefore, the agricultural development should be integrated with other programs such as off-farm employment, transmigration, and family planning programs.

Consumption of rural basic needs simulation model

The population and unemployment dynamics are useful towards the formulation of the rural resource management with emphasis on community forestry development. It is assumed that an increase in population leads to an increase in amount of food, fodder for their livestock, fuel for domestic cooking, and timber for handicraft and construction activities; therefore there is a need to develop the rural basic needs consumption model. The assumptions of the consumption of the rural basic needs model include:

a) The total quantity of food, timber, fuel, fodder consumed varies proportionately with population growth;
b) The food consumption is computed in terms of rice consumption (rice equivalent of maize and cassava) which is the staple food of the rural community. At the beginning of model time, the average per capita rice consumption is 0.35 kg/day. However, per capita consumption of rice will decline to 0.25 kg/day due to the improvement in consumer incomes and the introduction of other alternative foodstuffs such as fast-foodstuff; the other important foodstuffs are maize and cassava expressed in rice-equivalent units;
c) In order to obtain a good comparison of the food production and the food consumption, the production of maize and cassava are converted into the rice equivalent units. 1 Kg of the maize production is equivalent to 0.5 Kg of husked rice, and 1 Kg of cassava production is equivalent to 0.3 Kg of the husked rice, and 1 Kg of the unhusked rice is equivalent to 0.7 Kg of husked rice;
d) The consumption of fuel is computed in term of fuelwood as the standard requirement. For future prediction purposes, at the beginning of simulation period, the per capita fuel wood consumption (PFWCI) is 0.5 m^3/capita/year and it is assumed to decline to 0.25 m^3/capita/ year. This is because the role of fuelwood to fulfill energy requirement of the people is gradually being substituted by agricultural wastes, kerosene, gas and electricity, however, the majority of the poor people still depend on wood.
e) Given the present increase in income of the people, the per capita consumption of fuelwood will decline. In this model the consumption of fuelwood is assumed to fall to (PFWCF) a scenario of 0.2 m^3/year by the year (t+n) of the model time;
f) Similarly, the demands for timber (TBC) in m^3/year is computed too, and at the beginning of simulation period (1995) the per capita timber consumption is (0.04 m^3/year (PTBCI) and the trend is assumed to flattened thereafter, and the per capita timber consumption is expected to stabilize at an arbitrary figure of 0.01 m^3/year (PTBCF) by the year (t+n) of the simulation time;
g) There is food sufficiency and a shortage is compensated with a lower export, while a surplus is sold off for cash incomes. The data got from the regency statistics shows that the export of food out of the region occurs even in years of famine because the rural farmers require cash incomes for their essential domestic purchases and payment of school fees for their children.

TABLE 4: Scenario data required in the consumption (demand) of basic needs model

Aspect	1995	(t+n) 2010
i) Per capita consumption of husked rice (kg/day)	0.35	0.25
ii) Per capita consumption of fuelwood (m3/year)	0.5	0.25
iii) Per capita consumption of timber (m3/year)	0.04	0.04

3.5.3. Production simulation program

To anticipate the increasing people's needs, it is necessary to predict the production of basic needs from the existing community forest management system using the computer simulation methods. In this study, the simulation production outputs computed from the community forestry system are food, fodder, fuelwood, timber, and job-opportunities. By comparing the people's consumption and the total output from the agricultural/tree (forestry) sub-system, the surplus or shortages of those basic needs of the people can be estimated. Based on the model results, the community forestry development are then be planned based on the situation, and with reference the primary goal of improving the system's contribution to the people and regional development. The information needed to run the production model are the landuse types and their extent, and the current productivity level of each landuse type to produce food (rice, corn, and cassava), fuel and timber, and their productivity level in the future period (scenario). The extent of each landuse is shown in Table 5.

TABLE 5: Data of acreage of each landuse type for production model

Land use	Coverage (ha)
1. Irrigated rice-field	2,540
2. Rainfed rice-field	5,527
3. Homegardes	24,172
4. Dry-land cultivation	82,193
5. Mixed farms	81
6. Plantations/Estates	55
7. Private forest	11,750
8. State owned forest	13,212
9. Roadsides	274

Source: UBOS, 2002.

Given the purpose of this study, the region is classified following the existing landuse pattern adopted in the region. The land-use is aggregated into two categories, namely, the agricultural lands which include: Irrigated rice-fields (not very significant); Rainfed rice-fields; Dry-land cultivation; Homegardens; Mixed farms, and Estates; and the communal or public lands which include: Road (RDS); River banks (RB).

3.5.3.1. Food-crop production simulation model

The landuse classes in the agricultural and private lands are represented in hectares whereas the extent of the roadsides is represented in kilometers. The possible land-use types in this model are illustrated in Table 6.

Table 6: Matrices of source of the food and wood production program

Land-use typ	Rice	Maize	Cassava	Wood fuel	Construction wood
1. Irrigated rice	R				
2. Rainfed rice	R	M			
3. Dryland farms	R	M	C		
4. Homegardens			C	FW	TB
5. Mixed farms			C	FW	TB
6. Private forest	R	M	C	FW	TB
6. Estates				FW	TB
7. Road				FW	TB

The main crops cultivated on the dry-land farms are maize, cassava, beans and vegetables. For easy conversions into the rice equivalent, these traditional foodstuffs are simplified into maize and cassava production. On the dry-land fields, small quantities of the rice are also grown. The estates produce commodities such as the coffee, the cocoa, and the tea. However, the fuelwood or sometimes also timber are available as by-products from this type of land use. With the increasing scarcity of the fuel and the timber production due to forest deterioration, that small role of the estates towards fuelwood and timber production makes the estates an important land-use in the production model. The role of the roadsides and the canal banks is to contribute a greater amount of fuelwood as compared to homegardens and mixed-gardens. In the future, the role of road sides, river and canal banks to produce fuelwood and timber is expected to increase. Although, the primary purpose of tree planting along the road sides and canal banks is to provide shade and to act as wind-break (conservation and environmental function). The growing trees on river banks are important for land-slide protection purposes and very often they also provide additional source for fodder production, especially during the dry season where grass is scarce, and therefore, its role is vital to feed the cattle during the declined supply of fodder.

Based on the fact that there are difficulties in developing the production functions, the future agricultural production is expressed by the scenario value based on the recent past. The agricultural production is based on the following assumptions:

a) On dry fields, the trees are grown, and unlike on the rice-field cultivation where the artificial fertilizers are applied extensively, the sustainability of the dry-land cultivation depends on tree management practices;

b) The crop yield is an estimator of the soil productivity and soil productivity declines if harvested crops are not substituted by inputs or artificial fertilizer;

c) The rain-fed rice-fields are also cultivated by beans, groundnuts, peas and maize;

d) The existing land-use types are considered as state parameters, assuming there are no significant changes between those classes along the model time;

e) Soil fertility is maintained or recovered by management practices and the level of crop production depends on different factors and the production function can be stated: Crop production (p) = f (soil, landscape, climate, inputs, management);

f) The productivity will increase over time, due to technical improvements.

A serious difficulty occurs in estimating the production level of small branches or twigs which is actually very important to the local population as a source of energy from their agricultural lands and private forests. From the quantity point of view, this product plays a very important role in the fulfillment of fuel requirement of the local people. The potential productivity level of twigs (small branches) in stand density and its actual production level is determined by the human activities to utilize it.

Initially, the productivity for each class is determined based on the current level. However, for some land use types the information are not available, at least not reliable enough for example timber and fuelwood productivity of dry-field cultivation, mixed farming, plantations/estates, and roadside. This information is estimated based on the past trends, the experiences from the other land use types as well as comparison with figures from the other regions.

FPROD = SPRO + TEGALPRO

where :

SPRO = irrigated rice production (ton);

TEGALPRO = production of dryfields (ton);

At least with the currently available data, it is difficult to develop a classical production function. The "technology-improvement" and increase of productivity is incorporated as an arbitrary function expressed as the control variable and for the case of the rice production calculation:

SPRO = SAWAHAREA* SAWAHPRY

whereby:

SAWAHAREA = area of sawah in hectare.

SAWAHPRY = productivity of sawah cultivation (ton/ha)

SAWAHPRY = f (TIME)

The dynamics of productivity are beyond the scope of this study, however, the productivity index represents the actual productivity as proportion of the potential productivity of the dry-fields (tegal). The dry land production is assessed similarly :

DLFPRO = DLFAREA * ACTPRO

where:

DLFAREA = dry field area (ha);

ACTPRO = productivity of dry fields (ton/ha)

Two significant different assumptions are made compared to the estimation of production on drylands.

ACTPRO = POTPRO * PRFRACT * PRINDEX ———————— (23)

whereby :

POTPRO = potential productivity of dry-fields (ton/ha)

PRINDEX = fraction of potential productivity which is "active" (number)

PRFRACT = fraction of area used for food cropping (number)

POTPRO = f (TIME) (year)

The potential productivity of dry fields is given as an arbitrary function of time. The food-crop production on the dry fields is in reality difficult to assess, this is because of fluctuations in rainfall, cultivated areas differ from year to year and a reference is difficult to establish because of the mixture of crops. The variable productivity fraction indicates the interaction and competition between tree growing and production of good crops in the Hoima districts dry land production systems. Under a tree canopy, there are possibilities for production of other crops. However competition between trees and food crops will limit production of food-crops.

The exact relationship between the physical, the chemical soil properties, and productivity of agricultural crop is difficult to assess. The agricultural productivity itself is determined by factors such as the landscape, the climate and the management. In the model, the actual productivity of dry fields is given in the proportion of the potential productivity which is active.

PRINDEX = PRINDEX(t-1) - dt * DEP + dt * REC .... (28)

whereby:

PRINDEX = fraction of potential productivity which is "active" (number)

DEP = depletion of productivity in fraction of prindex (number)

REC = recovery of productivity in fraction of prindex (number)

The total depletion is derived from depletion rate and productivity index.

Dep = Deprate * Prindex (t-1) .... (29)

deprate = rate of depletion in % actual productivity (number)

The rate of depletion is a scenario parameter whereby the soil fertility is recovered or retained through better tree management practices. The application of the artificial fertilizer is obvious, but also the biological conservation practices can have a significant effects for example the crop residue management, the tillage management, and the agroforestry.

2.8 Wood production simulation model

The wood is the prevailing source of fuel (energy) in Hoima district. Data from the regional statistics office states that 85% of all consumed energy is used for domestic cooking. The energy need in Hoima is fulfilled by the wood and residues of the agricultural products (89% depend on biomas) as cooking fuel (UBOS, 2002). The assumptions that underlie the wood production model are:

a) Fuelwood and timber are obtained from the homegardens, mixed-gardens, estates, and the road-sides;

b) Fuelwood production increases with time because there is enough underutilized land;

c) Timber production increases by very small percentages, this is because of the long time period required for a tree to mature into timber;

d) The trees are cut for wood as a source of fuel and for timber and the total energy needs are satisfied through harvesting of trees. The energy is derived from the wood and from the agricultural wastes. The consumption of the crop residues as source of energy is given as a percentage of the total energy consumption;

e) The soil productivity declines if harvested crops are not substituted by inputs such as manure and artificial fertilizer. In case this assumption is true, then the soil has a certain richness which decreases with agricultural production, but increases concurrently by manuring;

In this model, opportunity is given to determine a minimum number of trees to be retained. It is assumed that a low productivity do induce tree planting to increase the soil production capacity. It is assumed that if subsistence needs for food are not completely met, then the competition between tree growing and food crop production is perceived as not profitable and trees are cut to decrease the competition. To maintain the tree cover, the harvested the trees must be replaced by tree planting.

PLANT = RPL + NE

whereby:

RPL = regular planting activities expressed (m3/ha/year);

NE = added planting as a function of productivity (number);

NE = f (Productivity index)

The regular planting standard is assumed to be 1 m^3 annually. To the rural farmers, the planting of trees is primarily an activity aimed at soil conservation. It is assumed that below a certain limit of productivity (productivity index), the planting of trees is the only viable remedy to the soil depletion problem in Hoima district. The growing stock on agricultural lands is estimated by the formula:

$$Wood_t = wood_{(t-1)} + dt * inc * cut + dt * plant - dt * xtrcut$$

whereby:

Wood = standing stock of wood on dry farming fields (m3/ha);

Inc = annual increment of wood (m3);

cut = annual harvest (m3);

plant = annual tree planting activities expressed (m3);

Xtrcut = emergency cut of trees (number)

The standing stock of wood is related with the tree-canopy cover and this is important variable in the assessment of competition between the growing of trees and the growing of food crops. As a result, the initial standing stock and tree canopy cover are model parameters, and the tree-cover is subsequently derived from the growing stock.

CUT = (ENER * (1-RR) / WEN+ (POP * TN)

where: Cut = harvest of wood (m^3)

ENER = fuel need (ton);

RR = amount of energy needs through crop-residues (number);

WEN = energetic value of wood (ton/m^3)

TN = need for timber (m^3/capita);

Given : Cut < = Inc + (wood - (tamin * ta)) (ha/year);

where: Wood = volume of standing stock (m^3);

Inc = increment (m^3);

tamin = minimum treecover (m^3);

ta = tegalan area (ha)

As stated before, wood is the prevailing source of fuel in Hoima district and it is assumed that 85% of all consumed energy is used for cooking. The total energy needed is almost completely met by wood and residues of agricultural products. Approximately 90% of the total population uses biomas as cooking fuel (UBOS, 2002). The regional needs are derived from food consumption, for each unit of consumed food a certain amount of energy is required for preparation.

The assumption is that the trees are exclusively cut for fuel and timber. The total energy needs are satisfied through harvesting of trees. The energy can be derived from wood and from agricultural wastes. The uses of residues however has an effect of the productivity of agricultural fields. It is assumed that the depletion of productivity to be twice as high when systems with no recycling of residues is compared with a system in which residues are recycled and the take away of residues is balanced by application of artificial fertilizer. Not all functions of organic matter are restored, but mulch helps to prevent slacking of the soil.

In this study, great importance is attached to the determination of the minimum number of trees and a low productivity will induce tree planting to increase production capacity. If subsistence food needs are not completely met then competition between tree growing and food-crop production are perceived as not profitable and therefore the trees are also cut in order to decrease the competition with food-crops.

The inter-linked relationship between the people sub-system, the consumption and production of the rural basic needs, and the general agricultural and tree growing development is established using the system analysis approach. The tree growing helps to maintain the stability of the rural system, and in case the tree growing sub-system is ignored then the rural development system will only consist of positive feed-back loops that means that less progress will be achieved to reduce the number of poor farmers and increase the regional incomes.

In the wood production system, problems are not totally be deliberated, therefore, in tree (forestry) management problems do not have answers in the ordinary sense. For this reason, the problems and their solutions will be represented in models, this is because models generate possible alternatives and not answers.

Presently, the main objective for the tree farming system is to obtain maximum forest outputs because for a long time there has been shortages of food, fodder, fuel, timber, and employment opportunities among the rural dwellers. With the full implementation of community forestry based on the farmers participation, the required maximum forest outputs can be obtained. The farmers make decisions concerning silvi-cultural aspects of the tree growing on their farms.

2.10 Animal population, its growth and migration

The accounting functions for livestock population can be written as:

$$LIVPOP_{t+1,a} = LIVPOP_{t,a} + LIVBR_{t,a} + LIVPUR_{t,a} - LIVSOLD_{t,a} - LIVSLT_{t,a} + LIMP_{t,a} - LEMP_{t,a}; \quad (8)$$

Where: LIVPOP = Number of livestock during the beginning of the next period

t = Time periods

a = Livestock type (cattle, buffaloes, goat and sheep).

LIVBR = Livestock borne (net of livestock died)

LIVPUR = Livestock purchased

LIVSOLD = Livestock sold

LIVSLT = Livestock slaughtered, with a fixed value of zero for cattle

LIMP = Livestock immigration into the area

LEMP = Livestock emigration out of the area

2.11 Study sample and model data collection

The input for the bio-economic model comprising of socio-economic, ecological and vegetative biomass data were collected for the Hoima district by using appropriate sampling methods.

Both primary and secondary data (on prices, growth rates, human and animal population, vegetative biomass, past soil erosion rates.) were collected from published records and through a socio-economic survey, using both participatory and interview methods. Detailed input-output data on crops, livestock and livestock products and forest activities (fuel wood collection and timber harvesting) were collected for the modelling exercise. It should, however, be mentioned that these data were obtained from only the sample households in the village.

Bio-economic model was calibrated and run for a period of 20 years (2000 to 2024) under four alternate scenarios, namely, (1) base scenario (BASE) run with the existing data, (2) introduction of improved agricultural technology (TECH) where yields of major crops, namely, paddy, maize and wheat were assumed to increase by 5% per year, (3) reduction in population growth (POPG), where population growth rate is halved to 0.67% per annum, and, (4) increase in the prices of major agricultural crops (PP), where prices for paddy, maize, wheat were assumed to increase by 5% per year.

Increasing population reduces the size of landholdings and the landless out-migrate in search of off-farm employment in the urban areas. The other positive loop shows that population density reduces the size of landownership which consequently leads to farm unemployment and incites families to out-migrate to urban areas. The negative loops show that the community forestry system is a goal-seeking system trying to find equilibrium. The unemployment will be countered by out-migration, secondly the unemployed labour-force will be absorbed in the urban employment until such a time of saturation. The unemployment implies that many will shift to the urban areas for employment to earn incomes which they bring back to develop and buy land in Hoima district and as such regional incomes increase.

In relation to these population-caused problems, the population sub-system has got connections with other parts of the rural system through:

a) Fuel consumption is directly related with food consumption, for a certain amount of food a related amount of fuel is needed for preparation of the food;

b) Food consumption is, in a subsistence production system, directly related with production. Higher rates of production will induce a higher consumption rate and export of "surpluses".

3.0 RESULTS AND DISCUSSION

We estimated the value of fuelwood based on the labour time used to collect it. This was probably a minimum value as it did not include any scarcity value of fuelwood nor was it related to the cost of the next best alternative fuel source. The average family surveyed collected 65 kg per week of fuelwood and used both adult and child labour. Valuing the time spent in fuelwood collection by the appropriate wage was equivalent to Ushs 2,750 which is approximately Ushs. 42 per kilogram. In terms of solid, air-dried fuelwood, this is equivalent to Ushs. 32,340 per m^3 (770 x Ushs. 42). We assumed that there were alternative productive opportunities available for adults and children at the specified wage rate. If such opportunities were not available or the true opportunity costs were higher or lower than assumed, the shadow price of fuelwood changed.

The mean annual consumption of fuelwood was estimated at 4,400 kg which implies that per capita fuelwood consumption in the study area was approximately 330 kg/user household, or 53 kg / capita per annum. The mean fuelwood consumption rate of 692 kg/capita of wood per annum in this study compared satisfactorily with figures elsewhere. For example, in South Africa as summarized by Shackleton (1993), who reported a mean fuelwood consumption of 687 (49 kg/capita annum) across 12 studies. Kalumian and Kisakye (2004) reported a lower consumption figure 485 kg per capita. On a household basis, the amount of 4,400 kg/household per annum for Hoima is comparable to 4,300 kg in parts of South Africa (Dovie *et al.* 2004). This similarity in figures represents a source of vital information for comparing analyses of the fuelwood crisis and comparing adaptive management strategies towards sustainable management of forests and allied natural resources.

3.1 Fuelwood farming option

Tree farming was evaluated using a net present value (NPV) approach. The costs of tree-farming were derived from the data given in Table 1. The various activities occurred in different years which resulted into different labour requirement and consequently varied costs and benefits for each year. For each activity, the labour requirements were converted to Uganda shillings and input costs (seedling, fertilizer) were also calculated.

The operational costs were calculated from activities 12, 13, and 14 in Table 1. Similarly, input (material) costs were only seedling and fertilizer and all these costs occurred in years 1 and 2. Labour occurred in all years. In year 1, labour is required for land clearing and preparation (35 man-days), seedling planting (35 man-days), fertilizer application (7 man-days), and weeding and brush removal (10 man-days). This adds to 87 man-days valued at Ushs. 321,900 (87 x Ushs. 3,700). In the succeeding years the following amount of labour is required:

TABLE 2: Direct costs and benefits from tree farming (Ushs/ha)

Year	Labour	Operating costs Materials	Operational cost	Benefits	Fuelwood Net benefits	NPV (15% discount rate)
1	321,900	442,700	208,00	0	-764,600	-114,690
2	77,700	102,000	36,300	0	-216,000	-32,400
3	14800	0	0	0	-14,800	-2220
4	14800	0	0	0	-14,800	-2220
5	577,200	0	14,800	3,030,000	2,438,000	365,700
6	14800	0	14,800	0	29,600	4440
7	14800	0	14,800	0	29,600	4440
8	14800	0	14,800	0	29,600	4440
9	547,600	0	0	3,030,000	2,482,400	372,360
Total	1,598,400	544,700	95,500	60,60,000	3,999,000	599,850

The fuelwood was harvested in years 5 and 9 and it was discovered that pure stands yielded about 123 m^3 per hectare; but, even with replanting, there was some mortality (approximately 25 percent). Therefore, the fuelwood harvest total was estimated at 92.25 m^3 (123 x 0.75). Harvesting required 1.6 man-days of labour per m^3; total labour required was thus 147.6 man-days (92.25 x 1.6). The man-days were then converted to Uganda Shillings at the stipulated wage rate, Ushs. 3,700 per day. We, however, recognize that no costs were assigned to tools used for harvesting or fieldwork.

Recent public surveys (MWLE, 2002b) indicate that it is common practice for schools in the rural areas to request pupils to carry firewood to school. How often this is done varies from one school to another. It could be daily, weekly or monthly depending on the demand. There are also cases of where pupils pay for firewood as part of their school fees (MoFPED, 2003). In the urban areas, and in some boarding schools, the situation is different. Most of the schools buy firewood (ranging between Ushs 50,000 – 800,000/per term of schooling) depending on the firewood demands of the school. Many poor communities derive their incomes from charcoal burning and trade.

Although, the present national resource basis for fuelwood is not known precisely and the few existing studies about the local fuelwood situation mainly have been consumption surveys rather than supply-and- demand studies, there are several reasons why the future fuelwood production capacity of the district is in doubt (Kalumian and Kisakye, 2001). First, the production capacity of most public forest lands is threatened by conversion to non-forest land uses. Second, the future demand for fuelwood will rise in response to population growth and increases in tobacco production. Third, there exist several plans in Hoima district for the establishment of large-scale projects such as charcoal-using iron blast furnaces, which will increase the demand for fuelwood tremendously (MWLE, 2002b). As planned, these projects are assumed to provide their own wood energy resources from associated plantations. Because the slow progress in planting and high rates of mortality, however, it seems reasonable to expect that commercialisation will result in at least some degree of substitution of fuelwood from the existing traditional fuelwood sector to this new modern fuelwood sector (Jacovelli and Caevalho, 1990).

The benefits of fuelwood farming are the value of wood produced (direct benefits) and environmental quality benefits of maintaining land under forest cover or reforesting land. Such benefits may include reduced erosion or increased forest production from "protected" forest lands previously used for fuelwood collection. The direct benefits occur in years 5 and 9. Each, year, 92.25 m^3 of fuelwood are harvested. Converted to kilograms and valued at fuel wood's shadow price, the value per hectare in each year is Ushs. 3,030,000.

When net benefit in each year are discounted at 15% and summed, the NPV is found. In this case, it is Ushs. 599,850 per hectare (Table 2). If the environmental quality benefits are included, the positive NPV becomes even larger. The environmental quality benefits consist solely of effects on land quality, valued at Ushs. 65,700 per hectare. These benefits begin in year 2 and continue for each succeeding year. Discounted, these benefits increase the NPV by Ushs. 256,000 for a total NPV of Ushs. 3,999,000.

If only direct costs and benefits are included (Table 2), the NPV at a 5 percent discount rate is Ushs. 2,302,000 per hectare of tree farms. With a 20% discount rate, it is Ushs. 460,000 per hectare. At higher discount rates, the NPV will turn negative since the large initial expenses involved in tree stand establishment outweigh future benefits. The inclusion of other environmental benefits not quantified in this study would increase the NPV. Since most of the costs are incurred in planting: additional coppice rotations (> 10 years) would increase the NPV because the harvest benefits are much larger than the costs of harvesting and weeding.

Fuelwood is collected by the people and used directly to meet their energy requirements. Fuelwood collection is an activity that is largely informal, and the economic value associated with it is not easy to capture. However, it is generally agreed that households consume a lot more fuelwood than what is actually recorded in the formal sector. Falkensberg and Sepp (1999) estimated fuelwood production in the informal sector to be about eight times higher than in the formal sector. Given that over 85% of Uganda's population is rural (UBOS, 2002), fuelwood contributes to the livelihoods of the majority of Ugandans. Moreover, the poorest 35% of the population who live below the poverty line cannot afford buying fuelwood, and hence rely heavily on collecting from the natural forests and trees to meet their energy requirements. The use of fuelwood depends more on the locality than on the level of household income, as the households in urban and semi-urban communities tend to prefer charcoal for cooking. This is because urban consumers find charcoal to be a cheaper source of fuel. In 2000 for example, the cost of cooking with electricity was roughly double that of cooking with charcoal and paraffin was about 3 - 4 times higher than that charcoal (MWLE, 2002a).

3.2 Improved woodstove option

A new woodstove design could improve fuel efficiency from 8 to 20% (MoEMD, 2002). Fuelwood consumption would therefore decrease since now 0.4 as much wood would supply the energy required to cook a given amount of food. The wood fuel savings is Ushs. 40 kg (0.6 x 65 kg). At a price of Ushs. 43 per kg, this is a saving of Ushs. 1,700 per week (Ushs. 89,900 /year). Over a 9-year period the annual benefits of the new stove are Ushs. 89,960. The costs are the Ushs. 21,000 purchase price and Ushs. 25,000 for administrative costs. Both costs occur in year 1. The NPV over nine years at a 15% discount rate is about Ushs. 390,000. The introduction of improved stoves appears to be very attractive based on fuel savings alone. If the new stove's fuel efficiency were only 10%, the annual fuel saving falls to Ushs. 36,000, and the first-year cash flow is negative. The NPV over nine years falls to Ushs. 8,800 which is one third of the previously calculated amount.

We assumed that the decreased fuelwood demand reduces the cutting of forests for fuelwood production, we therefore used the previously assigned values these benefits (Ushs. 65,700 per ha) and determine what part of a hectare would not be cut if the new tree farming areas (25m^3 x 770 kg/m^3 = 19,250 kg/ha), the annual wood fuel benefits of about 2,111 kg are equal to 0.11 of an hectare or about Ushs. 7,400 in annual environmental benefits per improved woodstove. Other studies (Moyini and Muramira, 2002; Tumuhimise and Kutesakwe, 2003), have, however, proved that natural forest yields are substantially lower than managed tree farms and therefore the environmental benefits from each kilogram of wood saved are greater. In addition, there are other benefits from each kilogram of wood increased production of other forest products when fuelwood needs are reduced. If the natural forest's annual increment were only 5.75m^3/ha, the environmental benefit would be Ushs. 31,000 per year. At a 15% discount rate, the total NPV for year 2 to 9 amounts to Ushs. 121,000.

3.3 Kerosene stove option

The kerosene stove is assumed to eliminate the need for fuelwood; 65 kg valued at Ushs. 300 per year. Annual fuel costs are Ushs. 134,900 (2.25 litres x 1153.5 x 52 weeks). In addition, the stoves cost Ushs. 17,200 and there is a one time administrative cost to the government of Ushs. 24,700 per stove adopted. The direct benefits and costs are very close. In the first year, net benefits are negative Ushs. - 32,350. In subsequent years, the net benefits are marginally positive (Ushs. 8,800). If environmental benefits are included, however, this option becomes more attractive. Reasoning as before for the woodstove example, the annual land quality benefits would be Ushs. 11,360 per ha which increased the NPV by Ushs. 44,400.

We faced problems associated with the option to use commercial fuels such as kerosene. This is because Uganda imports most of its kerosene using scarce foreign exchange. The price of kerosene has greatly increased therefore, making the fuelwood the most feasible alternative for the rural poor households. A new stove may last longer than nine year hence increasing the NPV or it may require repairs and replacement parts whereby lowers the NPV.

3.4 Model simulation results

The tree cover in the forest areas declined by 6% in the BASE scenario, 4.8 % in POPG scenario and 4.7% in TECH scenario, indicating an overall trend of forest degradation in the Hoima district (Fig. 3) under each of these scenarios. Reduction in the population growth rate (represented by POPN model scenario) and introduction of improved farming methods (TECH model scenario) did, however, to a somewhat reduced forest degradation rate. Reduction in the population growth rate would reduce demand for wood fuel and timber. Increase in crop yields due to introduction of technology would increase the household income, and hence, activities such as the sale of fuel wood and timber will decrease. In both cases, the rate of tree removal will be lower than in the base scenario.

The price increases for the major crops (PP scenario) reversed the degradation process and actually indicated regeneration within the study area during the latter part of the simulation period. This option seems to be more effective in mitigating forest degradation rate by enabling an increase in income from crops, and thereby reducing pressure on sales of fuel wood and timber, to meet the household requirements. Furthermore, enhanced income form existing farm production would mean that less land would need to be brought into cultivation, hence less deforestation.

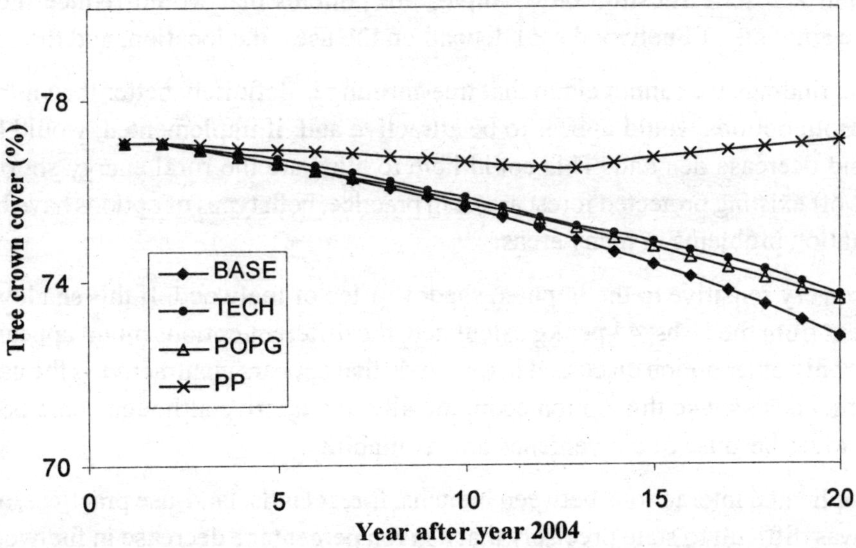

FIGURE 3: Projected tree crown cover (%) under alternate scenario in Hoima district, Uganda

Scarcity of fuelwood is one of the key factors that motivate farmers in adopting rotational woodlot technology (Jacovelli and Caevalho, 1999). As long as fuel wood could be collected without paying for it, farmers always have little incentive to plant fuelwood producing trees. Shackleton (1998), however, reported that high fuel wood demand stimulate tree production that this is only the case when there is a fuel wood crisis. Thus, the high cost of fuel wood may motivate farmers to establish woodlots (Tumuhimise and Kuteesakwe, 2003).

The results of this study have shown the importance of trees in supplying domestic fuelwood. Wood and charcoal remains the predominant cooking fuels for households in rural areas with growing population, less innovative energy utilization technologies, increasing price for agricultural produce, During the period of 2004 to 2020 the fuelwood consumption is predicted to increase by 2.2% annually, which is about the same percentage as the increase in average amount of 0.5 - 0.6 m^3 fuelwood consumption per household each year (MWLE, 2002a).

4.0 CONCLUSION

Environmental degradation, rural poverty, food shortages, wood scarcities and heightened concern for energy supplies have all contributed to an increased awareness of the importance of forests and trees for the well-being of rural people. With the growing impact of deforestation and land deterioration, new approaches are called for. This study has shown that tree-farming is one of the possible approaches to increase the supply of fuelwood (energy), while the woodstoves and kerosene substitution are policies that reduce the demand for fuelwood. This helps to alleviate the rural energy shortage and take some pressure off existing protected forest areas. The tree-farming illustrates one of the possible approaches to how the supply of fuelwood (energy) can be changed, while the woodstoves and kerosene substitution examples are policies that would reduce the demand for fuelwood. The amounts of Fuelwood sold depend on the uses, the location, and the sources.

Based on these findings we cannot claim that tree-farming is definitely better than introducing new woodstoves. Both options would appear to be attractive and, if implemented, would both increase fuel supply and decrease demand. This could help to alleviate the rural energy shortage and take some pressure off existing protected forest areas. In practice, both types of options have been hindered by implementation problems in many areas.

Our results are very sensitive to the imputed shadow price of fuelwood. If this shadow prices were greatly different from the Ushs. 44 per kg calculated, the different options might appear more or less attractive. The only other option discussed in this study that appears unattractive is the use of kerosene stoves. High fuel costs make this option economically unattractive although some people will opt for kerosene stoves because of convenience and availability.

Given the complicated interactions between humans, forest lands, land-use practices, and fuelwood collection, it was difficult to state precisely that a given percentage decrease in fuelwood collection resulted in a proportionate improvement in forest land cover or decrease in deforestation. Regardless of the exact environmental consequences of deforestation, there is general agreement on several facts: there is a rural energy shortage; fuelwood is the most common rural fuel, and its use will increase for both continued household use and for new commercial purposes; deforestation has decreased the production capacity of fuelwood; and reforestation of denuded lands may assist to ensure a future resource base for fuelwood production, as well as to control environmental degradation.

The price of kerosene has greatly increased therefore, making the fuelwood the most feasible alternative for the rural poor households. A new stove may last longer than nine year hence increasing the net present value or it may require repairs and replacement parts whereby lowers the total benefit to the poor farmers.

The necessity to plant trees on private fields probably will become more important in the future in satisfying the needs of the fast growing population, therefore, an understanding of the factors that determine the decisions about tree growing of the farmer will prove important. The social problems emerging from population development and traditional agricultural expansion seem to remain unsolved constraints for rehabilitation and community forest development is in relation with land status and property rights on its resources.

ACKNOWLEDGEMENT

Authors are grateful to the staff of the Directorate of Natural Resources, Hoima District Administration for the technical support. They also thank the anonymous referees and editors for their comments. An early version of this paper was presented at the Fall 2006 Meeting of the Norwegian Society for Development Research, Oslo.

REFERENCES

Andersson D. and Fiswick, R. (1984). Fuelwood consumption and deforestation in African countries. World Bank staff Working paper No. 704. World Bank, Washington DC.

Dovie, B.D.K., Witkowski, E.T.F. and Shackleton, C.M. 2004. The fuelwood crisis in southern Africa – relating fuelwood use to livelihood in a rural village. GeoJournal 60:123-133.

Falkenberg, C.M., and Sepp. S. (1999). Economic Evaluation of the forest sector in Uganda; Forest Sector Review. Ministry of Water, Lands and Environment. Kampala, Uganda.

Jacovelli, P. and Caevalho, J. (1999). The private forest sector in Uganda – Opportunities for greater involvement. Forest Sector Review, Ministry of Water, Lands and Environment. Kampala, Uganda.

Kalumian, O.S and Kisakye, R. (2001). Study on the establishment of a Sustainable Charcoal Production and Licensing System in Masindi and Nakasongola Districts. EPED Project. Ministry of Water, Lands and Environment. Kampala, Uganda.

MoEMD (2002). The Energy Policy for Uganda. Ministry of Energy and Mineral Development. Kampala, Uganda.

Moyini, Y. and Muramira, E. (2002). The Cost of Environmental Degradation and Loss to Uganda's Economy, Reference to Poverty Eradication. Policy Brief No.3 IUCN.

MWLE (2001). Forest Sector Review. Ministry of Water, Lands and Environment. Kampala, Uganda.

MWLE (2002a). Ministry of Water, Lands and Environment. The Uganda Forestry Policy 2001. Kampala, Uganda.

MWLE (2002b). National Biomass Study, Technical Report. Ministry of Water, Lands and Environment. Kampala, Uganda.

Sankhayan, P.L., Gurung, N. Sitaula, B.K. and Hofstad, O. (2003). Bio-Economic Modelling of Land Use and Forest Degradation at Watershed Level in Nepal. *Agriculture Ecosystems and Environment* 94: 105-116.

Shackleton, C.M. (1998). Annual production of harvestable deadwood in semi-arid savannas, South Africa. *Forest Ecology and Management* 112:139-144.

Tietema, T. (1993). Biomass determination of fuelwood trees and bushes of Botswana, Southern Africa. *Forest Ecology and Management* 60(3-4):257-269.

Tumuhimise J and Kuteesakwe, J. (2003). Sustainable Charcoal Production and Licensing System in Masindi District. Ministry of Energy and Mineral Development. Kampala, Uganda.

UBOS (2002). Population Provisional. Uganda Bureau of Statistics. Government of Uganda. Kampala, Uganda.

Energy & Sustaniable Development

4

Optimizing Green Energy: Recipe for Sustainable Development

Jitendra Behari
School of Environmental Sciences, Jawaharlal Nehru University,
New Delhi 110067, India
E-mail: jbehari@hotmail.com

Energy is a key issue in the present day scenario which ensures the survival of life on planet earth. In an effort to ensure this all possible sources of energy have been exploited and as dictated by increasing population, are in rising demand. Renewable (green) and non renewable sources of energy both have been used to meet these needs. Renewable energy has found favour with the environmentalists and the scientists alike and the latter has been implicated for causing a rise in pollution (air, water, soil, noise, electropollution and electronic waste) levels. Electronic waste is a new environmental menace, which also generates materials like lead, cadmium, mercury etc. These pollutants cause not only health problems but also leads to green house gas effect, global warming and subsequently the climate change. The last of all these is a pointer toward the possibility of a major catastrophe and if steps are not taken immediately it may threaten the very survival of the planet earth. It is therefore an urgent demand to develop newer technologies, optimize energy generation, distribution and its efficient use, define the sustainable life style and implementing it globally. The problem as such has a global dimension but the solution has to be initiated locally and an integrated approach involving the total system needs to be evolved. In the present context, it is planned to present a picture in its totality, keeping Indian scenario in focus.

KEYWORDS: *Renewable and non-renewable energy, Pollution, Green house gas effect, Electronic waste.*

1.0 INTRODUCTION

The energy (security) is a key parameter indicating the development of a country and is intimately linked with the state of its economy. It is evident from the fact that about 2 billion people spread over Latin America, Asia and Africa, who are below the poverty line have no access to modern energy supply. Therefore a development process has to be set in coexistence with the environment. As an indication of this, the term 'sustainable' was coined during 1970's to imply that any human activity for its subsistence and betterment in the life style should be in equilibrium with the natural processes to maintain the environmental quality. However, the load of the increasing population and the demand for a better life style has disturbed this balance, necessitating a need for a greater exploitation of the available resources. Already we have on this planet more number of people than can be spatially sustained. The problem is more in developing countries than the developed ones. In fact the demarcation between the developed and developing countries is largely because of this, if not completely. Development is considered sustainable, if it meets the need of the present without compromising the ability of future generations to meet their own needs (WCED, 1987). While the relationship between poverty and energy available is well understood, its link with sustainability is one of recent concern. Unmindful exploitation of natural resources has led to a rise in pollution levels (natural and manmade) beyond the tolerable level of the presently available means to absorb these. Sustainable development has become an integrated approach embracing economic, social and environmental issues. Use of non renewable energy is permissible but it must be used after full estimate of its impact assessment and should be eco friendly. However, this has not happened in many developing countries and may also be in developed countries to attain the faster economic growth.

In India, the situation is rather precarious. With a rural population of over 741 million people and the UNDP HDR reporting human poverty value of 31.4% for 2004, India ranks among 95 developing countries on poverty levels (Srivastava and Rehman, 2006). In this context, inequitable access to clean energy is a major bottleneck toward achieving sustainable development. In India 30% of the total population do not have the access to potable drinking water and 40% do not have adequate health care facilities and can only be addressed through the provision of clean energy. Moreover, the agriculture and macro enterprise sector that together employ more than 30% of the population can become more productive and efficient through the provision of clean and sustainable energy options. Hence, in order to address the needs for sustainable development, it is necessary to examine the constraints related to rural energy and find appropriate solutions that have bearing across all sectors of development, which is also key to national development.

2.0 MAIN BODY

2.1 Sources of Energy

Basically the energy resources can be divided into two parts: primary and secondary. First of the two are those that are found or stored in the nature. Commonly found in this category are coal, oil, natural gas and biomass (such as wood). Other primary sources are: nuclear energy from radioactive materials, thermal energy stored in earths interior, wind energy and potential energy due to earth's gravity. Other stores of energy such as: firewood, cow dung and agriculture wastes are also sources of energy but are often not considered for energy accounting. All the present day use of energy is directly or indirectly derived from these sources.

These energy resources are further subdivided into two parts: renewable and non-renewable.

Renewable include solar power, geothermal, tidal power and hydroelectric power. A characteristic of these power generating methods is that they can be harnessed without the release of much harmful pollutants in the environment.

Non-renewable sources of energy include mostly fossil fuels. These sources not only produces pollution but also get depleted with time. Coal and other fossil fuels, which have taken three million years to form, are likely to be depleted in less than hundred years. We have consumed about 60% of the available resources. For sustainable development, we need to adopt energy efficiency measures. Today about 85% of primary energy resources comes from non-renewable and fossil sources (coal, oil etc.).

In order to achieve sustainable development, we need to observe energy conservation and energy efficiency. Energy conservation is achieved when growth of energy consumption is reduced, measured in physical terms. Energy conservation can therefore, be the result of several processes or developments, such as productivity increase or technological progress. On the other hand, energy efficiency is achieved when energy intensity in a definite product, process or area of production or consumption is reduced without affecting output, or consumption is reduced without affecting output or comfort levels. Promotion of energy efficiency will contribute to energy conservation and is therefore an integral part of energy conservation promotional policies. It is often considered as resource base. It provides additional economic value by preserving the resource base and reduces pollution. For example by adopting more efficient means of electricity transmission one can reduce the power loss by as much as 20% and by replacing traditional light bulbs by compact fluorescent lamps will require only 1/4th of the energy to light a room. Pollution levels will also be correspondingly reduced. Energy efficiency in general implies greater output for the same input. While there is an upper limit for efficiency in each process, we have long way to go to achieve the maximum and anyway this is an ongoing process.

2.2 Bioenergy

Biomass is the dominant source of energy in rural regions of most developing countries for cooking, water heating, etc. The energy required for domestic consumption is mostly met by biomass, with fuel wood claiming the dominant share (woody). It is estimated that this domination will remain around 52% till 2031-2032, with the remainder burden being borne by fossil fuel-based energy carriers (Ravindranatha and Balachandra, 2009). Non-woody biomass comprises agricultural and agro-cultural residues and animal, municipal and industrial wastes. The problem facing the use of these materials is on how much of these materials can be routed through the modern technologies to reduce the burden of pollution and make the system more efficient. The advantages of these modern bioenergy technology is that they use significantly less amount of biomass resources as input for providing higher quantity as well as quantity of energy services for more number of people, compared to the conventional processes. They are environment friendly; reduce indoor pollution significantly while improving the living standards for rural poor. It is possible that use of such technologies will result in climate change mitigation.

At present, coal is the mainstay of power generation in India (50%), through the burning of coal. With the estimated reserves of 221 billion tonnes (Government of India, Ministry of coal, Annual Report 2001-2002) and with limited availability of oil and gas, and even with poor quality of coal, it is imperative that the dominance of coal as a source of energy will continue for quite sometime.

India has a large hydro potential and only a part of this potential has been exploited. All issues relating to harnessing this resource need to be addressed and its full potential to be exploited expeditiously. Figure 1 shows the contrast between renewable and finite energy supplies.

3.0 NUCLEAR POWER

Nuclear power is a well established source of nuclear energy, the foundations of which were laid in our country immediately after independence, thereby committing ourselves to its peaceful uses. An immediate edge of this technology over others stems from the following facts: while other fossil fuels (wood, coal—even of a poor quality, oil etc) are presently available, they produce pollution that is now assuming a gigantic dimension. Out of all the sources of energy, one that is based on use of nuclear power is important and is likely to make significant contribution to electricity supply in future. A program to introduce 1000MW fast breeder reactor is likely to be in place soon. Recycling of plutonium obtained from the reprocessing of spent fuel gives us a very large energy resource. A

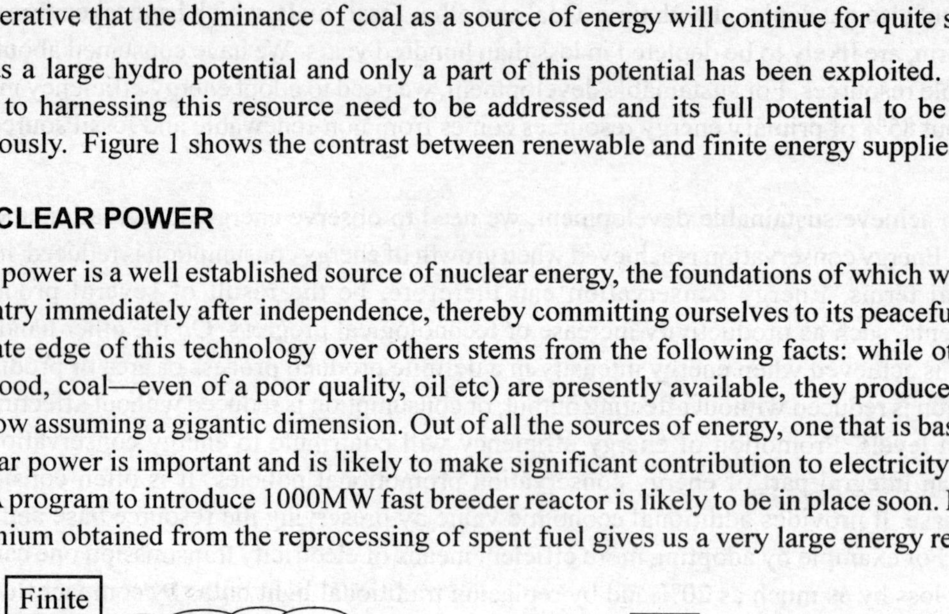

FIGURE 1: Contrast Between renewable (green) and finite (brown) energy supplies Environmental Energy flow ABC, harnessed energey flow DEF. (Adapted from Twidell and Wein (2006), Renewable Energy Resources

TABLE: 1

S. No.	Source/System	Estimated Potential	Achievement as on 30 September 2008
I	**Power from renewable**		
A	**Grid-Interactive renewable Power**	(MW)	(MW)
1.	Wind power	45 195	9521.80
2.	Bio power (agro residues and plantations)	16 881	656.60
3.	Bagasse congeneration	5 000	993.83
4.	Small hydro power (up to 25 MW)	15 000	2220.99
5.	Energy recovery from waste (MW)	2 700	55.25
6.	Solar photovoltaic power	—	2.12
	Sub Total (A)	**84776**	**13450. 59**
B.	**Captive/combined heat and power/distributed renewable power**		**(MW)**
7.	Biomass/cogeneration (non-bagasse)	—	136.70
8.	Biomass gasifier	—	102.21
9.	Energy recovery from waste	—	31.07
	Sub Total (B)	—	**269.98**
	Total (A+B)	—	13720.57
II.	**Remote village electrification**	—	5379 villages/hamlets
III.	**Decentralized energy systems**		
10.	Family-type biogas plants	120 lakh	40.32 lakh
11.	Solar photovoltaic systems	50 MW/km^2	120 MWp
	i. Solar street lighting system	—	70 474 nos
	ii. Home lighting system	—	434 692 nos
	iii. Solar power plants	—	697 419 nos
	iv. Solar power plants	—	8.01 MW_p
	v. Solar photovoltaic pumps	—	7148 nos
12.	Solar thermal systems		
	i. Solar water heating systems	140 milion m^2 collector area	2.456 million m^2 Collector area
	ii. solar cookers		6.37 lakh
13.	Wind pumps		1342 nos
14.	Aero generator/hybrid systems		723.00 kW
IV	**Awareness programmes**		
15.	Energy parks	—	516 nos
16.	Aditya solar shops	—	269 nos
17.	Renewable energy clubs	—	521 nos
18.	District advisory committees	—	560 nos

large source of uranium and thorium which exists in the country is capable of making this a possibility (Kakodkar, 2002).

Given this scenario, a large number of villages and those in semi urban areas, which are without and with short supply of electricity, are likely to be immediate beneficiary. This method of generating energy has the advantage that does not contribute to the generation of gases that contributes toward pollution.

An additional support to this emerges from the fact that Indian industry has now reached a level of maturity, whereby it can provide support to the increasing infrastructure demands. In India the generation of energy from various sources is presented in Table 1.

4.0 RESULTS

Above scenario of energy production has resulted in the following:

4.1 Environmental Pollution

Since the rate of energy consumption is much higher than it was ever before and demand is on the increase, there is a vast negative balance between source depletion and replenishment by natural processes. This has also resulted in uncontrollable rise in pollution due to air, water, soil and noise besides electropollution, making the very survival of life on planet earth difficult. A major cause among these is air pollution.

4.2 Greenhouse gas

The major greenhouse gases include CO_2, CH_4, NO_X (Table 2). An increase in the level of CO_2 is causing a temperature rise, leading to global warming and subsequent melting of glaciers. This results into climate change - a phenomena of utmost concern. This has serious economic and social dimensions. Climate change is linked to CO_2 emissions. This is suggestive of sharing growth among nations, which has serious economic overtones.

TABLE 2: Green house Gas

GHG (CO_2 equivalent) 1994	1,228,540 Giga gram per year (Gg)
Per capita GHG (1994)	1,3 tones
Main constituent of GHG (1994)	CO_2 (65%), CH_4 (31%), N_2O (4%)
Main contributors of GHG	Energy Sector 61%. Agriculture 28%, Industrial process 8%, waste 2%, Land use and land use change and forestry (LULUCF) 1%

Source: MoEF, (June 2004), India's Initial National Commitments to the United Nations Framework convention on climate Change, Govt. of India.

Electronic waste (e-waste) has become another environmental menace. India generates around 1,050 tonnes of electronic scrap each year, which is likely to increase. This also generates toxic materials like cadmium, mercury, etc. making e-waste a major health hazard.

The emerging scenario has led to a tremendous pressure on survival of species types (plants and animals). The number of such species is registering a significant decline, thereby posing a serious threat to the sustainability of ecosystem itself. A demand for green energy is now better understood, the quantum and the methodology of its exploitation is under constant review. The better use of solar, the ultimate source of green energy is now widely discussed, and increase in its efficiency usage remains a matter of continued investigation. Government of India has taken a major initiative by committing to produce 5% energy by solar and wind (Sharma, 2009).

Any attempt to increase the resources for energy generation is to increase the pressure on the clean environment and therefore the issue is one of constant concern. It is suggestive that an immediate task would be to increase the efficiency of the presently adopted means. As an example, the loss due to electricity transportation in India is about 20%, which can be reduced to generate a corresponding extra power. A public awareness to utilize resources with greater care would play an important role. A suitable fresh area may be assigned for tree plantation, acting as carbon sink. In fact, carbon credits may be assigned on a unit basis and any extra use may have to be heavily levied. Most of such efforts are contributory towards maintaining a clean environment. This implies that whatever may be the form of energy production the end result is one of generating and controlling the pollution, and providing maintenance to the population at large. The problem is interwoven, daunting, not well understood and solutions-sought but the slide is yet unstoppable.

Another related problem is the ever increasing demand of water by the industry and by the increasing population, leading to water scarcity. This is not only related to food security but also to the climate change and ultimately the survival of the planet. A related issue is how the regulated supply of water from the dam can improve ground water and surface water. It is again suggestive that the distribution of water may have to be rationed.

The decline and degradation of natural resources such as land, soil, forests, biodiversity and ground water, resulting from current unsustainable use patterns are likely to be aggravated due to climate change in the next 25-50 years. Besides many sub-continents, India is also facing severe land degradation and freshwater scarcity problems.

Increasing pollution has caused stress on every component of living system and a solution to this is no where on the horizon. It is therefore imperative to look at the issue from a holistic point of view. It is one of paramount importance in the third world and in the developing countries for their inaccessibility to the emerging technologies. An efficient use of existing resources along with their sustainability is interrelated along with critical aspects relating to biodiversity.

5.0 DISCUSSION

5.1 Remedial Measures

Even with more efficient burning of fossil fuels, an enormous quantity of CO_2 will still be emitted from fossil–fired plants over the coming decades and contribute further to global warming. A novel approach to the problem is carbon sequention (also called carbon capture), in which carbon dioxide is captured before it enters the atmosphere. One idea is to isolate the CO_2, liquefy under pressure and can then be sealed. However, carbon sequention is an expensive technology.

Pollutants from burning forests, crop wastes, and fossil fuels can form aerosols that disperse worldwide (aerosols are tiny liquid and solid particles suspended in the air). Many of these are damaging to health but the overall effect of these aerosols is to reduce the rise in temperature by shading the earth. The levels of aerosols are likely to increase than those of CO_2, due to emission controls, the effect of CO_2 on temperature will be exacerbated. One possible important negative feedback effect suggested is that global warming might cause a drying out of the upper levels of the troposphere. The decrease in water vapour will reduce the greenhouse effect and hence reduce the temperature rise.

Renewable energy concerns include energy security, fossil fuel pricing, climate change, air pollution, supply efficiency and other issues that are uniquely poised to be address. Like plants, solar cells turn light into energy. Plants do this inside vegetative matter, while cells do it in a semiconductor crystal doped with impurity atoms. Current solar cells cannot efficiently convert the incoming light into usable energy because some of the light escapes back into the air. Additionally, sunlight comes in a variety of colours and the cell might be more efficient at converting bluish light while being less efficient than reddish light. Hence, there is an inherent element of uncertainty in this approach.

The nanoparticle approach seeks to partially remedy problems of efficiency. This lies in creating a tiny electrical disturbance called "surface Plasmon". When light strikes a metal it set up waves on the surface of the metal. These waves of electrons then moves like ripples on the surface of a pond. If the metal is in the form of a tiny particle, the incoming light can make the particle vibrate, thus effectively scattering the light. At certain colours, the scattering process is particularly strong. Use of nanoparticles causes the incoming sunlight to scatter more fully, and inside the solar cell. Secondly, by changing the size of nanoparticles, light capture can also be improved (Catchpole and Polman, 2008).

The enhancement of biomass production and its utilization is yet another possibility of improving the scenario. The problem with renewable energy is their status cost, dependence on subsidies, incentives, besides uncontrollable environmental parameters. Renewable energies are often suitable for distributed generation/decentralized energy systems. In order to mainstream renewable energies in India, it is necessary to focus on capacity building. Unfortunately, at the present level of technological development, it is not possible for central power stations and yet is expensive due to high capital costs and their low availability factor. These resources also suffer from seasonal variations.

While new and fresh approaches to discover energy sources will continue, the approach to achieve higher efficiency has gained immense importance. This has led to control of high level of CO_2 emissions .The industrial sector accounted for some 41% of global primary energy economy demand and approximately the same share of CO_2 emissions.

6.0 CONCLUSIONS

While there are many ways to evolve a methodology for sustainable development, some inherent component of strategy can be visualized. Short term and long term strategy for energy use and storage which can result in sustainable development are:

- Efficient generation and transmission, and distribution and end use can reduce the emission of greenhouse gases (GHGs). A better idea will be to adopt renewable sources of energy. Along with this, an assessment of actual needs for the kind of population (their survival needs) is also of paramount importance. A detailed data compiled by The Energy and Resource Institute (TERI) study from 32 villages in the part of Jharkhand revealed that there is a vast variation in the agriculture load requirement. There is no linear relationship between the load requirement and the number of households, as the agriculture loads is dependent upon the local land holdings and crop patterns. The data indicate that the average agriculture load per household ranges from 0.075 to 0.185 kW, hence the former can be taken as the bare requirement. Hence, electricians need to have these figures in mind to meet their demands in a comprehensive manner (TERI, 2003).
- Adoption of forest conservation, reforestation and sustainable management practices can contribute to conservation of biodiversity, water shed protection, rural employment generation. This will lead to subsistence of forest dwellers and carbon sink enhancement.
- Efficient fast and reliable public transport systems such as metro-railways can reduce urban congestion, local pollution and green house gas emissions.
- Reduction of natural gas flaring
- Improving energy infrastructure
- Privatization (domestic land and foreign) of energy sector

6.1 Carbon Credits

This is a concept used internationally to provide an incentive to the developing countries to reduce green house gas (GHG) emissions, by shifting to cleaner technologies. Under this protocol, each company that shifts to cleaner technologies obtains to its account, one credit point per ton of CO_2 emissions. This credit point that company obtains is carbon credit, which is a right to emit one tonne of CO_2. The protocol imposes target commitments upon countries, which in turn, set emission quota on companies in their country, in order to fulfill their quota (Misra and Rangwala, 2009).

By extension of the above, those companies whose emissions are above the permitted quota buy carbon credits from companies who have excess credit to their account. Hence, protocol imposes target commitments upon countries, which in turn, set emission quota on companies in their country. Allowing such credit to be bought and sold makes it possible for companies to achieve their target commitments. These carbon emission certificates may be purchased and sold like any other trading commodity. Since this is going to be an expansive proposition, puts pressure on companies to follow environmental norms.

Carbon Trading in India: India is one of the leading generators of carbon credits through clean development mechanism (CDM) projects and has a great scope in emission trading. The forecasted credit trading in India is expected to touch US Dollar 100 Billion by 2010. At present, the total number of registered projects is over 300, which accounts for about 1/3 of the total CDM projects registered with the United Nations Framework Convention on Climate Change (UNFCC). The total issued credits with India stands at about US Dollar 34 Million.

There are many ways to pursue sustainable development strategies that contribute to mitigation of climate change. An example is leadership in energy and environmental design of buildings, which uses 69 point criteria to award a certificate at platinum, gold and other levels to buildings. Criteria for evaluation include: sustainable sites, water efficiency. An integrated approach involving all these issues is thus required to be pursued to attain the goal. A comprehensive view is presented in Figure 2.

6.2 Climate change and sustainable development

Sustainable development has become part of all climate change policy discussion at the global level, particularly due to adoption of Agenda 21 and the various Conventions resulting from the UNCED-1992. The generally accepted and used definition as given by the Brundtland Commission is development that meets the needs of the present without compromising the ability of future generations to meet their own needs (WCED, 1987). Sustainable development has become an integrating concept embracing economic, social and environmental issues. Sustainable development does not preclude the use of exhaustible natural resources but requires that any use be appropriately offset. This concept is not acceptable to many developing countries since it seems to disregard their aspirations for growth and development. Further, sustainable development cannot be achieved without significant economic growth in the developing countries (Goldemberg *et al.*, 1996).

Three critical components in promoting sustainable development are economic growth, social equity and environmental sustainability. The question often asked is, should the current economic growth (GNP, employment, etc.) be sacrificed for long term environmental conservation? Policy makers in developing countries often perceive a trade off between economic growth and environmental sustainability. However, there is a growing evidence to show that environmental conservation for sustainability of natural resources is not a luxury but a necessity when considering long term economic growth and development, particularly in the least developed countries. The decline and degradation

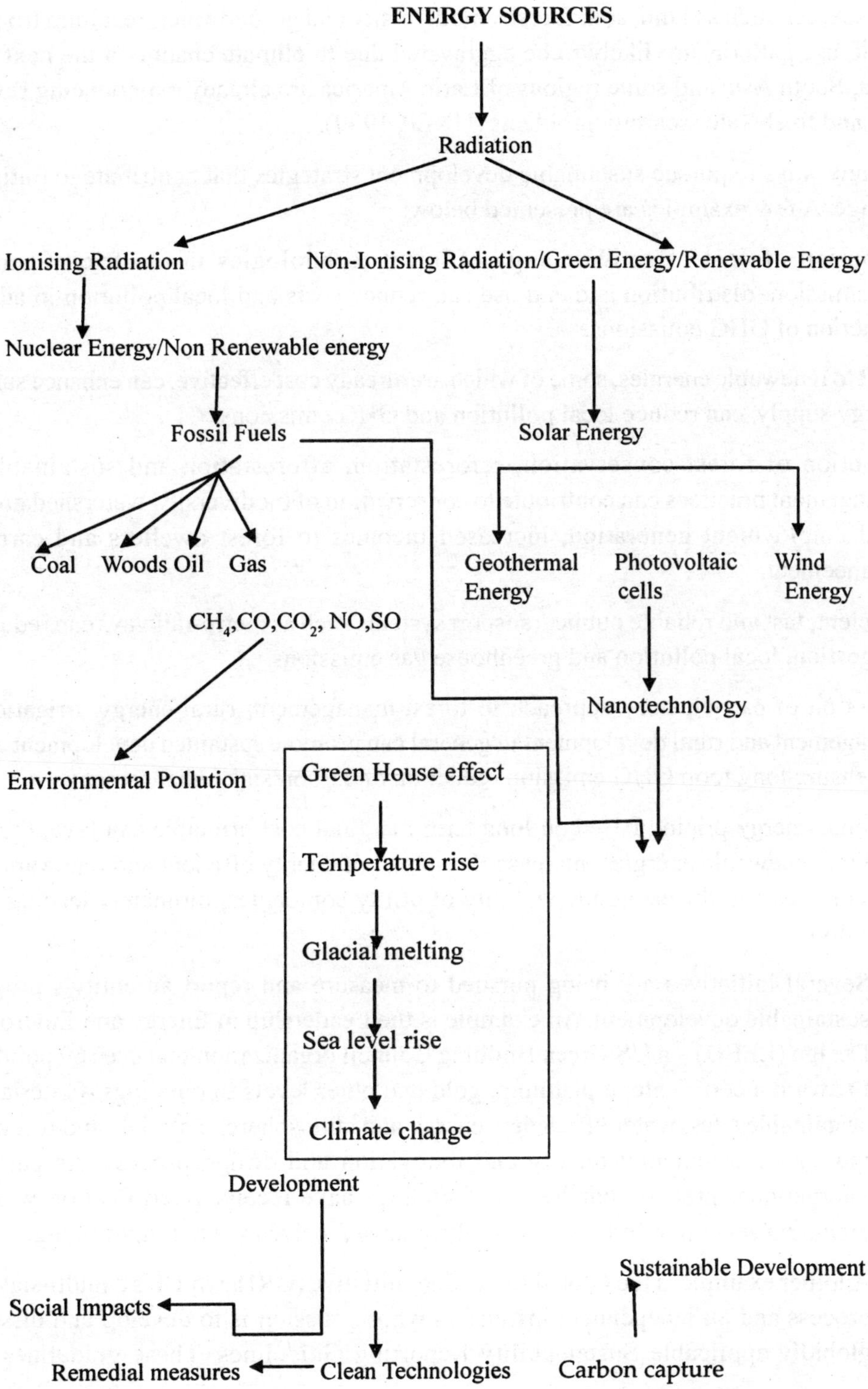
ENERGY SOURCES
Radiation
Ionising Radiation
Non-Ionising Radiation/Green Energy/Renewable Energy
Nuclear Energy/Non Renewable energy
Fossil Fuels
Solar Energy
Coal
Woods Oil
Gas
Geothermal Energy
Photovoltaic cells
Wind Energy
CH4,CO,CO2, NO,SO
Nanotechnology
Environmental Pollution
Green House effect
Temperature rise
Glacial melting
Sea level rise
Climate change
Development
Sustainable Development
Social Impacts
Remedial measures
Clean Technologies
Carbon capture

of natural resources such as land, soil, forests, biodiversity and groundwater, resulting from current unsustainable use patterns are likely to be aggravated due to climate change in the next 25 to 30 years. Africa, South Asia and some regions of Latin America are already experiencing severe land degradation and freshwater scarcity problems (UNEP, 1999).

There are many ways to pursue sustainable development strategies that contribute to mitigation of climate change. A few examples are presented below:

- Adoption of cost effective energy efficient technologies in electricity generation, transmission, distribution and end use can reduce costs and local pollution in addition to reduction of GHG emissions.
- Shift to renewable energies, some of which are already cost effective, can enhance sustainable energy supply, can reduce local pollution and GHG emissions.
- Adoption of forest conservation, reforestation, afforestation and sustainable forest management practices can contribute to conservation of biodiversity, watershed protection, rural employment generation, increased incomes to forest dwellers and carbon sink enhancement.
- Efficient, fast and reliable public transport systems such as metro-railways can reduce urban congestion, local pollution and greenhouse gas emissions.
- Adoption of participatory approach to forest management, rural energy, irrigation water management and rural development in general can promote sustained development activities and ensure long term GHG emission reduction or carbon sink enhancement.
- Rational energy pricing based on long term marginal cost principle can level the playing field for renewable energies, increase the spread of energy efficient and renewable energy technologies and the economic viability of utility companies, ultimately leading to GHG reduction.

1. Several initiatives are being pursued to measure and report an entity's progress on sustainable development. An example is the Leadership in Energy and Environmental Design (LEED) – a US Green Building Council organization that uses 69 point criteria to award a certificate at platinum, gold and other levels to buildings. Criteria include sustainable sites, water efficiency, energy and atmosphere, materials and resource use, indoor environmental quality and innovation and design process. As part of this international process, hundreds of buildings have received certification worldwide, including several in India some of which have received the platinum rating.
2. Another example is the Global Reporting Initiative (GRI), which is a multi-stakeholder process and an independent institution whose mission is to develop and disseminate globally applicable Sustainability Reporting Guidelines. These guidelines are for

voluntary use by organizations for reporting on the economic, environmental and social dimensions of their activities, products and services. Started in 1997, GRI is an official collaborating centre of the United Nations Environment Programme (UNEP) and works in cooperation with UN's Global Compact.

The motivation for using the above types of reporting criteria is diverse. In a recent evaluation of GRI, 85% of the reports addressed climate change and 74% of respondents identified economic reasons and another 53% gave ethical reasons for reporting their company's performance to GRI. India's ITC Limited, for example, has won a platinum LEED rating for its Gurgaon (Haryana, India) building and also reports its sustainable development performance to GRI as a carbon positive corporation, i.e. it sequesters more carbon than it emits.

6.3 Climate change at global level and possible implications for developing countries:

The first commitment period of the Kyoto Protocol ends in 2012. Given the relatively short period to its termination, participating countries have been engaged in several dialogues within the UNFCCC auspices and elsewhere about post 2012 commitments on emissions reductions and adaptation measures. The discussion at these dialogues ranges from mandatory economy wide targets to sector specific ones on all countries, to bilateral and/or multilateral agreements to voluntarily reduce GHG emissions. Industrialized countries, except notably US and Australia, already have agreed to adhere to economy wide targets and they are keen to continue such an approach post 2012. Others have proposed sector based approaches that require adoption of voluntary carbon intensity targets for the energy and major industry sectors in all countries. Key questions include: how are sectors defined, how does the voluntary target setting process unfold, are there separate benchmark targets for new and existing facilities within a sector, when and how are reductions generated that can be sold. Sectoral benchmarks as part of a country target, is another issue to be resolved. Studies (Worrell and Price, 2005; Galitsky *et al.,* 2005) have tested the use of tools and voluntary approaches for benchmarking energy efficiency and carbon intensity in a variety of industrial sectors in both industrialized and developing countries, and these could form the basis for setting verifiable sectoral targets. A key to making a sector based approach attractive to developing countries is to create financial incentives to adopt such a target. A combination of technology finance and CDM/trading revenues could serve as one basis for making such targets attractive to developing countries.

Addressing adaptation in post 2012 international climate regime could be done through the use of insurance based approaches, mainstreaming and innovative financing mechanisms, to at least partially off set the effect of environmental damage. However, this exercise is parasitic in nature and does not contribute to generation of primary sources. Anyway the implications for developing countries with nascent insurance industries, however, need to be examined.

Over time, as indicators and measurement tools become available, the pursuit of sustainable development is moving out of academic world and being put into practice increasingly by institutions and private industry. The trend is likely to be strengthened globally as nations come to recognize the limits on access to and development of natural resources.

While in general, most people agreed that this is an important aspect of adaptation response, its implications for on the ground actions need to be addressed.

For developing countries, enhancing the economic well being of their citizens remains an urgent and pressing goal. To the extent the new climate architecture would be perceived as a barrier to this. For coming decades, the GHG emissions per citizen from most developing countries would remain significantly below those in industrialized countries. For most developing countries, this is the century when majority of their citizens are likely to first experience economic prosperity. It would be an exciting time and experience to rationalize the economic well being with the climatic change scenario.

3. We can also solve the problem of carbon emission with the help of algal farm by recycling the carbon. The simple, direct method of greenhouse gas mitigation is the removal of CO_2 from stack gases, followed by long term sequestration of CO_2 by microalgae ponds. Large raceway algal farms could be located adjacent to power plants. The bubbling of flue gas from a power plant into these ponds provides a system for recycling of waste CO_2 from the burning of fossil fuels. According to one estimate, 1,000 tonnes of CO_2 can be sequestered by Algae/acre/year. The concept of coupling a coal fired power plant with an algae farm provides an elegant approach to GHG mitigation of the CO_2 from coal combustion into a useable liquid fuel. Thus, microalgae technology can extend the useful energy we get from coal combustion and reduce carbon emissions by recycling waste CO_2 from power plants into clean burning bio-diesel. Microalgal-based carbon sequestration technologies can, in principle, not only cover the cost of carbon capture and sequestration but also produce environment friendly bio-diesel.

Large scale plantation of Jatropha plants to generate bio-diesel has also been an attractive proposition but has been associated with several difficulties. A few have been the use of large cultivation area and the consumption of valuable water resources. The generation of waste will add to additional burden on pollution. However, government has in principle agreed to mixing of a proportion (5-10%) of the diesel with the petrol.

Recent soaring oil prices, diminishing world oil reserves and the environmental deterioration associated with fossil fuel consumption and green house gases emission have generated renewed interest in using algae as an alternative and renewable source of energy. Algal biofuels are likely to have negligible impacts on the World's food supply than conventional biofuel producing crops. Microalgal Bio-diesel is technically feasible and carbon neutral renewable alternatives. Transition to algal biomass can deliver development and climate co-benefits. Use of the biorefinery concept and advances in raceway ponds engineering will further lower the cost of production. It is the only renewable bio-diesel that can potentially completely displace liquid fuels derived from petroleum. If the microalgal biorefinery concept can be adapted to a country like ours, it could become a highly distributed source of fuel oil and perhaps make us self reliant and improve our economy many folds. In the present scenario, algal biomass is a key link between energy, local environment and climate change and further research are necessary to unlock full potential of algae (Khan, 2008).

REFERENCES

1. Catchpole, K.R. and Polman, A., 2008. Plasmonic solar cells. Optics Express, 16 (26), pp. 21793-21800.
2. Climate Change, 2001. Impacts, Adaptation and Vulnerability, Summary for policy makers and technical summary of the working group II report, Intergovernmental panel on climate change (2001), Geneva, Switzerland, IPCC.
3. Galitsky, C., Worrell, E., Radspieler, A., Healy, P., Zechiel, S. and Fetzer Vineyards, 2005. BEST Winergy Guidebook: Benchmarking and Energy and Water Savings Tool for the Wine Industry. LBNL 3184.
4. Goldemberg, J., Squitieri, R., Stiglitz, J., Amino, A., Shaoxiong, X. and Saha, R., 1996. Introduction: scope of the assessment. In: Climate Change 1995: Economic and Social Dimensions of Climate Change, Contribution of Working Group III to the Second Assessment Rebridge University Press Cambridge.
5. India's Initial National Communications to the United Nations Framework convention on climate Change, 2004. Ministry of Environment and forests, New Delhi.
6. Kakodkar, A., 2002. Nuclear power in India: An inevitable option for sustainable development of a sixth of humanity. World Nuclear Association Annual Symposium, Londan, pp. 1-7.
7. Khan, S.A, 2008. Algae: A novel source of renewable energy and carbon sequestration, Akshay Urja, 2(2), pp. 14-18.
8. Mishra, S. and Rangwala, K., 2009. Carbon credits –In Response to climate Change. International Legal Consultants, pp. 1-4.
9. Ravindranath, N.H. and Balachandra, P., 2009. Sustainable bioenergy for India: Technical economic and policy analysis (In Press).
10. Report of the World commission on Environment and Development (WCED), 1987. Out common Future, Oxford University Press, New York.
11. Sanghi, A., 2003. Workshop on improving electricity access and role of market based subsidy mechanisms, organized by public private infrastructure advisory facility (PPIAF), 22-23 September 2003, New Delhi.
12. Sharma, C., 2009. Times of India, February 29.
13. Srivastava, L., Rehman, I.H., 2006. Energy for sustainable development in India:Linkages

and strategic direction. Energy Policy 34, pp. 643-654.

14. The Energy and Resource Institute (TERI), 2009. Strategies for electrification of villages inaccessible to the grid in the state of Jharkhand. TERI, New Delhi.

15. Twidell, J. and Wein, T., 2006. Renewable energy resources, pp. 8.

16. United Nations Environment Program (UNEP), 1999. Global Environmental Outlook, Oxford University Press.

17. Worrell, E. and Price, L., 2005. An integrated benchmarking and energy savings tool for the iron and steel industry. Int. J. Green Energy, In Press.

5

"Maurice Ile Durable" Green Mauritius: Towards Sustainable Development

Doonunjoy Dassaye
Ministry of Renewable Energy and Public Utilities,
Port Louis, Mauritius
E-mail: ddassaye@mail.gov.mu

Mauritius, a small island and developing state is working on a number of energy policies and strategies to be able to achieve sustainable development. The Maurice Ile Durable Fund (MIDF) has been set up on the 19 of June 2008 under the Finance and Audit (Maurice Ile Durable Fund) Regulations 2008, in line with programme-based budgeting to achieve Government vision and goals for sustainable development. Accordingly, the overarching objective of the MIDF is to pave the way for Mauritius to move towards a 'green' island in the next twenty years by enlisting the collaboration of all stakeholders. The policies and strategies are directed towards the preservation of local natural resources with a view to adapting to climate change. Projects are financed to explore and harness all potential for local sources of renewable energy and to reduce dependency on imported fossil fuels. In addition, promotion of energy savings through the "Maitrise de la Demande en energie" is also encouraged by regulating the efficiency of new electrical appliances used by households and the imposition of appropriate code of practice for industrial and commercial utilization of energy. Programmes are being financed to reduce consumption of fossil fuels, achieve greater efficiency in the use of energy in enterprises, offices, homes, public sector, and transportation sector and in hotels. Schemes are financed to encourage innovation by households as well as by business to produce their own energy requirements and for the sale of any surplus at a premium. The MID Fund also finance projects and programmes to support efforts to protect the environment through recycling of waste, to encourage more efficient use of energy and to increase reliance on renewable energy. A programme is presented for research and analysis pertaining to the development of renewable source of energy and consumption trends and to ensure environmental sustainability. Energy management programmes through networking with local and international partners is underway. The supply side strategy is underpinned on the use of renewable energy. Mauritius has been using bagasse, the byproduct of the sugarcane industry. Mini-hydro potential is being exploited. A major wind-park project is well under way to improve the energy mix of the country. Use of solar water heaters for substitution of electricity and gas for domestic water heating has been enhanced. Actions are taken to allow households to generate their own electricity from either solar or wind and sell any excess to the national grid.

KEY WORDS: *Policies, Strategies, Renewables, Sustainable, Development, Recycling and Conservation.*

1.0 Introduction

The price of oil has gone up from around $40 a barrel some three years ago to $136. The petroleum import bill of Mauritius has risen from Rs6.5 billion in 2000 to Rs 22.3 billion in 2007. The share of petroleum in the total import bill has gone up from 12 percent to 18 percent over the same period and keeps on rising. It is clear; the world has entered an era of expensive energy. As a country that relies on around 80 percent of its energy on imports, Mauritius is extremely vulnerable. Though the price of oil has gone down, Mauritius has to reduce its reliance on fossil fuel. Given that oil reserves are expected to last some 40 years or so as predicted by many experts. Mauritius has developed an energy policy and strategy for the medium to long term that would reduce fossil fuel reckoning with the constraints that the country has in terms of its geographical location, availability of other less volatile and better distributed reserves of fossil fuel.

This paper provides a background of the institutional set-up of vision and goals for sustainable development in the energy sector in Mauritius, gives a description of the current electricity mix, the energy policy and initiatives that are necessary to allow the country to achieve its goal and objectives for a Green Mauritius.

2.0 legal and institutional framework in the energy sector

The Ministry of Renewable Energy and Public Utilities (MREPU) is responsible for the energy, water and wastewater sectors, and oversees the power utility, the Central Electricity Board (CEB). The Ministry of Business, Enterprise and Cooperatives oversees the petroleum industry, including the State Trading Corporation, which is responsible for the import of all petroleum products. The import and distribution of coal are not regulated. The principal legislations governing the sector include the: (i) Electricity Act of 1939 (amended in 1991); (ii) Electricity Regulations of 1939; (iii) Central Electricity Board Act (1964); (iv) Petroleum Act of 1970 (amended in1991); and (v) Consumer Protection (Price and Supplies) Act of 1998 (vii). The Finance and Audit (Maurice Ile Durable) Regulations 2008.

The Ministry of Renewable Energy and Public Utilities manager and ministers, has provided a Maurice Ile Durable Fund of a total budget of Rs 1.0 Billion to be supplemented by an annual amount of about Rs 300 million from Government in subsequent years. This is for promoting the use of renewable energy through the provision of appropriate financial incentives and grants.

3.0 Analysis – demand and supply in Mauritius

3.1 Supply

Mauritius has no fossil fuels and, therefore has to import approximately one million tons of petroleum products, comprising 380 cst and 180 cst heavy fuel oil, gasoline, diesel and LPG, and 650,000 tons of coal annually. The balance of supply comes from local renewable sources, namely bagasse, a by-product of sugarcane processing, and hydro.

3.1.1 Solar Energy

As a tropical island, Mauritius has a very good solar regime of an average potential annual solar radiation value of some 6 kWh/m². Government through the MID Fund has provided for an outright grant of Rs 10,000 for every solar water heater purchased by 31 December 2009 through the Solar Water Heater scheme operated by the Development Bank of Mauritius.

3.1.2 Wind Energy

Mauritius has a wind regime with an annual average speed of 8.1 m/s at 30m above ground level. A number of project proposals have been considered in respect to tapping energy from the wind on large scale.

3.1.3 Bagasse

Bagasse is the primary energy resource used by the sugar industry to meet all their energy requirements in terms of heat and electricity generation. Surplus power from bagasse is fed into the national grid. Consideration is being given to more efficient use of bagasse.

3.1.4 Hydropower

Hydropower potential can be considered as fully tapped in Mauritius. Two mini-hydro projects at Midlands and La Nicolière are on-going and are expected to be completed by 2010. Both of them would generate a total of 750 KW. Presently the total combined installed capacity for hydro power amounts to 59 MW, and there is competing demand for the existing water resources.

3.1.5 Electricity Demand

Total electricity generation in 2008 is as follows:—

Hydropower	:	107.5 GWh
Thermal & Heavy Fuel Oil	:	761.6 GWh
Thermal & Kerosene	:	6.6 GWh
Coal	:	998.7 GWh
Bagasse	:	366.4 GWh

It has been forecast that in 2025 about 4500 GWh of energy would have to be generated to meet the demand, with a peak power demand of more than 700 MW. There is a tendency for coal based energy to replace heavy fuel oil, thus reducing the country's dependence on oil.

4.0 Prospects and challenges

4.1 Government's energy policy

The energy policy for the short and medium-to-long-term of Government is as follows: Firstly, availability, security and diversity of supply; secondly, affordability; thirdly, energy conservation; and fourthly, reform of the power market and industry.

In addition, the strategy of Mauritius is to replace imported fossil fuel by local renewable sources and to better manage demand. We plan to double the share of renewable sources of energy to produce electricity to around 40 percent within the next decade. This is achievable by increasing the megawatts from renewable sources and the megawatts from energy savings.

A Maurice Ile Durable (MID) Fund with a provision of Rs 1.3 billion has been set up to kick start the Fund which will support efforts to protect the environment through recycling, to encourage more efficient use of energy and to increase reliance on renewable energy. Hence, this will build the energy independence whilst allowing Mauritius to do its share in the fight against global warming. The Fund will mobilize resources from taxes, Government subsidies, development partners, carbon credits, and the private sector including airlines offsetting their carbon footprints.

The Fund, under the aegis of the Ministry of Renewable Energy and Public Utilities, supports action in three crucial areas namely:

(i) Supply side;

(ii) Demand side; and

(iii) Institutional Framework.

The supply side:- All potential for local sources of renewable energy and their use to replace imported fuel is being explored. Mauritius has renewable sources of energy on land, in the air and in the sea. The country is exposed throughout the year to the South East Trade Winds and hence the exploitation of the wind potential is a priority for development, both in Mauritius and in Rodrigues. It is to be noted that Government is also relying on Technical Assistance from other countries such as France and India, whose assistance would be crucial in the development of the sector.

In this context, the Ministry of Renewable Energy and Public Utilities, in collaboration with the Public Private Partnership Unit of the Ministry of Finance and Economic Empowerment launched a PPP project at Bigara for 25 to 40 megawatts (MW) of wind energy on a BOO scheme which is expected to cost about Rs 1 billion and yield 50 to 60 million kilowatt hours (KWh) representing 3 percent of electricity consumption. The Bigara site has been chosen given its very good wind potential. A Transaction Adviser is being appointed to advise and assist the PPP Unit and the MREPU in the preparation of the Request for Proposal.

4.2 Electricity from bagasse

Power generation from the sugar industry in Mauritius, in particular from bagasse, can be improved if there are fewer mills, but bigger in terms of cane handling capacity. Given the phased reduction of the guaranteed price of sugar with the European Union (EU), the sugar industry is compelled to accelerate its centralization programme as elaborated in the Multi Annual Adaptation Strategy (MAAS) 2006-2015 agreed with the EU for support in the context of the restructuring of the industry. In that context, potentially only four sugar factories will stay in operation in the medium to long term, one each in the North, East, West and South of the country. Active discussions are currently being held between the stakeholders on the strategy for power generation from bagasse and coal from the sugar industry.

In the future, Mauritius may have to resort to higher bagasse yield type of sugarcanes to further improve the contribution of bagasse-energy in its energy mix. Such a strategy can be beneficial to both the energy and agricultural sectors. Bagasse will, therefore, continue to play a vital role in the country's energy mix, not only in the short-to-medium, but also in the longer term. In that context, the Mauritius Sugar Industry Research Institute (MSIRI), an institution specialized in sugarcane research, is playing an important role in developing the right varieties of sugarcane. A 50% increase in the amount of bagasse available from an increase in the fibre content of cane for example would allow the generation of approximately 900 GWh of electricity from bagasse annually that is about 40% of current total consumption.

4.3 Electricity from hydro

As indicated earlier, the hydropower potential in the country is nearly fully exploited. However, two mini hydro power stations are being constructed at an existing dam for supply of potable and irrigation water, where an outlet pipe had already been provided during the construction of the Midlands dam. Electricity will be generated when water will be released to a downstream reservoir via the La Nicolière feeder canal. A similar mini-power station will be constructed on the feeder canal, where there is sufficient head to install a hydro turbine. The total cost of the two projects is estimated at Rs 70M.

4.4 Electricity from wind

Government policy is to substitute as far as possible imported oil for power generation, and wind power development is considered imperative. A Memorandum of Understanding has been signed with the Government of India for renewable energy development in Mauritius, under which Mauritius, *inter-alia,* can benefit from technical assistance in the various fields of renewable energy. In this context, Government relies on the assistance of Government of India.

The Ministry of Renewable Energy and Public Utilities in collaboration with the Public Private Partnership (PPP) Unit of Ministry of Finance and Economic Development (MOFED) will launch a RFP to develop a wind farm at a site called Bigara in the central area of the country with very good potential on a Build Own Operate scheme with the installation of bigger wind turbines for a capacity of 25-40 MW. A transaction adviser has been appointed to advise and assist the PPP Unit and MREPU in the procurement process till the commissioning of the wind farm. Given the initial high costs involved in the development of renewable energy projects.

As regards Rodrigues, the CEB has proceeded with the installation of two 275 kW wind turbines at Grenade estimated to cost Rs 40 M. Over and above the two units, with the financial assistance of Government five additional units will be installed estimated to cost about Rs 100M. The objective of the CEB is to increase the number of wind turbines at Grenade to ten in the medium term.

4.5 Solar energy

As at to date, 25,000 households are equipped with solar water heaters. A study carried out by the Ministry of Renewable Energy and Public Utilities has shown that the main barrier for further market penetration of solar water heaters is its high price. However, in view of the rising energy prices, Government has put a scheme with a direct grant of Rs 10,000 to households from July 2008 for the purchase of a solar water heater through the MID Fund.

Nonetheless, to increase further the use of solar energy in the country from a long-term strategic perspective, it is compulsory for promoters of development projects to make maximum use of solar energy for water heating in buildings and photovoltaic systems for outdoor lighting by introduction of appropriate conditions in the EIA license for new building projects. The target group includes new hospitals, hotels and IRS projects. Moreover, private households will also be encouraged through appropriate incentives to install photovoltaic panels to meet their own electricity needs and to supply any excess to the CEB grid. A grid code is under preparation.

4.6 Municipal wastes and landfill gas

The CEB and a private promoter is finalising the setting up of a 20 MW incinerator that would burn about 300,000 tons of municipal waste to provide some 144 GWh of electricity annually. In addition, the CEB is proposing to purchase about 2 GWh of electricity generated from a 3.5 MW plant that would use landfill gas.

4.7 Electricity from coal

Coal utilization is a major carrier in the country's energy mix for two reasons as follows: (i) it allows bagasse to be a reliable source for year-round electricity generation and not to remain idle during the off-crop season. (ii) Coal is the only other fossil fuel with reserves larger than oil, in addition to their even distribution around the globe. Given the huge investment in the developed world to make coal a still cleanlier and environment friendlier fuel, Mauritius would be able to use such available technologies in the marketplace in its new strategy.

4.8 Energy conservation/demand side management

The Ministry of Renewable Energy and Public Utilities has launched a comprehensive and sustained energy savings and energy efficiency campaign for the public at large. This is with a view to creating public awareness of the impacts of energy consumption on natural resources as well as the demands that consumption patterns place on electricity infrastructure. The campaign would also involve the inculcation of a culture of energy savings among school children, which would bear its fruits in the medium-to-long term. The ministry, with the help of funding agencies, is setting up an energy efficiency unit.

Mauritius is faced with a highly variable daily demand pattern: system load averages about 200 MW during overnight periods, to reach nearly 370 MW during the evening peak hours between 18.00 and 20.30 hrs. The CEB must have sufficient production capacity to meet these peak hour demands, yet much of this capacity remains idle for long periods each day and on weekends. In the national interest, regular appeals are made to industrial and commercial customers and the public at large to limit their electricity consumption during these peak hours. If customers shift their electricity consumption to hours outside this peak period, the maximum demand will be lowered and system generation, transmission, and distribution assets will be more efficiently used. For example, the CEB has been encouraging industrial customers to modify their processes so that production cycles use a multi-shift or batching mode in which electricity-intensive activities are performed outside of peak hours, and to curtail non-essential activities such as automatic water pumping, electric water heating, and washing machine use during peak hours. Furthermore, the CEB is looking into the possibility to introduce time-of-use tariffs, which would encourage off peak consumption.

The use of compact fluorescent lamps (CFL), also known as energy saving lamps, has been promoted as one of the conservation measures to reduce the use of fossil fuels for electricity generation. In this context, the CEB has procured one million CFLs for resale to its customers at half the price. Each customer is allowed to purchase a maximum of 3 CFLs. The project reconciles several objectives namely the savings of energy, reduction of lighting load, reduction of GHG, the reduction of the evening peak demand, lowering the electricity bills of customers and finally securing carbon credits from the market.

A GEF funded project on *Removal of Barriers to Energy Efficiency and Energy Conservation in Buildings* is being implemented over the next four years. During the first year of the project, the drafting of an Energy Efficiency Bill including secondary legislation to make mandatory the labeling of energy intensive domestic electrical appliances. The energy efficiency bill will be ready by mid 2009. This will give the opportunity to potential customers to choose between equipment, which generally have a low price at purchase but are not energy efficient, and energy efficient appliances which can be more expensive at purchase but have low running costs as they are energy efficient. The aim is to give customers the opportunity to make the judicious choice of their appliances.

Summer time has been adopted in Mauritius on a pilot basis whereby the clock has been moved forward by one hour for the period 26 October 2008 to 29 March 2009. Besides reducing energy consumption, the summer time has reduced peak demand by some 15 MW. The summer time will become a permanent feature if it proves successful.

The Ministry of Environment in collaboration with UNEP is actively developing a five-year project called 'Development of National Programme on Sustainable Consumption for Mauritius'. Resources use Efficiency with focus on energy is one of the five components of the programme. In this context, a number of energy conservation projects and initiatives will be identified for implementation over the next five years.

4.9 Transportation Sector

The transportation sector which accounts for about 50% of the consumption of imported oil products poses more challenges in terms of fuel substitution. Such a situation arises from the dominance of travelling by motor vehicles for personalized travelling and buses for mass transportation.

Government has a declared policy of substituting 20% of the gasoline utilized in cars by ethanol produced from molasses, a by-product from sugar processing. Such an initiative has benefits on two fronts in that it would reduce the reliance of the sector on oil with attendant benefits to the economy, and secondly it would help to boost up the revenue of the sugar industry that would offset part of the loss of revenue from the sale of sugar to the EU.

Government would proceed in phases with the E10 project at first, which comprises 10% gasoline substitution. Positive results were obtained on the use of E10 in twenty five cars over a period of three months. The bio-fuel was available at only one filling station. Under Phase II, it is proposed to make E10 available at 6 filling stations with appropriate incentives for users to shift to the bio-fuel mixture. Furthermore, the sustained high price of oil is likely to make hybrid cars more popular in Mauritius in the coming decade.

5.0 STRATEGIES AND PROJECTS UNDER THE MAURICE ILE DURABLE FUND

5.1 Medium to Long Term strategies

Medium to long term strategies, although evoked at various levels, are yet to be formulated and phased in as the options are and will still be heavily dependent on a number of key determinants, such as technological advancement in the field of renewable energy and energy conservation, and the pace at which the technologies will become affordable to the population at large. At this stage, the MIDF can consider short term strategies where consensus has more or less already been obtained at the national level.

A Strategic Business Plan has been prepared to describe the operational framework for the MIDF over the next three year-period including projects which are to be approved for either partial or total funding.

5.2 Approved Strategies

Three specific strategies are already outlined in the MID legislation, namely -

(a) Solar water heater scheme being implemented by the Development Bank of Mauritius,

(b) Replacement of conventional bulbs by CFL, being implemented by the CEB, and

(c) Bus Modernization Programme, being implemented by the Ministry of Public Infrastructure, Land Transport and Shipping

In addition, the MIDF will provide support to a few national endeavours which have received prior approval the Government. These are;

— the co-financing with the French Government of the joint Université Technologique de Compiègne and University of Mauritius assignment, to develop the 'Maurice Ile Durable' concept, including specialized courses on Energy Auditing. This exercise will constitute the basis for national consultations scheduled in June 2009.

— the funding of public awareness campaign for the implementation of summer time on a pilot basis from 26 October 2008 to 29 March 2009.

The above projects are considered as approved strategies as the scope and modalities for implementation have already been sufficiently described in the legislation.

5.3 Strategies of immediate national interest

The Legislation combined with measures announced in the June 2008 Budget Speech also provided additional strategies that have already reached a marked level of acceptance among relevant stakeholders, and the underlying rationale are unlikely to change in the forthcoming years. The implementation of such strategies are imperative and be pursued as strategies of immediate national interest. Their objectives have overall wide-reaching impact and are not disputed. These include projects aiming at sustaining activities for energy saving and energy efficiency practices, as well as recycling of solid waste.

5.4 Strategies under consideration

As regard the other projects mentioned in the legislation and Budget Speech, regular updates would be required as the status of each project depend on changes from the project idea to the end product or major policy decisions.

The MIDF Committee holds the view that these projects, together with other new project proposals received from the various stakeholders to date, are to be termed strategies under consideration. In fact, annual review of the Strategic Business Plan is expected and new initiatives should progressively be considered and included as priorities if considered to be of national interest.

Pending the national consultations to be held in 2009 and in accordance with the requirements of the MID Regulations, consultations with relevant Ministries and specific stakeholders have been engaged by the MIDF Committee since August 2008. To date, the relevant ministries and some non governmental institutions have submitted proposals requiring some support from the MIDF.

The MIDF strategies and projects will, therefore, encompass the various projects mentioned above and will make provision for the inclusion of new strategies as and when these will be received from the stakeholders. Strategies or projects that require in-depth technical expertise and advisory services prior to final examination would also be considered by the MIDF. In addition, the Ministry of Renewable Energy and Public Utilities is expected to firm up some of the initiatives falling under this category.

6.0 CONCLUSIONS

Government is promoting the use of local and renewable sources as much as possible to move away from fossil fuel. Initiatives are already being put in place for enhancing the energy potential with regard to bagasse and wind technology with a view to achieving the target of 40 % of the country's electricity supply within the next decade, inclusive of hydro power. Cultivation of new varieties of sugarcane with higher fibre content to generate more bagasse for conversion into electrical energy is being envisaged. In addition, with the introduction of a grid code over the next six months or so, Government will encourage households to generate electricity from either photovoltaic panels or wind turbines for their own consumption and the sale of any excess to the national grid.

A Maurice Ile Durable Fund with wider references has been set up. The object of the new Fund, for which Rs 1.3 billion have been earmarked, is to step up efforts to protect the environment through recycling, to encourage more efficient use of energy and to increase reliance on renewable energy.A number of measures are being introduced in the context of MID. Thus Summer Time has been introduced on a pilot basis. A new grid code is also being charted whereby Government will encourage all households as well as businesses to produce electricity with emerging technologies for renewable energy and guarantees that CEB will buy their renewable energy at a premium. A new solar water heater loan scheme by the DBM allows for a grant of Rs 10,000 for every solar water heater purchased through that scheme until December 2009. Other strategies for energy production and efficiency use include:

- Setting up an Independent Utility Regulatory Authority for the operation of the electricity market.
- Launching of a PPP project at Bigara for 25 to 40 megawatts (MW) of wind energy at an estimated cost of Rs 1 billion.

- Setting up of two new hydro units by the CEB at Midlands Dam and La Nicolière.
- Developing the Landfill Gas potential at Mare Chicose and using the gas for power generation.
- Burning solid waste for electricity production.
- Exploring the use of cane field residues mixed with bagasse as boiler fuel.
- Establishment of an Energy Efficiency Unit
- The use of energy savings lamps for street lighting and for households.
- Measures to reduce pollution due to traffic and to encourage greater use of public transport.
- Strengthening the institutional and legal frameworks relating to the energy sector.

BIBLIOGRAPHY

1. Budget Speech 2008-09, www.gov.mu.
2. CEB Annual Report 2007, ceb.intnet.mu
3. Digest of statistics, Central Statistical Office, www.gov.mu/portal/site/cso.
4. Energy Policy for Mauritius: Analysis, Projects and Challenges – Dr. P. M. K. Soonarane
5. The Outline National Energy Policy 2007-2013

ABBREVIATIONS

CEB	Central Electricity Board
CIF	Cost, insurance and freight
CER	Certified Emission Reduction
CEST	Condensing-extraction steam turbine
CFL	Compact Fluorescent Lamp
CSO	Central Statistical Office
Cst	Centistokes
DBM	Development Bank of Mauritius
E10	10 % Ethanol mixture with gasoline
EU	European Union
GEF	Global Environment Facility

GHG	Greenhouse Gases
GWh	Gigawatthour
kW	kilowatt
LPG	Liquefied Petroleum Gas
MAAS	Multi Annual Adaptation Strategy
MOFEE	Ministry of Finance & Economic Empowerment
MREPU	Ministry of Renewable Energy and Public Utilities
MRC	Mauritius Research Council
MSIRI	Mauritius Sugar Research Institute
MW	Megawatt
PPP	Public Private Partnership
RFP	Request for Proposal
Rs	Mauritian Rupees
UNEP	United Nations Environment Programme

6

Sustainable Utilization of Energy for Wealth Creation and Development: The Nigerian Scenario

B. O. Solomon*, R. S. M. Gidado, A. A. Akinrelere, N. C. Ezebuiro, G. A. Etuk-Udo and O. A. Adetunji

National Biotechnology Development Agency, 16 Dunukofia Street, Area 11 Garki, Abuja, Nigeria *E-mail: bosconsult@yahoo.com

The needs for sustainability in energy resources have driven research considerations and development options into bio-energy sources. Biodiesel, bioethanol and biogas have the highest investments in public and private interests. The abundant feedstock for these energy sources is a huge attraction for investment in technology optimization processes and product re-design for efficient conversion and utilization for transportation, manufacturing and domestic use. In Nigeria, overwhelming endowment in biomass, ranging from waste materials (municipal and industrial waste) to high energy plants (Jatropha, Neem, Cassava, Sugarcane, Maize) required for this innovative phase of global renewable energy supply is a huge advantage. Diversification of Nigeria's energy options would ease the current burden of erratic power supply, which has affected social development, production efficiency, poverty eradication and increased environmental pollution. The desire to create a balance in the energy supply mix is driving policies towards sustainability principles in energy development and consumption investments. The policies' overall aim is to reduce internal dependence and pollution arising from hydrocarbon based energy resources; stimulate new set of green industries; increase employment and diversify export. Policy implementation strategies involve awareness and capacity building in private- and public sector players and development of economic leverages in energy investment. Strong components of this strategy are technology identification and development, skill transfer and legislation to drive compliance and investment in sustainable energy, and development of market environment for the renewable/bio-energy products. The drive is necessitated by gross media contamination in hydrocarbon production sites, low employment capacity in oil related operations relative to GDP contributions, and global warming concerns. This paper presents institutional arrangements and efforts designed to develop different technologies, regulation, economic and market incentives to drive a sustainable strategy in the utilization of energy for wealth creation and development in Nigeria.

KEY WORDS: *Bio-fuels, renewable energy, utilization, diversification, production efficiency, investment, pollution.*

1.0 Introduction

The current energy situation in the world requires a rethink in terms of production and use. The need for sustainable sources of energy is creating new possibilities for research and business, new dimensions in legislation and new regulatory challenges. These contemporary issues are driven by the need to satisfy the anticipated rise in future energy requirement without continuing the wanton destruction of natural resources associated with energy exploitation. This is a huge task: achieving any tangible success in this will require alternative sources of energy. Fortunately, the concept of sustainability is gaining much political and economic impetus. However, the policy and development mechanisms to drive a sustainable energy use pattern for Nigeria is still preliminary.

Nigeria has begun a shift to renewable energy. This shift requires technology, time and policy. As in most developmental issues, the need for this shift is obvious and draws mainly from principal unfavorable antecedents such as: massive hydrocarbon pollution of water and soil following petroleum operation and rising uncertainties in the international oil market. Perhaps a remote internal cause for this shift is unconnected with the low contribution of petroleum operations to employment, relative to petroleum contribution to gross domestic product (GDP). Piecing these issues together gives a gloomy picture from an economic aspect: high unemployment rate, inflation and concentration of resources within a privileged few. Environmentally, the continuous loss of water and agricultural land in the Niger Delta region is a multi-dimensional threat with overwhelming social, political and economic crisis in the region. The list is endless. The threat is real. And most of all, the problems of the current energy regime will outlive the regime.

A peep into time suggests that we may be contending with the containment of devastations of natural resources following petroleum exploration and exploitation. If this is true already, then the need for a change to environmentally friendly and clean/renewable energy alternatives is behind schedule. Notwithstanding, acknowledgement of the problem and efforts at solution are desirable.

The strides are bold and the efforts elaborate. The Government of Nigeria is poised to make huge investments in bioenergy in both private and public sectors. The strategy is a three prong mechanism comprising awareness, technology and legislation to drive investment interests towards renewable energy. Policy is particularly essential in this because the challenge of the focus shift is capital and production intensive. Again, as with the petroleum era, Nigeria is overwhelmingly endowed with the resources for this great change. Massive arable lands suitable for some of the worlds known energy crops such as jatropha, neem, cassava, sugar cane and elephant grass attest to the nation's natural endowment for renewable and sustainable energy. This paper is divided into biomass resources of Nigeria; policy mechanisms for promoting bioenergy options in Nigeria; bioenergy options and their technologies.

2.0 Biomass Resources of Nigeria

Biomass refers to living and recently dead biological material that can be used as fuel or for industrial production. It excludes organic materials which have been transformed by geological processes into substances such as coal or petroleum that are generally referred to as fossil fuel (ECN, 2008). The biomass resources of Nigeria can be identified as wood, forage grasses and shrubs, animal waste and waste arising from forestry, agricultural, municipal and industrial activities, as well as aquatic biomass. The biomass energy resources of the nation have been estimated to be significant. Some non-food crops biomass are displayed in Table 1.

TABLE 1: Non-food crop biomass

S.N.	Source	Products	Function	Distribution (Nigeria)
1.	Jatropha	Biodiesel	Biofuel and Bioenergy	> 26 states (mostly Northern states)
2.	Neem	Biodiesel	Biofuel and Bioenergy	>30 states (mostly Northern states)
3.	Castor Seed	Biodiesel	Biofuel and Bioenergy	> 25 states (mostly Northern states)
4.	Elephant Grass	Bioethanol, Biobutanol		Biofuel and Bioenergy All over Nigeria

Biomass resources employed in energy generation are largely agricultural plants and animal wastes. It can be converted into energy by thermal and biological processes. The thermal gasification of biomass results in the production of a low grade fuel called producer gas, which is a carbon monoxide-rich gas. Biological processing of biomass results in liquid fuels such as bioethanol and biodiesel while biological gasification results in the production of biogas- a methane-rich gas that is used for cooking and heating. There is justification for the use of biomass in the energy mix of Nigeria; some of the reasons as stipulated in the Energy Commission, 2008 publication include:

i A greater number of Nigerians, about 80% of the population, live in the rural or semi-urban areas, and they depend solely on fuel wood for their energy needs. Indeed, less than 20% of the total households in Nigeria are connected to the national grid or diesel generators.

ii Biomass energy technologies can be installed in each household or community no matter how remotely located thereby eliminating the cost of installation of a distribution system, metering and networking problems. Also energy self-sufficiency and independence is assured.

iii Some blended bioenergy have better fuel characteristics than fossil fuel. These fuels confer on fossil fuel better octane and cetane number when blended. The use of gasohol, a blend of alcohol and gasoline in internal combustion engines in such countries as Brazil as a means of minimizing the green house effect illustrates this significance.

Biomass in the form of fuel wood contributes about 37% of the total energy demand, and the energy choice for the vast majority of rural dwellers and the urban poor in Nigeria. Currently, the country consumes about 43.4 x 10^9 kg of fuel wood annually. The average daily consumption is about 0.5 to 1.0 kg of dry fuel wood per person (Energy Commission, 2008). However, the resource base is under pressure from both human activities and natural factors such as drought. The rate of consumption hardly matches the rate of deforestation. This suggests urgency in action to save forest wood as source of energy.

The biomass energy resource base of Nigeria is estimated to be about 144 million tonnes per year. The biomass resources of Nigeria consist of wood, forage and shrubs, animal wastes and wastes arising from forestry, agricultural, municipal and industrial activities as well as aquatic biomass. Nigeria's land area is about 79.4 million hectares of which 71.9 million hectares can be considered to be arable. This shows a huge potential for the production of biomass since an estimated 94% of Nigerian households are engaged in crop farming.

Nigeria's aggregate annual crop production of 93.3 million tonnes of major crop yields far more quantity of straws, chaff, leaves and other biomass materials. This expanse of arable land holds promise for cropping of energy crops such as Jatropha for biodiesel production. These energy crops are not edible and as such cannot affect the food chain. Nigeria also produces an estimated 285.1 million tonnes of manure from her livestock population of 245.9 million, which can yield about 3 billion cubic meter of biogas annually. This is more than 1.25 million tonnes of fossil fuel oil per annum. Other possible biomass resource base includes aquatic plants such as water hyacinth and municipal wastes, both of which constitute major environmental problems (Energy Commission, 2008).

3.0 Nigeria's Renewable Energy Options

Bioethanol, biodiesel and biogas are the major renewable options for Nigeria. Each has its favourite range of crops or biomass for commercial production. The need to balance energy requirement and food security is complicating the choice of feedstock for Nigeria's move towards sustainable energy resources. Striking a sustainable balance between supply of feedstock such as cassava and sugar cane use as food raw material and demand for these products as feedstock for energy is one that requires multidimensional analysis of production conditions. These are real concerns that must be factored into realities such as population growth and possibilities of low agricultural output. The challenge is to find ways of ensuring continuous production above estimated requirements for food; developing other non food varieties of the plants; and identifying other non food crops capable of commercial-quantity yield of bioenergy.

Bioenergy, solid, liquid or gas fuels produced from biomass for transport, heating and electricity generation. They are produced from agricultural and forest products and the biodegradable organic portions of industrial and municipal wastes. Generally, bioenergy are renewable energy resources derived from recent biological materials, which distinguishes them from fossil fuels that are derived from long dead biological materials. There are various types of bioenergy such as bioethanol, biodiesel, biogas, and other fuels made from biomass. Presently, bioethanol, biodiesel and biogas are receiving global attention.

Bioethanol is an alcohol made by fermenting and distilling simple sugars. It can be produced from any biological feedstock that contains appreciable amounts of simple sugar or other materials that can be converted into simple sugar such as starch or cellulose. Bioethanol can be blended with gasoline in any proportion up to 10% (E10) without the need for engine modification.

TABLE 2: Profile of Biomass Energy in Nigeria

Fuel	Common use	Usage rate	Limitation	Negative impact	Potential
Biogas	Cooking, heating, refrigeration and lighting	Very low	Lack of awareness	None	Medium
Bioethanol	Gasoline operated internal combustion engines for transportation, lighting and other applications	Very low	• Difficulty in purification • Limited availability of feedstock due to its competing use as edible crops • Low level of awareness	• Depleting food to the population	High
Biodiesel	Diesel operated internal combustion engines for transportation, lighting and industrial process heating	Very low	• Lack of energy crops • Low level of awareness	• Deforestation and its attendant consequences	High

Source: Energy Commission Nigeria (2008)

TABLE 3: Average Fermentable Sugar Content of Some Saccharine Feedstocks

Feedstocks		Average Fermentable Sugar Content (%)
Fruits		
	Grapes	15.0
	Bananas	13.8
	Apples	12.2
	Pineapples	11.7
	Pears	10.0
	Peaches	7.6
	Oranges	5.4
	Water melons	2.5
	Tomatoes	2.0
Sugar beets		15.0
Sweet sorghum14.0		
Sugar cane		10.0 - 15.0
Molasses		50.0 - 55.0

Source: Energy Commission of Nigeria (2008)

TABLE 4: Bioethanol Feedstock and their Annual Yield

Crop	Annual Yield (Litres/Hectare)	Greenhouse gas savings (% vs Petrol)	Comments
Miscanthus	7300	37 - 73	Low input perennial grass. Ethanol production depends on development of cellulosic technology.
Switchgrass	3100 – 7600	37 - 73	Low inout perennial grass. Breeding efforts underway to increase yields. Higher biomass production possible with mixed species of perennial grasses.
Poplar	3700 – 6000	51 - 100	Fast growing tree. Completion of genomic sequencing project will aid breeding efforts to increase yields
Sugar Cane	5300 – 6500	87 - 96	Long season annual grass. Newer processing plants burn residues not used for ethanol to generate electricity. Only grows in tropical and subtropical climates.

Sweet sorghum	2500 – 7000	No data	Low input annual grass. Grows in tropical and temperate climates, but highest ethanol yield estimates assume multiple crops per year. Does not store well.
Corn	3100 – 3900	10 - 20	High input annual grass. Only kernels can be processed using available technology. Development of commercial cellulosic technology would allow stover to be used and increase ethanol yield by 1,100 - 2,000 l/ha

Source: Energy Commission of Nigeria (2008)

Biodiesel is most commonly made by chemically altering organic oils through the use of alcohol in the presence of catalyst. This process known as transesterification reaction brings about the breakdown of the oil molecules and replaces the glycerine portion of the molecules with alcohol molecules. The glycerine falls to the bottom and is drained off resulting in biodiesel.

TABLE 5: Oil Production Averages of Some Crops

Plant	kg Oil/ha
Oil palm	5000
Coconut	2260
Jatropha	1590
Rapeseed	1000
Peanut	890
Sunflower	655
Soybean	375
Hemp	305
Corn	145

Source: Reily and Paltsev (2007)

TABLE 6: Gas Production at the Various Temperatures

Temperature range	Advantage(s)	Disadvantage(s)
Psychophilic	None	• Bioconversion is slow and incomplete • Longer detention time is required • Heating of the digester is required
Mesophilic	This temperature corresponds to the ambient temperature	Longer detention time may be required to enable complete conversion of the

Temperature range	Advantage(s)	Disadvantage(s)
	of the Tropics and as such no heating is required thus reducing cost of production	available carbon
Thermophilic	• Higher rate of gas production • Allows heavier organic loading • Lower detention time • Enables the use of comparatively smaller size digesters • Digestion is much more sanitary than digestion at the other temperatures because of the few pathogens that can survive at this temperature • Mechanical transport and handling of the digester is easier because the slurry is less viscous	Digestion is easily upset at this temperature

Source: Energy Commission of Nigeria (2008).

Biogas is a methane rich gas that originates from bacterial attack on organic wastes under anaerobic condition. The waste type ranges from organic municipal to agricultural waste. The organic wastes are degraded by various types of bacteria such as *acidogenic* and *methanogenic* bacteria, under suitable and stable temperatures, living behind digested slurry known as biofertilizer. Other gases are produced such as Methane ($CH_{4)}$, Ammonia ($NH_{3)}$, Carbon dioxide ($CO_{2)}$, Hydrogen Sulphide (H_2S), Water vapour (H_2O), etc. All these gases make up biogas. Many factors affects biogas yield and the most prominent are temperature, pH, and feedstock type. Others include loading rate, toxicity of substrate and carbon-nitrogen ratio.

4.0 BIOENERGY POLICY

The potential of bioenergy is enormous but the challenge is tasking. Driving the change will certainly require strategies, most of which will require government intervention. This makes policy a possible instrument for this challenge. Consequently, the government of Nigeria is adopting certain policy measures to trigger the initial take-off of the change.

4.1 Objectives

The objective of the energy policy is to firmly establish a thriving bioenergy industry utilizing biomass as a means of improving the quality of automotive fossil-based fuels in Nigeria. The Policy shall link the agricultural and the energy sector, with the underlying aim of stimulating development in the agricultural sector (Energy Commission, 2008).

Specific to automotive use, the policy will allow:

The Minister of Petroleum Resources makes appropriate regulations pursuant to his powers under the Petroleum Act whereby bioenergy blends shall become available for sale and use within Nigeria.

Nigerian National Petroleum Corporation (NNPC) guarantee off-take of bioenergy produced within the country under agreed contractual terms.

In broad terms, the policy aims to

- promote job creation, rural and agricultural development and technology acquisition and transfer;
- provide a framework which is capable of attracting foreign investment in the bioenergy industry;
- streamline the roles of the various tiers of government in order to ensure an orderly development of the bioenergy industry in Nigeria; and
- involve the oil and gas industries in more purposeful development of other sectors of the nation's economy.

4.2 Policy on Environment

4.2.1 Creating Market Demand for Bioenergy

The Federal government of Nigeria approved the blending of bioenergy as a component of fossil-based fuels in the country as required for all automotive use. This blend shall involve the process of upgrading fossil-based fuels.

It is envisaged that from the take-off of the programme, full national implementation shall be achieved within 10 years.

4.2.2 Market Entry

Market entry shall be through registration of bioenergy plants manufacturing fuel ethanol or/and biodiesel with the possible addition of a co-generation power plant by the Department of Petroleum Resources (DPR).

Organisations/investors interested in bioenergy production in the country shall be duly registered with the DPR.

4.2.3 Industry Classification

For the purpose of this policy, investment in the bioenergy industry shall be treated as an agro-allied activity and shall benefit from the incentives put in place to foster the development of the agro-allied industry, in addition to other incentives under this Policy.

4.2.4 Industry Structure, Bioenergy Producers and Out growers

- Policy will support the emergence of an industry in which substantial portion of feedstocks used by bioenergy plants will be produced by large scale producers and outgrowers.
- Bioenergy operators will be encouraged to have resources in their organizations to support the operations of out growers.
- Bioenergy producers are required to establish documented, transparent, fair and auditable contractual terms with bioenergy feedstock out growers; long term contracts are encouraged.
- Bioenergy producers are required to use auditable feedstock, weighting equipment and methodologies that are prescribed by the Department of Weights and Measures of the Ministry of Commerce and Industry.

5.0 BIOENERGY TECHNOLOGIES

5.1 Bioethanol

Bioethanol is simply ethanol ($CH_3.CH_2OH$), a clear colourless liquid derived from agricultural sources, as distinct from petrochemical sources. It is a renewable liquid fuel which is the principal fuel used as a petrol substitute for road transport. It is mainly produced by simple sugar fermentation process, although it can equally be manufactured by the chemical process of reacting ethylene with steam. The main sources of sugar required to produce ethanol come from fuel or energy crops. Presently, there is an on-going research and development into the use of municipal solid wastes, which compose principally of cellulosics, to produce ethanol fuel.

5.1.1 Bioethanol Production and Technology

Certain materials (biomass) require less processing than others. Generally, small scale production from the saccharine materials is the easiest and most economical in terms of labour and energy consumption. However, starchy materials usually produce the most alcohol on a weight per weight basis, and cellulosic materials are the cheapest because most of them are currently regarded as wastes.

Bioethanol is produced through the conversion (fermentation) of sugars to ethanol. The biomass that exists as complex sugars (polysaccharides) are however first broken down into fermentable sugars through a process of chemical reaction called hydrolysis (Figure 1). The simplified fermentation reaction equation for the simple 6-carbon sugar, glucose, is:

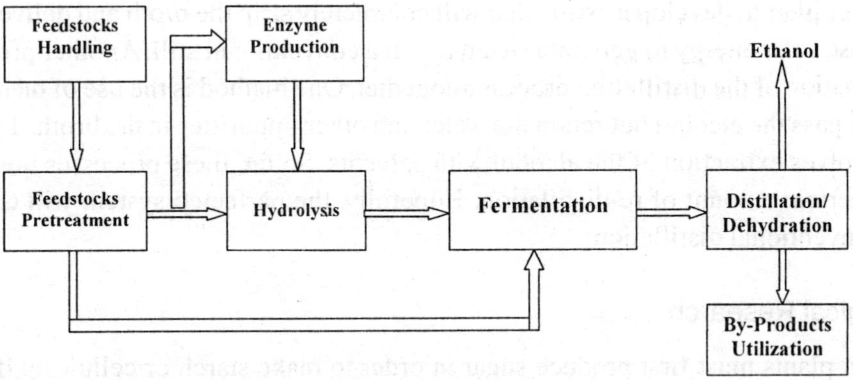

FIGURE 1: General Scheme of Bioethanol Production

$$\underset{(Glucose)}{C_6H_{12}O_6} \longrightarrow \underset{(Ethanol)}{2CH_3CH_2OH} + \underset{(Carbon\ Dioxide)}{2CO_2} \quad (1)$$

5.1.2 Emerging Technologies in Bioethanol

5.1.2.1 Immobilized Enzymes

One of the most important areas of current research concerns the enzymes used to convert starch to sugar. Enzymes are expensive and are a major expense in starch conversion. Using present technology, enzymes can only be used once. They go into solution in the mash and are all but impossible to recover for reuse. The current research which began in 1950s involves immobilizing the enzymes. Instead of being allowed to go into solution, the enzymes are fastened or "immobilized" on an inert substrate. The mash is then passed over a bed of these immobilized enzymes. The end result is that the enzymes do not go into solution and they may be reused many times at a considerable saving. This technology is not yet fully developed for starch conversion, but should be available in the near future.

5.1.2.2 Cellulose Conversion

Cellulose is the most plentiful and cheapest potential source of fuel alcohol. Attempts are being made to develop new enzymes to economically convert cellulose to glucose. The problem, as mentioned earlier, is freeing the cellulose from the lignin. Strains of organisms, such as those that cause dry rot in wood are being studied. The objective is not only to extract the cellulose, but to allow easy recovery of the lignin. Lignin, in turn, makes an excellent fuel with which to run a 'still'.

5.1.2.3 Alternatives to Distillation

The study is in three basic areas. First is the development of large, efficient solar distillation apparatus.

The approach is either to develop a "still" that will completely strip the broth and deliver high proof alcohol, or to use solar energy to generate steam to run a conventional still. Another promising area involves elimination of the distillation process altogether. One method is the use of membranes that will selectively pass the alcohol but retain the water and other impurities in the broth. The third area of research involves extraction of the alcohol with solvents. So far, these processes involve the use of heat and a certain amount of re-distillation. Hopefully, the perfected system will consume less energy than conventional distillation.

5.1.2.4 Biological Research

It is known that plants must first produce sugar in order to make starch or cellulose. If there were some way to pick plants at the peak of sugar production, the manufacture of ethanol could be greatly simplified. In addition, crops with inherently high sugar content could be mutated to produce even more sugar and grown specifically for the production of alcohol.

There are certain bacteria, algae, and fungi that can ferment alcohol directly from cellulose without a conversion step. It is also within the realm of possibility that a plant could be developed to produce alcohol directly. Certain plants, such as mint, already possess the basic biological mechanism. Even the relatively simple development of a yeast strain that is more tolerant to alcohol would allow more concentrated mashes and a significant saving in distillation energy.

The full utilization of existing technology and the promise of new developments will make the production of alcohol fuel easier and more economical in the near future. However, as fossil fuel supplies dwindle, it will become increasingly important to utilize every shred of available material and waste in the production of energy.

5.1.3 Economics of Production and Utilization of Bioethanol

Bioethanol has a lower mileage compared to gasoline. At N300.00 per gallon, for instance, it takes about N396.00 worth of E85 (85% Ethanol and 15% gasoline) to drive the same distance as a N360.00 worth of gasoline. It is therefore clear that before bioethanol can be competitive and have any economic acceptance, the cost of production must be much lower than the current gasoline prices.

An important factor in reducing the cost of bioethanol production is the use of larger industrial facilities rather than smaller ones. By increasing the plant size, the investment per unit output of product falls off.

To further improve the economy of ethanol production, energy integration of the ethanol production to already existing plants, such as pulp and paper plants, and sugar producing factories, may be employed.

Apart from the cost of production technology, the cost of the feed stock, which represents about 40% of the total production cost, and that of production enzymes are two other important parameters in the production cost of bioethanol. The cost of feedstock and the choice of the feedstock to use wholly lean on the availability and the ongoing uses.

5.2 Biodiesel

Biodiesel is a renewable fuel manufactured from vegetable oils, animal fats and recycled cooking oils. It is the product obtained when vegetable oil or animal fat is chemically reacted with an alcohol in the presence of a catalyst to produce a new compound known as fatty acid alkyl ester. Methanol is the preferred alcohol although ethanol is also used. The use of ethanol produces ethyl ester that is produced at a lower conversion rate and it is difficult to recycle the excess ethanol in the production process. Biodiesel should be clear, although it may come in a variety of colours. Biodiesel colour does not in any way predict the fuel quality. Equation (2) expresses the production of biodiesel from methanol using sodium hydroxide as catalyst.

$$\underset{(100Kg)}{\text{Oil}} + \underset{(10Kg)}{\text{Methanol}} \xrightarrow{NaOH} \underset{(100Kg)}{\text{Biodiesel}} + \underset{(10Kg)}{\text{Glycerol}} \qquad (2)$$

Biodiesel can be used in different ways. 20% biodiesel can be blended with 80% diesel fuel (B20) for use in most compression ignition engines. B20 is the most commonly used blend because it balances property differences, performance, emission benefits and costs with the conventional diesel. Higher blended levels such as B50 or B100 require modification of the compression ignition (CI) engine.

5.2.1 Production of Biodiesel

The transesterification of triglycerides results in the production of biodiesel and glycerol. It is important to note that the process used for each treatment is different and plant design for one feedstock may not be able to satisfactorily treat the other. However, two-stage transesterification appears to be the norm in most modern plants

The transesterification process is achieved in four principal steps. The steps are:

Step 1: *Pre-treatment of Feedstock*

This is the pretreatment of feedstock to remove contaminants and components that will be detrimental to subsequent processing steps. Example is the removal of gummy materials from vegetable oil.

Step 2: *Transesterification*

Two basic events are involved in this process. These are the reaction process followed by the separation of the methyl ester and glycerol. These events are undertaken twice to enable the completion of the transesterification process by reducing the concentration of the glycerol in the second stage. The pre-treated triglyceride is reacted with alcohol (methanol) to form raw methyl esters and glycerol in the presence of a catalyst. The reaction is also pushed closer to completion by using an excess of methanol. The transesterification process can be undertaken using simple equipment such as buckets and drums. Commercial production of biodiesel will however require more sophisticated equipment.

Step 3: *Methyl Ester Purification*

This stage removes the excess methanol, catalyst and glycerol from the transesterification process. Methanol removed is recycled to the transesterification process.

Step 4: *Glycerol Purification*

This is a further purification step to remove substances such as catalyst and methyl esters that are contained in the glycerol in order to produce a higher grade of glycerol if economics dictate.

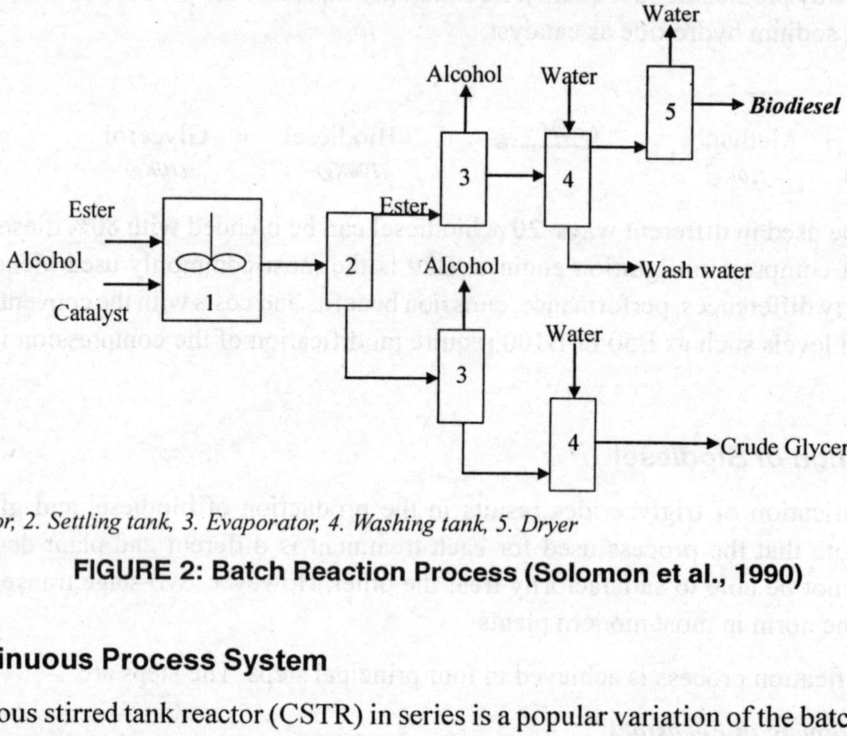

1. Batch reactor, 2. Settling tank, 3. Evaporator, 4. Washing tank, 5. Dryer

FIGURE 2: Batch Reaction Process (Solomon et al., 1990)

5.2.2 Continuous Process System

The continuous stirred tank reactor (CSTR) in series is a popular variation of the batch process. The Continuous Stirred Tank Reactor (CSTR) can be varied in volume to allow for a longer residence time in CSTR 1, to achieve a greater extent of reaction. After the initial product glycerol is decanted, the reaction in CSTR 2 is rather rapid with as much as 98% plus completion. Sufficient mixing of the input is essential in the design of a CSTR to ensure a constant composition. This facilitates the dispersion of the glycerol product in the ester phase. This leads to an extension in the time required for phase separation.

The intense mixing to initiate the esterification reaction can be done by use of pumps or mixers. Instead of allowing time for reaction in the agitated tank, the plug flow reactor is used. This is a tubular reactor in which the reaction mixture moves through the reactor in a continuous plug with little mixing in the axial direction. This is like a series of small CSTRs chained together. This is a continuous system that requires a short residence time - as low as 6 to 10 minutes - for near completion of the reaction. Figure 3 is the schematic diagram of the plug flow reactor.

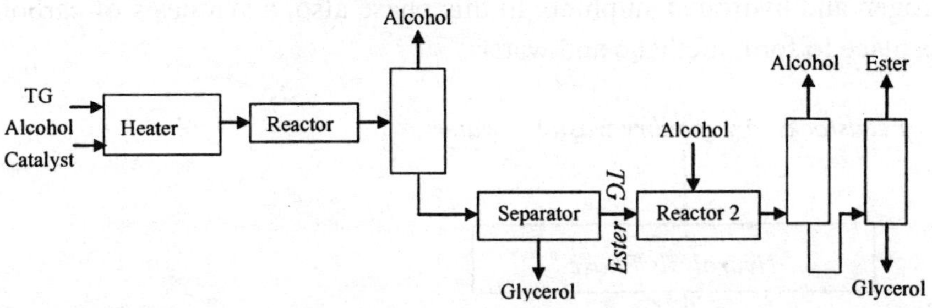

FIGURE 3: Plug-Flow Reactor System (Ojumu et al., 2003)

5.2.3 Economics of Biodiesel Production

The cost of producing biodiesel is influenced by the following cost elements:

1. Capital and operating cost of the plant
   - Cost of operating plant (machinery)
   - Services
   - Cost of product storage
   - Cost of buildings
2. Cost of feedstock used in the process
   - Cost of vegetable oil or waste oil
   - Cost of alcohol
   - Cost of catalyst
3. Sale of glycerol and other by-products provides secondary revenue to the biodiesel production which helps offset some of the costs of production.
4. The sale of the biodiesel produced also represents a revenue receipt.

5.3 Biogas

5.3.1 Production Process

Biogas is produced from a three-phase process, namely, the hydrolysis, acid-forming and methane-forming phases. In the hydrolysis phase, extra cellular enzymes secreted by acidogens, break down the complex organic material into simple, soluble molecules. These molecules are further broken down into volatile fatty acids (e.g. propionic and butyric acids), carbon dioxide, ammonia and hydrogen by acidogens in the acid-forming phase. In the methane-forming phase, methanogens or methane-formers convert the volatile fatty acid (VFA) into methane, carbon dioxide, carbon monoxide, nitrogen and hydrogen sulphide. In this phase also, a synthesis of carbon dioxide and hydrogen takes place to form methane and water.

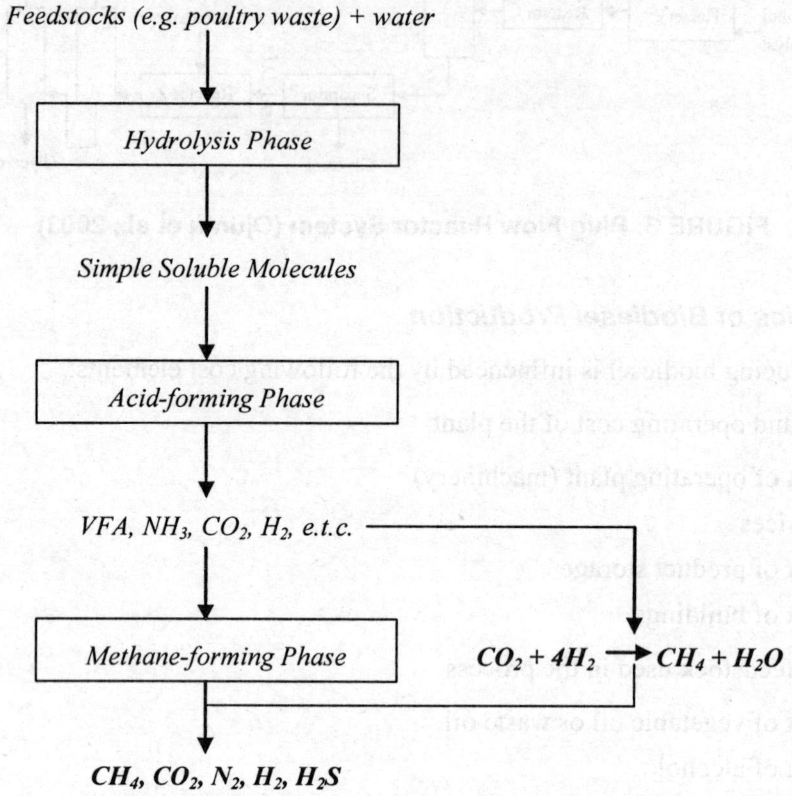

FIGURE 4: Flow Diagram of Biogas Process

5.3.2 Methods of Improving Biogas Production

The anaerobic digestion process can be improved by the pre-treatment of organic wastes, the mixing of different waste types, the addition of chemicals and use of media materials among others.

- The addition of certain chemicals such as nickel accelerates and increases gas yield.
- The use of mixed substrate, for example, 25% poultry manure and 75% brewery waste water.
- Pre-treatment of the feedstock such as milling to diminish the particle size, swelling in an

alkaline solution heating of the feedstock.

- Re-concentrating flushed waste (waste to which fresh waste has been added to increase its total solid concentration) and recirculating leachate from old digesters.
- Stirring or mixing of the digester content.
- Two-phase anaerobic digestion of waste. The two-phase concept consists of first digesting the waste in an acid phase. The gaseous and liquid products of this phase are separately conveyed into separate methane fermenters to enable biogas production.

6.0 ENVIRONMENTAL ISSUES ASSOCIATED WITH DIFFERENT TYPES OF BIOENERGY

All energy options have implications on the environment, land and agriculture. These are obviously less intense in terms of pollution but have dimensions that could be worrisome and impact mostly on agriculture and land use. Polaski (2008) puts some of these issues as follows:

6.1 Biodiesel

- When land is cleared, vegetable matters that absorb CO_2 will be removed, and there will be more CO_2 arising from emissions from the automobiles used for clearing thereby contributing to environmental pollution.
- Agrochemicals used in the farming of biodiesel crops contribute to environmental pollution.
- Agricultural residues left after isolation of the seeds (e.g. cobs, pods, stems etc) constitute environmental problems.
- During the production of biodiesel, oil spillage in the area is an environmental issue.
- Noise pollution arising from the processing of the biodiesel constitutes environmental pollution.
- Combustion of the biodiesel emits CO_2 and some other gases.
- The presence of oxygen in biodiesel improves combustion and therefore reduces hydrocarbon, carbon monoxide and particulate emissions but oxygenated fuel tend to increase nitrogen oxide emissions.
- Edible crops such as soybean could lead to food crisis when used recklessly for bioenergy production.
- Biodiesel is a good solvent that can dissolve certain types of paints, however, lower blends, up to B20, does not.
- Biodiesel has the tendency to dissolve the accumulated sediments in diesel storages and

vehicle fuel system tanks, however, lower blends, up to B20, does not.

- Dissolved sediments can block fuel filters and foul the fuel injection.
- Biodiesel may dissolve the fuel filter and the fuel host therefore; special materials should be used for the production of fuel hosts and fuel filters e.g. Teflon and Fibre glass.

6.2 Bioethanol

- When land is cleared, vegetable matters that absorb CO_2 will be removed, and there will be CO_2 arising from emissions from the automobiles used for clearing thereby contributing to environmental pollution.
- Agrochemicals used in the farming of bioethanol crops contribute to environmental pollution.
- Agricultural residues left after isolation of bioethanol crops (e.g. corn cobs, cassava and sugar cane peels etc) constitute environmental problems.
- On fermentation of bioethanol, carbon dioxide is released, large quantity of water used for hydrolyses and fermentation.
- During distillation, thermal pollution is encountered.
- Noise pollution is encountered during the crushing of raw materials for bioethanol production.
- Combustion of bioethanol emits carbon dioxide and some other gases
- Bioethanol is a good solvent and can dissolve certain types of paints, however, lower blends, up to E10, does not.
- Bioethanol has the tendency to dissolve the accumulated sediments in diesel storages and vehicle fuel tanks, however, lower blends, up to E10, does not.
- Dissolved sediments can block fuel filters and foul the fuel injection system, however, lower blends, up to E10, does not.

6.3 Biogas

- Land clearing: when land is cleared vegetable that absorb CO_2 will be removed, meaning that there will be more carbon dioxide in the environment.
- The automobile used in clearing will produce emissions that pollute the environment.
- Organic wastes used for biogas production contain pathogens that are harmful to living organisms therefore, proper handling should be ensured.

6.4 Mitigations

- Strengthening of the existing National Energy Policy and its Master Plan through an Act of

National Assembly parliament for sustainable production and utilization of bioenergy.

- Creation of public awareness on safety issues involved in bioenergy production and utilization.
- Intensification of research and development into bioenergy technology to curb emissions arising from bioenergy production and utilization.
- Involvement of the full participation of the community for the development of bioenergy production and utilization to avoid socio-cultural conflicts.
- Maintaining bioenergy standards of Standard Organization of Nigeria (SON) and Department of Petroleum Resources (DPR).

7.0 CONCLUSION

It is necessary to note that bioenergy is occupying an increasingly important position in the world's energy mix especially as a result of the current global fossil fuel experience and the global cry for environmental protection as clearly expressed in the millennium development goal (MDG). There is therefore the need for constant increase in research and development, and knowledge of the technological processes of bioenergy production in order to meet the global demand for bioenergy that is surely rising.

REFERENCES

Energy Commission of Nigeria (2008)

Gidado, R. S. M. (2008), Contributor: Biofuels Training Manual Publication of Energy Commission of Nigeria (ECN) 2008.

Ojumu, T.V., Attah-Daniel, B.E., Betiku, E. and Solomon, B.O. (2003) Auto-hydrolysis of Lignocellulosics under Extremely Low Sulphuric Acid and High Temperature Conditions in Batch Reactor. Biotechnology and Bioprocess Engineering, 8: 291-293.

Polaski S. (2008). Rising Food Prices, Poverty, and the Doha Round, Carnegie Endowment for International Peace.

Reily J. and Paltsev S. (2007). Biomass Energy and Competition for Land, Report No. 145.

Solomon, B. O., S. K. Layokun, P. K. Nwesigye and P. O. Olutiola (1990) Hydrolysis of Sawdust by Cellulase Derived from *Aspergillus flavus* linn Isolate NSPR 101: Beyond the initial fast rate period. JNSChE 9(1&2), 46 - 50.

7

Trends in the Sustainable Utilization of Renewable Energy Resources of Sri Lanka for Facing the Energy Challenge

J.G. Shantha Siri
National Science Foundation, Colombo, Sri Lanka
E-mail: shantha@nsf.ac.lk

Sri Lanka's current economic development policy framework is a consolidation of Pro-poor growth strategies based on private and public sector partnerships in economic and social development. The policy framework is developed on a market friendly, export oriented and competitiveness basis to sustain eight percent economic growth in the GDP during the medium term. The Millenium Development Goals are also incorporated in the economic development policy framework. Energy security has become an important factor in achieving the objectives articulated in the economic policy framework. Supply of reliable, affordable and clean energy during the medium term for industries, individuals and other sectors are extremely important in this regard. Manufacturing as well as service industry contributes a significant amount to the GDP of the country. Energy is the key input for all the sectors namely manufacturing industry, construction industry, hotels, restaurant and transport. Biomass, hydropower, and thermal power (fossil fuels) are the conventional energy sources in supplying energy. The utilization of renewable resources such as solar, wind, dendro, and mini-hydro are relatively small. The transport sector is entirely depending on petroleum based fuels such as gasoline and the diesel. The escalating prices of petroleum-based fuels in the recent past have resulted into a great burden to the economy. This has created a trend to explore alternative avenues to fulfil energy requirement with a view to tackling the energy challenges. The available options are the utilization of renewable energy sources and conservation of energy. The renewable resources are also important in reducing air pollution, mitigating adverse climate effects and the provision of clean energy. The improvement in efficiency in energy resource is also useful in facing the energy challenge. Several measures such as reducing system losses, use of low energy consuming equipment, saving energy through changing behaviours, etc. have been introduced and practicing at present. The solar power is the largest renewable energy source in terms of quantity. It is useful in providing electrification in rural people in remote areas where electrification through the national grid is difficult due to capacity limits and cost effectiveness. The utilization of large hydropower projects has been limited due to the carrying capacity limits and absence of hydropower sources. The future hydropower expansion will be therefore mainly composed of introducing mini hydropower schemes. The wind power is also a useful energy source in providing energy supply to the national grid in future. The Dendro power has become increasingly popular among the renewable energy sources at present. It utilizes biomass to generate electricity. The cultivation of fuel wood species in marginal/uneconomic lands will generate employment opportunities as well as additional income to the rural population which in turn address the rural poverty. The direct output of Dendro power is the supply of electricity which eases the facing of energy challenge.

KEY WORDS: *Renewable energy, Biomass, Biogas, Gasification, Hydropower, Dendro-power, Bio-fuel, Energy efficiency, Rural electrification*

1.0 INTRODUCTION

The world oil reserves are being depleted at an alarming rate and also consumption of petroleum based fuel has environmental implications. The annual demand (by domestic as well as commercial sector) for electricity is in an increasing trend in Sri Lanka. These factors had made people to shift their attention to sustainable utilization of renewable energy resources. Even though the biomass and hydro power are the major renewable sources of energy, Sri Lanka has the potential to harness wind and solar energy.

Three main energy sources in Sri Lanka are biomass (inclusive of fuel-wood), petroleum oil and hydro-electricity. Biomass accounts 48%, while petroleum and hydro electricity accounts 43% and 9% respectively. With the escalating oil prices in the world market and world trends of harnessing renewable energy resources, initiatives have been taken by the Government of Sri Lanka to utilize indigenous renewable energy resources.

The Ceylon Electricity Board estimates growth in the consumption of electricity to be approximately 8% annually in the next five to ten years requiring an additional 120-150 MW annually to meet this demand. While only 60% of households presently have access to electricity, this is expected to increase to 80% by 2010.

1.1 Energy demand growth

With the increasing demand for energy to drive the country's economic and social development, total primary energy demand is expected to increase to about 15,000 kTOE (thousand tonnes of oil equivalent) by the year 2020 at an average annual growth rate of about 3%. Electricity and petroleum sub-sectors are likely to record higher annual growth rates of about 7- 8%.

1.2 Government Policy on Renewable Energy Resources

Promoting Indigenous Resources: Indigenous energy resources will be developed to the optimum levels to minimise dependence on non-indigenous resources, subject to resolving economic, environmental and social constraints. It is planned to increase up to 10%, the share of electrical energy generated from Non-conventional Renewable Energy (NCRE) resources and supplied to the Grid by year 2015.

NCRE resources include small-scale hydropower, biomass including dendro power, biogas, solar power and wind power. These are the leading sustainable, non-conventional forms of renewable energy promoted in Sri Lanka for electricity generation into the grid. In addition, other NCRE resources such as wave energy and ocean thermal energy are also encouraged where appropriate. Commercial development of biomass will be encouraged and facilitated as a new rural industry, allowing rural poor to engage in fuel wood farming and participate in the mainstream economic activity by supplying electricity to urban load centres.

Specific new initiatives are included in this policy to

- expand the delivery of affordable energy services to a larger share of the population;
- improve energy sector planning, management and regulation; and
- revitalize biomass as a significant resource of commercial energy.

It has been declared by the government that *Gliricidia* (an energy crop) as the fourth plantation crop in Sri Lanka. *Gliricidia* is grown as an energy crop with the participation of rural community with a view to increasing the level of income of the rural community.

2.0 SUSTAINABLE USE OF RENEWABLE ENERGY FOR ECONOMIC DEVELOPMENT

Sustainable use of renewable energy has been given a due consideration by the government of Sri Lanka. The renewable energy sector is already included as an element in the energy policy. Government has planned to harness the potential renewable energy resources to maximum by implementing various strategies such as concessionary financing, Clean Development Mechanism, systematic planning, enabling new industrial activities, Research and Development on adopting new and emerging technologies etc.

The Ministry of Science and Technology has promoted establishment of Gliricidia energy plantations to supply fuel wood to the following industrial locations

- Kottawa (Dipped Products. Ltd)
- Walapane (Lanka Transformers Ltd.)
- Madampe (Harcarb Ltd. and Lignocell Ltd.)
- Badelgama (Recogen Ltd.)
- Gampola (Deenside Tea Factory)
- Horana (EnerFab Ltd.)
- Sapugaskanda (EnerFab Ltd.)
- Digane (EnerFab Ltd.

The government of Sri Lanka has also taken initiatives to encourage the private sector to establish the following energy generating facilities:

- A 3 MW Paddy husk fired power plant in Polonnaruwa
- A 2 x 3 MW Paddy husk/Gliricidia biomass power plants in Dehiattakandiya.
- A 5 MW Co-Generation biomass (Gliricidia) power plant at Thulhiriya.

Government has also constituted a National Committee through the National Science Foundation (NSF) on Energy and a Thematic Committee on Bio-fuel. The NSF is the premier scientific research funding institution of the country. The NSF will foster R&D in the area of renewable energy sector with the assistance of above committees to identify priority areas and monitor of funded research projects:

Four research projects in the area of Bio-fuel have already approved for funding by the NSF.

2.1 Projects implemented at commercial scale harnessing renewable energy resources

Wind farm project: Suzlon Energy Ltd., the world's fifth leading and India's largest wind turbine manufacturer, supplied wind turbines of 10 MW capacity to a project developed by Senok Wind Power Pvt. Ltd., Sri Lanka. The project, supplied with eight units of Suzlon's S64 – 1.25 MW wind turbines, will located at the Kalpitiya region of Sri Lanka.

This is the first project to feature megawatt-class turbines, and also first project to be implemented by a the private company, Senok Group. Sri Lanka's Senok Group is a diversified business conglomerate with businesses ranging from construction machinery, automotives, infrastructure development, tea, haulage to mining. The group has an established presence in the clean energy space with investments in mini-hydro projects.

Power generation with agricultural waste: Pelwatte Sugar Industries has also build the largest Dendro-power plant in Sri Lanka. The Pelwatte Sugar has applied to the Ceylon Electricity Board to build a 25 megawatt power plant which will be fired by sugar cane waste (bagasse), making it the largest Dendro power plant proposed in Sri Lanka. The company submitted the application after a recent cabinet decision to raise the limit of buying power from Dendro plants to 50 MW from the previous limit of 10 MW.

The special feature of this plant is that, it can be switched on and off the grid by the utility's system control unit.

Other Dendro power projects invested by private sector: There are more than a million households in Sri Lanka which are in need of off-grid power. According to Ceylon Electricity Board annual growth in electricity demand is 8 per cent which indicates that the national grid would require an additional 150MW a year. Traditional biomass currently accounts for nearly 52 per cent of the primary energy supplied in the country, and nearly 76 per cent of the country's population still depends on fuel wood and other forms of biomass for household cooking.

Dendro power is therefore the potential solution to Sri Lanka's energy needs. It is estimated that the national potential for Dendro power in Sri Lanka is about 4000 MW annually generating over 24,000 GWh. This is nearly four times the total hydropower potential of the country and is adequate to meet domestic electrical energy demand for many decades. Sri Lanka could therefore achieve a projected saving of Rs 60 billion on fuel imports in 2010 by replacing 50 per cent of future electrical power with Dendro power.

A member of the Hayleys Group, Puritas is the country's leading builder of sewage, water and waste water treatment plants and has gained recognition in the private sector as a constructor of Dendro power plants. All the technologies used in its plants are designed locally with the construction and fabrication subcontracted under Puritas' supervision. Puritas also undertakes maintenance contracts and repair of malfunctioning plants.

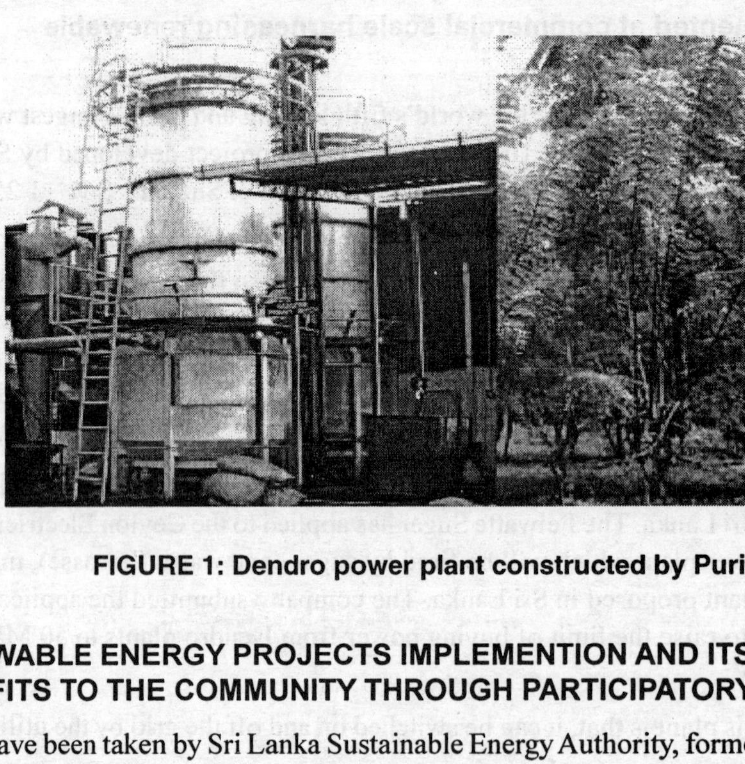

FIGURE 1: Dendro power plant constructed by Puritas

3.0 RENEWABLE ENERGY PROJECTS IMPLEMENTION AND ITS TANGIBLE BENEFITS TO THE COMMUNITY THROUGH PARTICIPATORY APROACH

Initiatives have been taken by Sri Lanka Sustainable Energy Authority, former Energy Conservation Fund to implement Bio mass energy development projects with the assistance of United Nations Development Programme (UNDP) with a view to enhancing the energy efficiency of small and medium enterprises (SMEs) by upgrading technology in the following areas associated with biomass

3.1 Efficiency improvement of biomass utilization in lime kilns

Replacement of fuel oil with biomass in brassware manufacturing industry

The technology used here is gasifier technology and this makes the process easy to control and thereby reducing losses. Pilot projects were carried out by Sri Lanka Sustainable Energy Authority in the Digana and pilimathalawa area and came out with successful results.

FIGURE 2: Kiln used in brassware industry

3.2 Quality improvement in agro produces processing with solar biomass dryers

Sri Lanka Sustainable Energy Authority in collaboration with the University of Moratuwa developed a dryer driven by biomass energy. The drier also uses solar energy when sunlight is available. Through this, solar energy can be productively used and quality improvement of products can be done through gradual drying operation using biomass. The designs and pilot projects carried out for pepper drying showed good results and the quality of the product could be maintained.

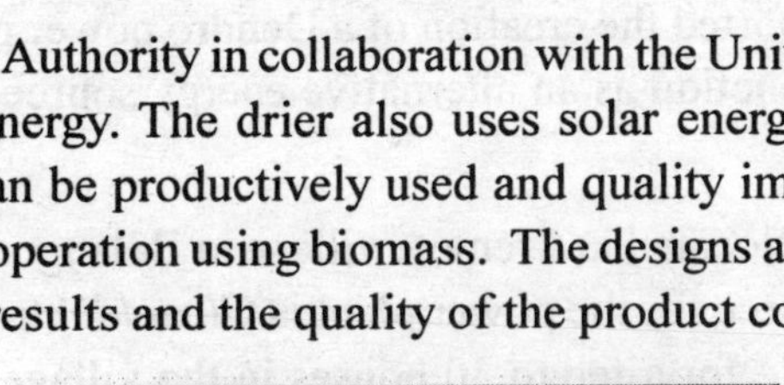

FIGURE 3: Solar assisted biomass dryer

3.3 Development of wood chippers for different volumes of biomass consumption

For efficiency improvement in the application of biomass, gradual feeding is a necessity, and this is long sticks of wood are used. Chipped wood therefore is necessary in this respect, Sri Lanka Sustainable Energy Authority developed wood chippers appropriate for chipping woods for this applications.

3.4 Energy Audits for Energy Conservation

Sri Lanka Broadcasting Corporation: The head office of Sri Lanka Broadcasting Corporation (SLBC) is situated at Independence Square, Colombo 07. The total annual electricity consumption of this is about 1.6 x 106 kWh and the average annual electricity bill was about 24.5 million rupees. According to the recommendation of the Energy Audit carried out by Sri Lanka Sustainable Energy Authority, the annual energy saving potential in SLBC was 27% and the corresponding annual cost saving was approximately 6.7 million rupees.

General Hospital – Kandy: A similar Energy Audit was done at General Hospital in Kandy. The total annual electricity consumption was about 3.88 x 106 kWh and the operating Maximum Demand was 1150 kVA. The average annual bill was about 59.9 million rupees. The saving potential of the recommendation was estimated as 17% of the total energy and the corresponding annual cost saving is approximately 10.36 million rupees.

3.5 Electricity from *Gliricidia* plants for domestic lighting in rural villages

Rural energy problem has been solved in some villages of Sri Lanka by providing electricity using Dendro-energy. Due to the harmful effects of climate change by certain fuels, Thurunushakthi Development Foundation supported the creation of a Dendro power plant. Dendro power, which is fuelwood-based energy, can function as an alternative energy source that has less negative effects on the environment.

The Dendro power project was located in a remote village in Balangoda district and will be used for domestic lighting. The fuel source for the power plant is *Gliricidia* trees which grow in the village. Dendro power will be available for a fee to 30 houses in the village. However, it will be free for certain community centres such as the temple, school, Grama Nildhari office and training centre for disabled children.

By using Dendro power, Thurunushakthi Development Foundation hopes to improve energy efficiency of *Gliricidia* and to make the community aware of alternative and more environmentally-friendly forms of energy.

3.6 Plug flow type biogas systems

Plug flow type biogas systems is an answer to the challenge of disposing organic waste including garbage & animal dung in an environmentally friendly manner. In January 2008 the Centre for Sustainable Technologies of the Indian Institute of Science, in collaboration with Practical Action of Sri Lanka, carried out an awareness seminar and hands on training for five Sri Lankan practitioners to enable the transferring of the technology.

Subsequently, three biogas systems were set up in Sri Lanka. In order to share this learning experience Practical Action organised a National Seminar on Plug Flow Type bio gas at the Sri Lanka Foundation Institute in Colombo on 26th January 2009.

4.0 SUCCESS STORIES AND EXPERIENCE OF SRI LANKA

4.1 Recogen plant- globally-unique charcoal manufacturing plant wins National Science & Technology Award-2008

Recogen plant is the world's only pollution-free charcoaling plant that converts the energy released in the process into electricity. It was recognised at national level by being one of the winners of the National Science and Technology Award given by the Ministry of Science and Technology in collaboration with the National Science Foundation.

The Recogen won the National Science and Technology Award in 2008. This award was under the category of 'Development of Eco-materials/Eco-friendly Processes for Industry'. The Recogen plant conceptualised and developed by Haycarb PLC and built by Recogen Ltd. The plant located at Badalgama converts coconut shell into charcoal in a mechanised process that uses 'Pyrolysis' (carbonisation in the absence of oxygen) and combusts the methane-rich gasses released, a major pollutant, to run a steam turbine that produces electricity for the national grid.

The process was designed as a backward integration to supply charcoal for Haycarb's activated carbon manufacturing operations. The patented process replaces the environmentally hazardous traditional charcoaling pits, captures and uses volatile gasses that would otherwise enter the atmosphere, to generate electricity and produces high-grade charcoal for its parent company. Recogen will by June 2009 represent an investment of Rs 1 billion after nearly 10 years of R&D and engineering operations. It will comprise of six charcoaling reactors that will produce 60 Tonnes of charcoal a day, generate 5 megawatts of power and earn the country Carbon Credits of 46,000 Tons a year under the Kyoto Protocol for the avoidance of methane and carbon dioxide emissions and the reduction of fossil fuel usage.

Activated Carbon Industry brings in more than Rs 2000 million per annum as foreign exchange to Sri Lanka, in raw material supply is guaranteed. Sri Lanka currently produces 45,000 Tons of coconut shell charcoal a year in about 300 open pits in the North Western Province. This process leads to air, water and soil pollution and is therefore a health hazard. It is also a waste of energy. The Recogen plant in its third and final stage will produce 19,800 Tons of pollution-free coconut shell charcoal, thus leading to gradual phasing out of the crude pit method, which releases an estimated 2000 Tons of harmful methane into the atmosphere.

4.2 Energy Efficient tea dryer wins National Science & Technology Award-2008

Tea industry is one of the major foreign earners for Sri Lanka. It is also the largest industrial fuel wood consumer in the country. The increasing cost of commercial energy and fuel wood is resulting into high cost of tea production.

Traditional fuel wood combustion in the tea industry is highly inefficient due to high moisture content and poor combustion of fuel wood. A research project was carried out by a team of researchers of National Engineering Research and Development Centre (NERD Centre) to improve energy efficiency and quality of made tea. A suitable combustor for accommodating wood pieces/chips or small sticks was developed together with a special feeder which is capable of controlling the feeding rate. An automatic system of feeding the fuel wood was incorporated which senses the temperature of hot air and control the feed accordingly.

After successfully producing a model at the NERD Centre, a pilot scale system was tested at a tea factory in Gampola. The promising results of the pilot study have shown that the overall system efficiency can be increased from 40% to 75%. Moisture content of fuel wood was reduced from 43% to 18%. Fluctuation of the temperature was reduced from ±20°C to ±3°C. There are about 600 tea factories in the country. According to the researchers, implementing this newly developed system in all the tea factories can save 150,000 MT of fuel wood annually in the tea industry, an equivalent cost reduction of SLR 1.2 billion. The above team of researchers were given the National Science and Technology Awards in 2008 under the category of "Contributions by Science and /or Technology in reducing use of energy".

5.0 CONCLUSION

Sri Lanka is blessed with the synergy of policy makers, researchers, technologists who can exploit renewable energy resources for the purpose of harnessing their potential so as to face the energy challenge effectively.

Renewable energy resources can be viewed as the most appropriate solution to mitigate the energy challenge. The operation and maintenance cost are very low compared to the conventional energy plants. To accelerate and utilize renewable energy sources, the following measures will be useful:

- Providing financial assistance to purchase machinery and equipment and encouraging private sector to invest in the renewable energy sector. This will facilitate generation of renewable energy to substantial amount.
- The power generated from the solar power is not strong enough when compared with that from the national grid. Therefore, there is need for technological improvement of solar power by providing assistance in research and development in the renewable resources to improve the quality and output. The Dendro power is becoming more and more popular but the utilisation at present is sub-optimal. Promotion of utilizing Dendro power will bring short and long term economic benefits by entering into carbon finance mechanism.

REFERENCES

"The biomass energy sector in Sri Lanka successes and constraints", Jayasinghe P., 2005

"An evaluation of the trends in the energy sector and potential for developing renewable energy", Perera K.K.Y.W., 1993

"National energy policy and strategies of Sri Lanka", Public Utilities Commission of Sri Lanka, 2006.

"Renewable energy development for electricity generation", Sri Lanka Sustainable Energy Authority

National Science and Technology Awards, special publication, National Science Foundation, 2008

www.most.gov.lk - Ministry of Science and Technology, Sri Lanka

www.nsf.ac.lk – National Science Foundation

8

Energy Efficiency as a Source of Income and Improved Livelihoods: The Case of GTZ Improved Stove Projects in Kenya and Uganda

Ms. Myra Mukulu
Programme Officer, Regional Energy Advisory Platform Eastern Africa (REAP-EA),
German Technical Cooperation (GTZ),
Nairobi, Kenya

The German Technical co-operation (GTZ) is a company owned by the German Government which serves as the main implementing agency of capacity building projects on behalf of the German Government. Sustainable develop-ment entails activities that bring a populace to a better state of living economically, socially, and politically whilst not causing any detrimental effects to the ecology or being short lived/temporary interventions in any way. GTZ is active in over 120 countries all over the world with offices in 67 countries. The nature of most projects is bilateral agreements with Governments in developing countries with a view to building local capacity for sustainable development.

GTZ AND ENERGY PROJECTS IN EAST AFRICA

GTZ has successfully implemented various energy projects on behalf of the German government as well as other clients such as the Dutch government and UNHCR in East Africa for over 20 years. These projects are in Kenya, Uganda, Rwanda, Tanzania and Ethiopia.

The Millennium Development Goals (MDGs) are the benchmark of realising sustainable energy projects for enhancement of community livelihoods worldwide and in the East Africa region. The MDGs are targets that were agreed upon by various countries at the millennium summit in the year 2000 where they committed to a new global partnership to reduce extreme poverty and setting out a series of time-bound targets, with a deadline of 2015.

These goals are to:

1. Eradicate extreme poverty and hunger
2. Achieve universal primary education

3. Promote gender equality and empower women
4. Reduce child mortality
5. Improve maternal health
6. Combat HIV/AIDS, malaria and other diseases
7. Ensure environmental sustainability
8. Develop a global partnership for development

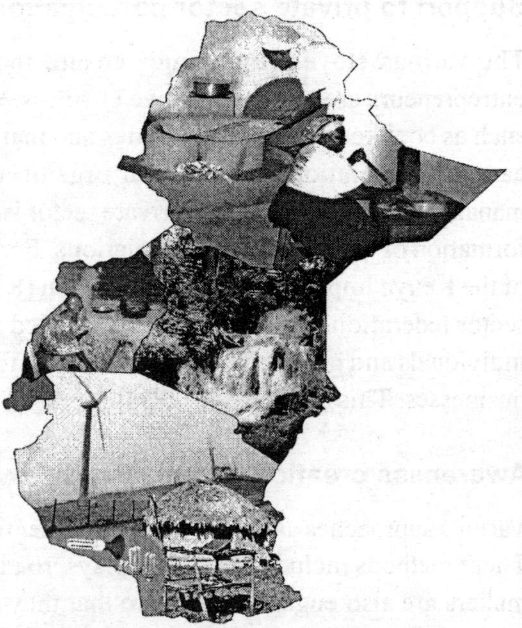

FIGURE1: A map of the EAC with various renewable energy and energy efficiency applications

GTZ projects promote the use of renewable energy and energy efficiency in the East Africa region. In Uganda, various activities are implemented such as solar PV for social institutions, advisory services to the Ministry of Mines and Minerals on energy, energy saving lamps, micro hydro projects for rural electrification, pilot projects on biofuels and promotion of improved cook stoves for household and institutional use among others.

GTZ implements similar interventions in Ethiopia except that the emphasis is on micro-hydro and solar power for rural electrification, grid densification and improved cook stove promotion. In the case of Kenya, biogas and improved stoves are promoted whilst in Rwanda; micro hydro and improved stove projects are implemented. This paper focuses on the improved stove projects in Uganda and Kenya since GTZ has a greater experience in these projects with very high success rates.

IMPROVED STOVE PROJECTS IN KENYA AND UGANDA

One of the energy efficiency projects that have proved to have notable impact on the rural population and the poor are the improved stove projects in Kenya and Uganda. The majority of the people in Kenya and Uganda (over 80% of the population) continue to rely on wood fuel to meet their basic energy needs, in particular for cooking (Stove bulletins 1-4). GTZ implements improved projects based on the recognition that wood fuel can be a viable and affordable renewable energy source if the use of firewood and charcoal is less than the biomass produced growth of forests. Through these projects, the stove programmes in Kenya and Uganda mainly promote fuel wood saving stoves as well as solar driers. The stoves are disseminated in a sustainable way through the following approach:

Support to private sector participation improved stove business

The various stove programmes ensure that a fully commercial approach is promoted where entrepreneurs are trained on stove business. Various aspects of business management are addressed such as bookkeeping, effective sales and marketing and quality stove production and control. In the case of associations training on organisational development, group dynamics and effective management is provided. The private sector is equipped to be independent and self-sustaining through formation of stove industry associations. For instance, in Kenya, GTZ supported the establishment of the Kenya Improved Stove Association (KEISA), which is now affiliated to the national informal sector federation. No subsidies are provided for stove activities except in very rare cases. However, individuals and groups are provided with information and assistance to obtain loans to finance their businesses. This addresses MDG1.

Awareness creation on improved stoves:

Various approaches are used to sensitize members of the public on improved stoves and their benefits. These methods include Market displays, road shows, exhibition stands at agricultural events. Policy makers are also engaged to ensure that they are keen on promoting improved stoves as a result of awareness of their benefits.

FIGURE 2: Stove awareness creations to firewood collectors by the road side n Kenya

FIGURE 3: Hon Akaranga, former Minister in the office of the president promotes improved stoves at a meeting in his constituency

1. **Appropriate technology:** Local expertise and raw materials are used in the design of improved stoves. For instance, in Uganda, the rocket mud stove is designed using affordable and locally available material such as grass and anthill soil. Likewise, training of producers does not require high technology facilities. This however does not imply that quality is neglected. Artisans are trained to ensure that what they produce is of high quality and GTZ is actively involved in stoves standards development in both countries.

2. **Encouraging all genders to participate in stove business, hence realizing gender parity is addressed:** Through GTZ interventions, men, women and youth groups are involved at all levels of stove production, including marketing and installation ensuring all genders benefit from the business. This addresses MDG3.

3. **Support to vulnerable groups:** Special attention is provided to disadvantaged and socially neglected groups such as people living with AIDS (PLWA) and widow groups by equipping them with stove making and stove business skills. This not only provides them with a means of a living through stove business making, but also improves their health since they strain less looking for firewood and are prone to less respiratory and eye diseases as a result of reduced smoke. PLWA are also able to cook faster and easier ensuring they eat healthy which is vital for their survival. This in line with MDG 6

4. **Special attention given to social institutions:** Schools, prisons, hospitals and other key social institutions are targeted as specific beneficiaries of improved cook stoves. This helps them save fuel wood costs as well as promote a healthier cooking environment for their staff. Some of the costs saved are ploughed back into the institutions to purchase critical materials or put up necessary infrastructure. For instance, in Kenya, Eluuya secondary school managed to save money by investing in a rocket-baking oven, which was later, used to construct a dining hall.

BENEFITS /OUTCOMES

GTZ's approach serves to realise tangible results for Kenya and Uganda's populace. The following are some of the tangible benefits accrued by some of the stove business people and users:

1. **Reduction in fuelwood expenses:** various businesses such as food processing and hotels save great amounts of fuel. For instance Eluuya secondary school had 550 students and invested in a fuelwood saving baking oven in 2006. They employed four people who bake 150 loaves of bread and 500 scones daily. The school gets fresh bread and saves Ksh 35,000 that it used to spend on bread. They also save up to 70% of the fuelwood expenses per school term. Several farmers have invested in soil sterilisers to treat soil for crops such as passion fruit. Operating passion fruit nurseries requires the soil be sterilized to ensure that all types of diseases and weeds are eliminated from the soil before it is used. Sterilization is normally done on a 3 stone fireplace where a 200-litre drum with water ¾ full is boiled for

2 hours to produce vapour. The vapour is allowed through pipes to pass through the soil in another tank before being released outside. Typically 16 drums are used in the process over 2 seasons in the year. On a 3 stone fire each drum consumes firewood worth Ksh. 500. Hence 16 drums x Ksh. 500 = Kshs. 8000 (USD 121) with the new installed rocket stove, sterilization for every drum they use firewood worth Ksh.150.Hence for 16 drums x Ksh. 150 = Ksh. 2400 (USD 36). This is to say that they save 70% on the costs for firewood, thus cutting down expenses on the production of seedlings and making the group more competitive. The same technology is being applied for mushroom substrate sterilization.

FIGURE 4: A rocket stove being used to sterilise soil

Prisons in both Kenya and Uganda have also invested in fuel wood saving stoves creating extra income that is used to improve prisoner welfare and other investments in the institutions.

2. **Jobs created:** Stove producers, installers and marketers in Kenya are making money from improved stoves. Stove business has created employment for 20 groups and 24 individuals as producers for Jiko Kisasa while there are 220 rocket stove builders, 75 groups and 13 individuals are working on marketing and installation of stoves with a total of more than 1,200 people involved. For Uganda, stove business has created 290 jobs in 2 years with an income of about €261,000 per year.
3. **Time savings:** Use of improved cook stoves provides for reduction in time for cooking

(hours/day), reduction in distances walked for firewood collection (kilometers) as well as reduction in time spent for firewood collection (hours/day). This time can then be used for studies by children who are typically involved in fuel wood collection. Women who are typically involved in cooking and firewood collection can also use time saved productively, for instance by working on their farms or engaging in income generating activities.

4. **Savings in firewood** (kg/day) help conserve local forests e.g. in Uganda over 200,000 tonnes of wood are saved per year, which is equivalent to spending almost €8,000,000 in afforestation.

5. **Reduction in illnesses related to indoor air pollution** (number of incidences) as well as risks to burns due to exposed fires e.g. in Uganda every second woman using improved stoves has reported suffering less eye irritation, burns or coughs.

REFERENCES

Ingwe A., Wanjohi and P. Kenya Stove bulletins 1-4

GTZ HERA stoves compendium

www.gtz.de/en

Alternative Fuels

9

Sustainable Energy Development In Myanmar

Lat Lat Tun
Materials Science and Materials Engineering Research Department,
Ministry of Science and Technology, Mandalay, Union of Myanmar
E-mail: lat.tun@gmail.com

The balance utilization of different energy types have all along been emphasized and increased production of commercial energy has been prioritized in order to replace traditional energy types and at the same time meeting the requirement of industrial development programme. Biomass fuels were produced as environmental management. Biomass resources are animal waste, agricultural waste, industrial waste, municipal waste, byproduct of forest, among others. The produced biomass fuels are producer gas, biogas, biodiesel, bioethanol from different raw materials. Biomass fuels reduce air pollution, greenhouse gas emissions, global warming and ensure environmental sustainability. Bio-fuel generation from different raw materials is also encouraged in rural areas. Use of bio-fuels prevents deforestation and at the same time it controls pollution which improves health and sanitation standards of the area.

KEY WORDS: *Sustainable energy, Biomass fuels, Producer gas, Biogas, Biodiesel, Bioethanol*

1.0 INTRODUCTIOIN

Union of Myanmar is geographically situated in Southeast Asia between latitudes 09° 32' N and 28° 31' N and longitudes 92° 10' E and 101° 11' E. It is bordered on the Southeast by Thailand and Laos, and on the Northeast by China and on the Northwest by Bangladesh and Indian States. The span from North to South is approximately 2050 km and 935 km from East to West at its widest extent. The area of Myanmar is 677,690 square km and its current population is about 56 million with annual growth rate of about 2.3%.

Apart from its wide latitudinal range, the country also has an altitudinal range from sea level to a peak of over 5,730 m. Consequently, a wide range of temperature exists. In the most parts of the country, it ranges between 25° C to 33° C during the rainy season, 10° C to 25° C during the cold season and 32° C to 38° C during the hot season. However, the temperature can reach peak of 43°

C during the hot season in the central dry zone. The rainfall ranges from less than 800 mm to 5,000 mm and is distributed over an average period of 5 months. In facts, Myanmar, a country of varied ecological manifestation, is basically an agricultural country.

Due to low level of population pressure and relatively low level of industrialization, the state of environment in Myanmar is comparatively better than many of its neighbours. However, there is a need to develop a national program of sustainable utilization of energy and biodiversity resources for wealth creation and development. Although modern guidelines for the preservation of environmental quality are being introduced in the country, a coordinated and integrated system is needed for managing the environment.

2.0 ENVIRONMENTAL QUALITY MANAGEMENT IN MYANMER

Current process of national development and planning in Myanmar does not provide a systematic means for the integration of environmental dimensions. Environmental conservation and protection programmes are confined to particular areas and sectors, especially where projects have been funded by multilateral development banks and bilateral donors. Environmental conservation and management in the forestry sector are given particular focus, but other sectors have yet to comprehensively integrate environmental concerns in the development process.

Due to relatively lower level of economic activities, the deterioration of environmental quality in Myanmar has not been fortunately serious. However, in some urban pockets, such as Yangon and Mandalay and industrial centres in satellite towns, pollution levels have started showing damaging trends.

It is anticipated that with the opening up and modernization of the economy and robust economic growth over a longer timeframe, environmental pollution such as water and air contamination, and hazardous substances and solid wastes are bound to increase, unless preventive measures are incorporated in the development process. There is also clear evidence that developing countries, in their early process of industrialization, attract investment in high pollution intensity industries. It is necessary to avoid such a situation through a precautionary approach and effective screening of investment proposals.

The Materials Science and Materials Engineering Research Department had conducted a number of research works in the fields of renewable energy (such as wind, solar, biofuels), electronic and magnetic materials. Recently, under the guidance of the Minister, bioenergy had been produced using technology and local equipments. These technologies, after scaling up to pilot plant and semi-commercial scales will be transferred to the interested private entrepreneurs. As the Ministry of Science and Technology has a large number of qualified scientists and engineers, it is now capable of introducing appropriate technologies and providing technical assistance to these industries which are now in need of raising the quality standards of their products.

3.0 POLICY DEVELOPMENT

The policy calls for harmony and balance between environmental developments through the integration of environmental considerations into the development process. National environment policy forms the basis for developing environmental strategies, environmental programmes and plans.

3.1 National Environment Policy of Myanmar

"The wealth of a nation is its people, cultural heritage, environment and natural resources. The objective of Myanmar's environment policy is aimed at achieving harmony and balance between these through the integration of environmental considerations into the development process to enhance the quality of life of all its citizens. Every nation has the sovereign right to utilize its natural resources in accordance with its environmental policies; but great care must be taken not to exceed its jurisdiction or infringe upon the interests of other nations. It is the responsibility of the State and every citizen to preserve its natural resources in the interest of present and future generation. Environmental protection should always be the primary objective in seeking development."

The policy was proclaimed through the Gazette in accordance with Notification No. 26/94 dated 5 December 1994, of the Government of the Union of Myanmar.

4.0 ENERGY DEVELOPMENT IN MYANMAR

There has been increasing interest in the use of biomass for power generation in recent years. This has been due to several reasons, the principal one being that the utilization of biomass can significantly reduce net CO_2 emissions. The other advantages of the use of biomass include the fact that it diversifies to power plant's fuel portfolio by adding a potentially less expensive non-fossil fuel. Furthermore, biomass firing can reduce NO_x and SO_2 emissions and it helps to dispose of a solid waste that would otherwise need to be landfilled. With the search for alternative renewable energy sources, bio-fuels are fast becoming a viable solution, as they are non-fossil fuels from a renewable agricultural source, resulting in cleaner combustion.

Biomass fuels were produced as environmental management. Biomass resources are animal waste, agricultural waste, industrial waste, municipal waste, byproduct of forest, and so on. The produced biomass fuels are producer gas, biogas, biodiesel, bioethanol from different raw materials. Biomass fuels reduce air pollution, greenhouse gas emissions, global warming and ensure environmental sustainability.

4.1 Producer Gas

The utilization for energy is considered for agriculture and wood waste. The very common utilization of waste can be used for fuel in electric power plant that can also be done by gasification. However, viewed from economic side, a direct combustion process might be the best choice for a relatively large scale plant while gasification technology may be economical for a small scale power generation.

Myanmar's nature, biosystem and biodiversity are unique and precious. Myanmar also enjoys plentiful resources such as rice, forestry resources and mineral resources. There are no laws enforced related with biomass, but all residue is used because of the lack in material and fuel.

FIGURE 1: Downdraft Gasifier

All the gasifiers are domestic, and the down-draft type, based on the combustion or gasification of solid fuel, and featuring a comparatively simple structure and low equipment cost. Wood chips of about 2-3" diameter and 4-6" length are generally used as the raw material. They are supplied from the upper feed port and layered in the furnace. The gasifying agent (air, oxygen) is steam from the bottom in a rising flow. The gasification reaction proceeds from the bottom towards the top. The other elements are combination of water scrubber, filter and gas engine. The output of the most gasifiers is 20-50 kW.

4.2 Biogas

In olden days, Biogas was used only for cooking. Today, it is an efficient way of generating electrical energy from animal waste, industrial waste, municipal waste and agricultural waste by using the anaerobic fermentation process.

The main objectives of biogas production are to

- get the sustainable gas production rate by controlling the fluctuation of the substrate quality and process parameters

- prevent the de-forestation
- provide the electrical energy in rural area
- reduce the fossil-fuel and fuel wood consumption
- utilize the recycling of agro animal residues as a bio-fertilizer
- contribute towards environment protection, sustenance of eco-system and conservation of bio-diversity.

The energy from National grid is limited, with only 15% of the population having access to electricity. More than 90% of village householder of those off-grid areas is un-electrified. As Myanmar is an agriculture-based country, there are so many cows in central Myanmar Region especially Mandalay, Sagaing and Magway Division. On the average, the dung dropping of a medium size animal is estimated at 10 kg per day capable of producing 0.5 m^3 biogas through anaerobic digester/ bio-gasification.

The Government had launched a program to propagate the biogas utilization since 1980 and in 1983. A total of 867 floating type biogas plant, family size digesters have been constructed and utilized in 134 townships in all 14 States and Divisions with highest figure in central Myanmar Region where fuel-wood is rarely used. But the high capital investment of floating type biogas plant installation, requirement of the intensive maintenance, and short life-time of the floating drum are its main drawbacks in the local level and thus less interest was given from the national attention.

Fixed dome type biogas plants, 50 m^3 in size, have been constructed from locally available materials and utilized since the beginning of year 2003 in Panan and Pankque villages, Kyauk Se District, Mandalay Division. All houses, monasteries and streets of those villages have been provided light with biogas for two hours in the early mornings and four hours in the evenings. Now this project has extended its effort to many villages in upper Myanmar.

FIGURE 2: Biogas plant

4.3 Biodiesel

Fatty acid methyl ester produced by the trans-esterification of fat and oil with methanol has the similar physical properties as mineral diesel oil in term of heating value, viscosity, cetane number, etc. and it is used as biodiesel fuel.

In this esterfication process, free fatty acid (the main cause for biodiesel acidity) is transformed to biodiesel (methyl ester). Triglyceride and fatty acid which are still bonded in triglyceride in the oil is converted to methyl ester via transesterification. By this process, oil conversion to biodiesel (without glyceroyl) can be achieved.

Bio-diesel can reduce air pollution. It causes no environmental pollution. It is not harmful to the health of human beings. Use of bio-diesel in engines directly can reduce the global warming and contribute towards environmental conservation.

In Myanmar, physic nut (*Jatropha curcas*) is now popular as a crop to produce biodiesel. Physic nut plays an important role in implementing the rural development. In 2006, a bio-diesel pilot plant was established under Ministry of Science and Technology (MOST) with the capacity of 30 gal / day.

Every state and division is pursuing the target of putting 500,000 acres under cultivation of *Jatropha curcas* in three years. Every state and division is cultivating *J. cursas* on unused land areas for the production of bio-diesel.

FIGURE 3: Biodiesel Pilot Plant

4.4 Bioethanol

The principle fuel used as a petrol substitute for road transport vehicles is bio-ethanol. Ethanol can be produced by both the chemical process to synthesize ethanol with ethylene as the raw material, of which the end-product is called synthetic ethanol. The end-product of the widely used biochemical process involving domestic agricultural products containing starch or sugar as the raw materials is called bio-ethanol. Theses raw materials include among others corn, cassava, sugarcane and molasses. Bio-ethanol fuel is mainly produced by the sugar fermentation process, although it can also be manufactured by the chemical process of reacting ethylene with steam.

Ethanol fermentation is an excellent biological reaction to convert biomass to liquid fuel ethanol. Fermentation refers to the process using yeast to convert sugar into alcohol or ethanol, while distillation is the process where the fermented ethanol is distilled under the atmospheric pressure. It forms as an azeotrope with water at around (96% by volume) concentration, which is called rectified spirit (hydrated ethanol) and cannot be separated from water above that point by simple distillation. For ethanol used as fuel, dehydration techniques must be used to enable hydrated ethanol (96% by volume) to become absolute ethanol or anhydrous ethanol (99.5% by volume).

Commercial ethanol production plant uses molasses as a feedstock. Output of molasses from sugar mill in Myanmar is 1.23×10^5 MT. The expected producing bioethanol was 1.65×10^6 gallons. 500 gal/day and 3000 gal/day capacity anhydrous ethanol plant has been constructed in 2004.

FIGURE 4: Anhydrous Ethanol Plant

5.0 CONCLUSION

Biofuels have been attracted much interests from the viewpoint of global warming and energy security. For adoption of biofuels technology, an effective promotional and dissemination strategy is required. Programmes on capacity-building to enhance the skill-level, awareness creation among local people for the use of biofuels and know-how in eco-system management, conservation of biodiversity and sustainable use of natural resources should be embarked upon. Strategies for removing scepticism and socio-cultural barriers of adoption of biofuel technology should be put in place. There is need to develop environmental standards for energy sources, guidelines and identify cleaner technologies for energy sector. Research and development in cleaner production processes from different sources including biofuels substitution should be promoted.

10

Biofuels and Energy Security in Vietnam

Ngo Thi Loan
Project Department, Center for Regional Research and Development,
Ministry of Science and Technology of Vietnam
Email: loancrd@yahoo.com, ntloan@most.gov.vn

Energy in general and fuel for international combustion engines in particular plays an important role as a motive force for the economic development. National security or economic security is closely relative to energy security. Use of clean renewable fuel, such as bio fuels to secure the energy security, economic development and environmental protection is always the first national policies in the sustainable development strategy. Biofuel is one of the most important alternative fuels for the transportation sector. This paper presents the biofuels research and development situation in Vietnam, as well as opportunities and challenges. Biofuels project goals, major measures for the development and utilization of bio fuels in Vietnam up to 2020 were also discussed.

1.0 THE BIOFUEL PRODUCTION SITUATION IN VIETNAM

1.1 Overview

1.1.1 Three burining global issues

- Ensuring the substitution of energy source when fossil fuels, especially mineral-oil, are becoming scarcity.
- The degrading of natural environment due to too much usage of fossil fuels and rapid urbanization.
- Almost 2 billion poor people lack clean water and model energies.

1.1.2 New awareness of utilizing energy

The current socio-economic development based on the exhausting fossil fuel is unsubstainable in terms of energy security and bio-environment.

Therefore, for the past few decades, many countries had set up policies which combined economic utilization of current energy sources and exploitation of clean energies such as nuclear energy, renewable fuels including biofuel.

1.1.3 What are biofuels?

- Biofuels are fuels devised from biological materials.
- They are different from oils and other fossil fuels; they are a form of bio-energy which is obtained from renewable energy source.
- There are two main catagories of biofuels:
  - — Biofuel (ethanol and methanol) produced from biomass (not from oil). The main materials for producing ethanol are all kinds of grain with starch, plants with cellulose, even woods and other agricultural wastes.
  - — Biodiesel, a clean energy, produced from lipid, animal and plant fats (soya bean oil, coconut oil, sesame oil, fist fat.).

1.1.4 Advantages of biofuels

✓ Biofuels are produced on the spot from rich and renewable plant sources. Possible to produce and supply at huge amount. They are ideal substitutions when the costs of mineral oils and gas are increasing.

✓ Biofuels do not contain toxic elements such as mineral oils, easy to decompose. Utilizing the biofuels helps reduce the greenhouse effect and other toxic gas emission, thereby reduces the expenses on environmental pollution. The emission quota can also be controlled through CDM projects.

✓ They are easy and convenient to use. Biofuels can be used together with other kinds of fuels, e.g. bioethanol and petrol can be used alternatively without the need of changing the engines or distribution networks.

✓ The biofuel production technology from animal and plant lipid is not as complicated as petrochemical refining technology and requires much less investment. Furthermore, it can be applied to mass production at different scales.

✓ The biofuel development helps to improve the agricultural economy, rural development and people's living standard because:

- ■ Besides providing food for people, cattle and poultry as well as providing materials for industries, agriculture also provides energy for the whole society.
- ■ It creates job opportunities for people, making the best use of uninhabited land.
- ■ It improves value addition to agricultural products and by-products.

2.0 THE CONSUMPTION AND DEMANDS OF ENERGY IN VIETNAM

The energy consumption in Vietnam, according to the Draft Resolution of National Energy Development Strategy of Vietnam up to 2025, is as shown in Tables 1- 5 below.

TABLE 1: Total volume of energy consumption in Vietnam

Year	1996	1997	1998	1999	2000	2001	2002	2003	2004	2005	Growth rate/ year
Coal (mill. ton)	9.8	11.4	10.7	9.6	11.6	12.9	16.04	19.3	27.3	32.4	10.4%
Crude oil (mill. ton)	8.8	10.9	12.5	15.2	16.3	16.8	16.9	17.7	20,05	18.5	16.4%
Natural gas (mill. Cubic m)	290	540	1018	1414	1580	1720	2161	3720	—	—	

Energy exportation
TABLE 2: Total volume of energy exportation (Million tons)

Year	1996	1997	1998	1999	2000	2001	2002	2003	2004	2005
Crude oil	8.71	9.64	12.15	14.88	15.42	16.73	16.60	17.05	17,14	19,50
Coal	3.64	3.45	3.16	3.26	3.25	4.29	6.05	6.30	7,261	11,62

Energy importation
TABLE 3: Total volume of energy importation (Million tons)

Year	1996	1997	1998	1999	2000	2001	2002	2003	2004	2005
Crude oil	5.89	5.96	6.85	7.42	8.75	8.99	9.97	9.94	11.05	11.45
Petrol	109	1.08	1.2	1.32	1.48	1.732	2.09	2.18	2.60	2.63
Diesel	2.79	3.10	3.53	3.48	4.13	4.08	4.43	4.65	5.48	5.89

Energy consumption (Ktoe)
Table 4: Total volume of energy consumption

Year	1996	1997	1998	1999	2000	2001	2002	2003	2004	2005
Coal	2692	3327	3302	3166	3223	3743	4017	4105	4851	4950
Oil products	4986	5001	5538	6222	6759	7456	8761	9639	10317	11864
Gas	21	20	19	19	18	15	19	18	21	100
Electricity	1150	1316	1524	1681	1927	2214	2586	2996	3437	3697
Total	8849	9664	10383	11088	11927	13428	15383	16758	18626	20950

Energy Consumption Forecast in Vietnam

The need for energy consumption of Vietnam is rapidly increasing, collerating with the rate of economic development and population growth.

TABLE 5: Forecast of Energy demand by fuel type in 2030

Year	2010	2015	2020	2030
Options	Low/Standard/ High	Low/Standard/ High	Low/Standard/ High	Low/Standard/ High
Coal	5430/5821/6020	6234/6743/7067	7122/7723/8289	9028/9790/10507
Electricity	6610/6922/7368	9937/10799/11933	13497/15355/17533	19988/24908/30164
Oil products	14365/17402/17785	18533/23898/25445	24910/32615/36363	42599/55621/64995
Gas	743/900/1900	1396/1800/2650	2291/3000/3230	3584/4680/4781
Total	27148/31044/33074	36101/43241/47095	47821/58693/65415	75200/95000/110447

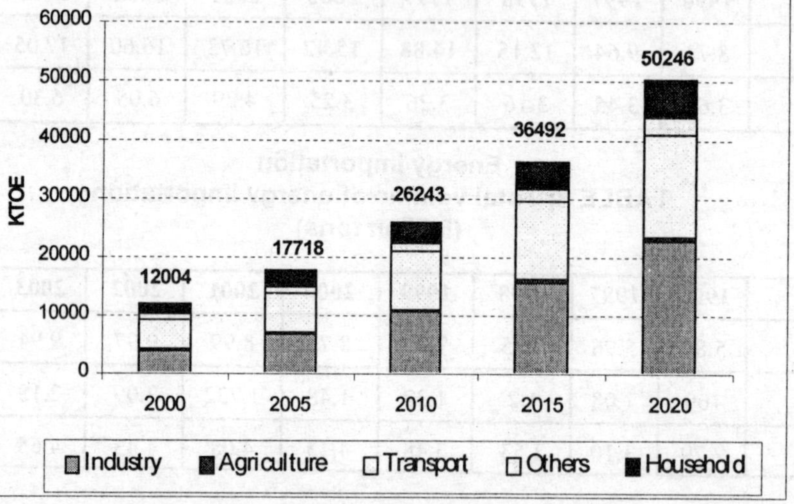

FIGURE 1: Energy Consumption Forecast by Sectors (according to the Draft Resolution of National Energy Development Strategy of Vietnam up to 2025)

2.1 The initial results of biofuel application research in Vietnam

— The Institute of Science and Technology of Vietnam already had basic researches on biofuel production and usage. Da Nang University of Technology and many other Research Institutes also conducted lots of researches about the production and application of biofuels

— APP company lead the state level independent research: "Research on biofuel production technology mixed with ethanol and some other compounds with plant origin"

— The Alcohol and Beverage Research Institute has implemented research using ethanol as a fuel substitution for some kind of engines. The Institute for Food Industry has been assigned to lead the state-level research on producing ethanol fuel from agricultural waste disposals by enzyme and microorganism technology.

— Sai Gon Petro, Lam Son Suger JSC, Agrifish and many other companies have conducted research, making the way for mass production at appropriate scale for some kinds of biofuel from fish fat, alcohol.

— Vietnam has potentials (peadology and climate condition) on some kinds of plant providing materials for alcohol production such as: rice, maize, cassava, sweet potatoes and sugarcane.

— Except for rice, the current production of the remaining kinds of food is just sufficient for current production and consumption. If alcohol production at large scale is developed, there will be a lack of material. However, at small and medium scale production, the material needs may be met, if the cultivating areas are expanded.

— The raw material source starch (cassava) and sugarcane have potential to grow in large area to provide enough material for alcohol manufacturing factory. However, use of new varieties with high productivity must be developed. New measure to increase the cultivation efficiency must also be taken.

2.2 Potential raw materials for biodiesel production from plant and fat oils

- ***Soya bean cultivation***

  — Total of 160,000ha was cultivated in the whole country in 2003 for the production of soya bean.

  — The total production of soya bean in the country was 225.3 thousand tons, 2.6 times as much as that of 1990.

  — Not sufficient for human consumption as food and livestock.

- ***Peanut cultivation***

  — Total area of peanut cultivation in the whole country increase from 200 thousand ha in 1990 to 243 thousand ha in 2003, at the average rate of 1.5%/year.

  — The country's average productivity reached 1.7ton/ha, increased by 0.6ton/ha (53%) compared to that of 1990.

  — This only met the consumption demand for human and livestock

- ***Aleurites moluccana/ A. cordata, Camellia sasanqua***

Aleurites moluccana and *Camellia sasanqua* are oil containing plants which are used in the state-funded upper basin forestation, protective forestation and greening wasteland and bare hills programs. The *A. moluccana* and *C. sasanqua* planting programmes in large scale are being implemented in the provinces of Nghe An, Thanh Hoa, Ha Giang, Lang Son and Quang Ninh. The programme will also be implemented in the Northern Midland Mountainous areas and other North central provinces.

— *Camellia assanqua* Cây So: In 2005, there was 20,000 ha of camellia cultivation with total production of 900 tons. By 2010, the total cultivation area is estimated to be 100.000 ha with production of 18,000-72,000 tons.

— Cây Trau: In 2005 the total production was 1200 tons. By 2010 cultivation areas is estimated to be 28,000 ha with total production of 12.600 tons.

- ***Sesame/ Jatropha curcas L***

Sesame *(Jatropha curcas L.)* is a potential plant with high biological energy. Vietnam has met several conditions to develop the sesame plant. With infertility land condition and tropical climate, sesame can grow fast and start to bear fruits after 6-12 months of cultivation.

The oil content of sesame is about 35 - 40%, therefore its oil productivity is quite high, from 2,500 to 3,000 litres of bio-diese/ha/year.

- ***Coconut plantation***

— Total plantation area in 2005 was appropriate 135,000ha with production of 920,000 tons and, average yield of 6.8 tons of coconut /ha.

— The two biggest coconut plantation areas are Mekong river delta and the Southern centre coastline with total areas of 105,000 ha and 20,000 ha, respectively.

— The coconut processing in Vietnam is still limited.

— The potential of developing and expanding the coconut plantation areas is really promising in terms of natural and peadological conditions in the key provinces, especially in the Southern centre coastal provinces where the coastal and wild land for coconut growing is immense.

- ***Cassava plantation (in 2005)***

— Total areas of cultivation was 423,000 ha

— Production was 6,646 million tons which was, twice as much as that of 1995

— Average yield was 15.7tons/ha

Regions that could expand plantation include:

— The northern mountainous midland: possible to expand by 60 thousand ha

— Northern centre: possible to expand by 21.8 thousand ha

— South centre coastline: possible to expand by 25.3 thousand ha

— Middland Tay Nguyen: possible to expand by 39 thousand ha

— Eastern South: possible to expand by 15.5 thousand ha

- ***Sugarcane plantation (in 2005)***

  — Toal area of cultivation was 266.4 thousand ha

  — Production:14,73 million tons (increased by 1,5 times compared to 1995).

  — Average yield: 55.3 tons/ha (quite low in comparision to the whole region and the world average yield about 100 tons/ha).

  — In 2006, there were about 37 sugar processing factories in the whole country with total capacity of 82,450 tons of sugarcane/year, honey yield was about 592 thousand ton which could produce almost 200 million litres of alcohol.

- ***Fish fat***

Total amount of catfish fat in Vietnam is 200,000 tons per year, mainly in the Mekong delta. Although these fats are sold to the engine oil and livestock food processing mills, the output and price is uncertain and the waste caused terrible pollution problem.

2.3 General assessment of oppotunities and challenges of biofuel development in Vietnam

Advantages

✓ Energy security and environment protection issues in Vietnam are becoming increasingly important and needs urgent and the careful attention of the government.

✓ Vietnam, a tropical agriculture based country, has good conditions to produce biomass of agricultural wastes.

✓ Vietnam had the opportunity of researching into the production and usage of biofuel and has obtained some important results.

✓ Vietnam had advantages of learning from the previous countries in research, production and usage of biofuel.

Disadvantages

✓ *Technology and equipment*

— Most of the alcohol production workshops in the country are now using the old backward equipment and technology with low capacity (under 10 million litres per year);

consuming lots of energy per product unit, overall productivity only reached about 80% compared to over 90% attained by developed countries. Vietnam only uses the conventional biomass material such as cereals, sugar etc. Therefore, the product cost is high.

— The animal and plant oil producing mills are using the backward oil extracting technology, so the production rate is low.

— The research to mix alcohol and animal and plant oil with mineral petrol technology is just at testing phase.

✓ *Human resource quality*

— The number of researchers and skilful technicians is quite low and their access to the modern technology to produce biofuel is commercially very limited.

— There are few highly qualified experts and they are not sufficient to catch up with the developing speed of modern technology.

✓ *Investment*

— The investment fund is limited, comes mainly from the state budget for researches and experiment production projects.

— Lack of investment focus and concentration.

— The investment rate for development and application of biofuel in Vietnam is rather low in comparison to the world's rate.

✓ *Research development and application*

— There is no cooperation between the closely related sectors in research and development (R & D) and application of these in the practical production. The current facilities have not been exploited effectively.

— The research target is not focus. Results of the researches are mainly of the laboratory types and experiment workshop scale-ups are not coming from the market demands.

— The biomass material resources are still produced using the same old varieties and cultivation methods, therefore productivity and quality is low. There is the need to have new and sufficient varieties as well as new cultivation processes to increase the productivity and quality of the raw material.

✓ *Legal system*

— There are no mechanisms and priority policies in investment, for human resource development, developing and improving quality of the raw materials, technology transfer, intellectual properties assurance and rights of investors to develop and apply biofuels.

— There are no regulations on environment protection that will encourage the use of clean energies.

— There are no standard systems for biofuel products.

✓ *International cooperation*

— There are few international cooperation programmes on R & D and application of biofuel. These results are not sufficient enough.

3.0 TARGET VIEWPOINT AND CONTENT OF THE BIOFUEL DEVELOPMENT PROGRAMME UP TO 2015, VISION 2025

3.1 Viewpoint

The development of biofuel to contribute to ensuring the energy security and environment protection is very important and should be given priority.

The biofuel development requires capital and technology investment. Therefore, investment should be focused with proper scales and synchronous.

The focus targets are:

— *Key products:* Set up and develop the biofuel processing and consuming schemes. By the year 2025, E5 petrol and B5 biodiesel would have been produced and used in the whole country.

— *Breakthrough measures:* Develop the national potential power (training the human resource and upgrading the technical and technological facilities) together with other proper policies to promote the internal power sources and the contribution of other economic sectors.

In the research and development of biofuel, the improvement of internal sources and learning of the world's experience and achievement should be combined to make the best application in the practical conditions of Vietnam.

3.2 Targets

3.2.1. General target

Develop the biofuel to partly substitute the conventional energy source, helping to ensure the energy security and environment protection.

3.2.2. Specific target

3.2.2.1. From 2006 to 2010 period

✓ Building the legal corridor to encourage the production of biofuel at industrial scale and using biofuel as substituted fuel.

- ✓ Building itinerary of using biofuel as substitute fuel in transportation and other industries and biofuel distribution pilot models in some provinces and cities.
- ✓ Accessing and mastering the technology of producing biofuel from biomass including: Having suitable mixture technology; increasing the productivity when transforming from biomass material to fuel; improve the public awareness about the importance ad benefits of biofuel.
- ✓ Training a group of professionals specialized in some major subjects relating to the production of biofuels; training technical workers to meet the demand of human resource during the development of biofuel.
- ✓ Building and developing the testing models for production and usage of biofuel with capacity of 100 thousand tons of E5 and 50 thousand tons of B5 per year, guarantee to provide 8% of the whole country's total demand for fuel oil and petrol with E5 petrol and B5 biodiesel.
- ✓ Approaching and mastering the technology of cultivating the high-productivity plant for biofuel production.

3.2.2.2. From 2011 to 2015 period

- ✓ Mastering and producing materials and additives for biofuel production.
- ✓ Developing and using biofuel, partly substitute the conventional fuel types. Expanding the scales of biofuel producing workshops as well as the distribution network (for transportation purpose and other industrial production).
- ✓ Developing to mass production of material plants with high productivity and good immunization to ensure adequate supply of biomass material for the biofuel production process.
- ✓ Successfully applying the modern zymotic technology to diversify the raw material sources for the biofuel transformation process.
- ✓ Building and developing the models for production and usage of biofuel nation-wide, guarantee to provide 20% of the whole country's total demand for fuel oil and petrol with E5 petrol and B5 biodiesel.

3.3. Vision up to 2025

- ✓ By the year 2025, the biofuel production technology of Vietnam would have reached the advanced level in the whole world. The ethanol and plant oil production meet 100% of the country's demand for fuel with E5 petrol and B5 biodiesel.

4.0 PROJECT CONTENT

1. Building mechanism, policies and legal document system relating to the research, production and usage of biofuel namely:
   — Investment encouraging policy; research and production promoting mechanism.
   — The technical standard system, the application itinerary (Vietnam National Standard TCVN, technical standard relating to the production and usage of E5, B5).
   — The compulsory regulations on environment protection and application itinerary for all users which encourage the use of biofuels.
2. Approaching and mastering the technology to mix petrol, condensat, nafta, mineral diesel with ethanol, biodiesel and additives to produce E5 petrol (95% mineral petrol and 5% ethanol) and diesel B5 (95% mineral diesel and 5% biodiesel). Operating the biofuel producing workshops with capacity of 100 thousand tons of E5 and 50 thousand tons of B5 per year. Developing the distribution network throughout the country, especially in major cities such as Hanoi and Ho Chi Minh City.
3. Approaching and applying the modern ethanol producing technology from different biomass sources.
   — Mastering the modern ethanol producing technology with high productivity, energy saving and qualify to be the biological fuel.
   — Upgrading the medium-scale alcohol producing workshops to large scale and synchronous to ensure enough ethanol supply for the E5 production.
4. Researching, technology developing and managing the production and usage of biodiesel from qualified animal and plant oils to mix with mineral diesel to form diesel B5. Developing the biodiesel producing workshops at suitable scales according to their raw material regions, ensuring the adequate supply of biodiesel for the production of B5.
5. Planning and organizing the raw material regions to produce alcohol, animal and plant oils (sugarcane, cassava, maize, oily plants, animal fat…), making the best use of the current land fund, at the same time considering the effectiveness of the land usage of each location, promoting the advantages of each material region. Selecting the sugarcane, cassava, rice and oily plant with high productivity, suitable with each planned land areas, especially the low quality and polluted land that can not be used for human food cultivation.
6. Approaching, researching and mastering the technologies to produce additives, chemicals necessary in order to mix biofuels and mineral fuel in order to ensure the fuel quality and safety to environment.

5.0 IMPLEMENTATION SOLLUTIONS

5.1 Mechanism and policies

- Changing the current mechanism and policies about taxes, giving priority to loans and land leasing, encouraging and creating favourable conditions for local and international enterprises, organizations and individuals to invest to biofuels manufacturing.

According to the Government Decree no. 24/2007/NĐ-CP dated 14 February 2007; investment in biofuel manufacturing is listed in the special investment incentives for the period of 2007 - 2015.

- During the construction of Vietnam standard and technical standard system for fuels, additives and biofuel products, enterprises are permitted to announce their own standards with reference to the standards of G7 countries.
- Develop and promulgate mechanism and policies to attract talents directly involved in research, development and production of biofuels in a sustainable manner.
- Encouraging and supporting the cross-breeding researches to create new microorganism inoculates and new varieties of industrial plants (including the genetic variations). Improving the cultivation process to produce materials with yield and quality meeting the demand for biofuel development.

5.2 Strengthening potentials

- Profoundly investing to upgrade the R&D institutes in biofuels; Supplementing and modernizing more advanced equipment and facilities for current laboratories in order to improve their capacity and qualification of expertises to approach and receive technology transfer, developing the biofuel to commercial products.
- Investing to fully equipped laboratories, ensuring sufficient strength for the purpose of researching raw materials, production and mixture process in order to effectively implement contents of this project.
- Giving priority to quick research methods to create products which can be potentially developed to mass industrial production.
- Importing some effective technology and developing to industrial production at large scale.
- Investing in breeding, creation and multiplication projects to have sufficient raw materials for production.

5.3 Human resource training

Types of training:

- Higher education: Talented students studying the technical subjects supporting the biofuel development program would receive encouragement and be funded by the state budget.
- Technician training: The state supports training and re-training programs to improve the qualification of technicians, skilled workers, meeting the need of biofuel development; helping the local people master some technologies so that they can establish small-scale raw material manufacturing workshops (ethanol and crude plant oil), meeting the diversified demands of each localities.
- Training, improving and receiving technology transfer: The state support technology transferring and receiving programmes.

Priorities being focused to the following training areas:

- Education on raw material plants production technologies (creating new breeds, cultivation methods)
- Education on modern fermentative technologies.
- Education on enzyme technologies
- Education on materials technologies
- Education on alcohol and plant oil production technologies
- Education on catalysts and additives production technologies

5.4 International cooperation

- Through the international cooperation channels, building up the mutual and multilateral cooperation programs with experienced countries in biofuel development in forms of cooperative projects, in which state budget would be used as Vietnam's reciprocal capital. This cooperation will help us to approach the new technologies as well as training for the human resource.
- Building up specific cooperative programs with experienced countries in biofuel development such as: Brazil, the US, Germany, Japan, India, China and Thailand.

5.5 Improving people's awareness

- Making campaign in the community about the important role and socio-economic effectiveness as well as environment protection of biofuel in the sustainable development.
- Step by step approach to establish and develop the biofuel market in Vietnam.

11

Uganda's Energy Crisis and its Quest for Biofuel Solutions

M. Bazira Henry
Water Governance Institute (WGI),
P. O. Box 23704, Kampala, Uganda
E-mail: bazirah@yahoo.co.uk; bazirah2@gmail.com; watergovinst@gmail.com

Uganda's population is currently 30.2million people. Over 90% of Uganda's population relies heavily on traditional bio-fuels for energy. These mainly consist of firewood, charcoal and crop residues. This dependency on traditional bio-fuels is exerting enormous pressure on natural resources resulting in the depletion and degradation of natural forest cover. The country has lost over 1.3million hectares of forest cover in the last 15 years as a result of human encroachment for settlements, farmland, commercial purposes and the search of fuel-wood. Currently, there is an increasing aspiration of investing in biogas, bagasse for co-generation and crops for biodiesel production. However, these aspirations and trends have not yet caught-on, because of the limited knowledge and technical skills among the populace to harness these resources.

For the last 5 decades, Uganda's electricity has predominantly been hydropower mainly generated from Owen Falls (Nalubaale) dam, and more recently (2002), from the Owen Falls Extension (Kiira) dam. This hydropower complex has a combined installed capacity of 380MW.This is in addition to approximately 10MW generated from several scattered small-scale hydropower dams across the country. This electricity, however, serves less than 10% of the households in the country mainly in the urban areas. It is riddled with technical and commercial losses of up to 41% that makes the generated power insufficient to meet growing demand, resulting in frequent power-cuts and load-shedding. This is further complicated by declining water available for power generation in Uganda's lakes and rivers due to over-abstraction and climate change.

In an attempt to contain the current energy crisis in the country, government installed 150MW of thermal power as a first-line remedy. However, this is very expensive and is exerting enormous strain on national treasury. About US$900,000/month is spent on thermal power. Use of thermal power is further disadvantaged by the significant carbon footprint associated with it.

The freely abundant solar, wind and geothermal energy sources in Africa are not given due consideration as alternative sources of energy for national development, yet they have great potential.

A new energy paradigm based on bio-fuel energy is slowly taking root, with countries in Africa and Uganda, in particular, desiring to free themselves from the bondage of imported fossil fuels and the need to generate revenues/incomes through growth of bio-fuel producing crops. While this seems to be ingenious, the manner in which it is being pursued leaves a lot to be desired. For instance, large chunks of natural forest estates (gazette & private) are being replaced with plantations (sugarcane, palm oil, sunflower, jatropha, etc) for biodiesel production, with little or no consideration of the social, economic and environmental consequences likely to accrue. Also, there is no clear evidence in Uganda that the other bio-fuel sources such as biodegradable plant and animal waste-based materials will be integral to this paradigm, yet they exist in substantial amounts. The few examples are based on sugarcane molasses and sunflower and Jatropha oils and are on trial basis. No Environmental Impact Assessment on bio-fuel production from these feed-stocks has been made public.

Countries, especially in southern hemisphere, need to be cautious in the manner in which they develop and utilize their energy resources without jeopardizing the very existence of their populations.

1.0 INTRODUCTION

Uganda is essentially an agriculture-based economy with a significant proportion of its national budget supported by external donor assistance. This is not an enviable situation. Every nation and individual aspires to be financially independent. However, this usually comes with a lot of sacrifice and investment. The question is not whether the sacrifice and investment should be made, but how much of that sacrifice, investment and the likely social, economic and environmental impacts Uganda is willing and capable of bearing as a nation and individuals.

Uganda has a population over 30million (CIA Factbook, 2008) on a land area of 246,038Km^2 with 18.2% of this land wetlands and open-water bodies. Less than 18% of the land is covered by natural forest cover. The natural forest cover used to be 44% of the land in colonial times, but has gradually been lost through encroachment for urban and rural settlements, industry, and agriculture and harvesting of forest products (timber, firewood, charcoal, medicinal plants & wild game) and now stands at less than 18% of the land cover. Therefore, the growing population is exerting enormous increasing pressure on the natural forest estate for settlements, food, energy and other products.

The paper highlights the energy situation in Uganda and discusses the challenges and opportunities of Uganda's quest for bio-fuel development. It is based on prevailing facts and review of literature available in the public domain.

2.0 THE ENERGY SITUATION IN UGANDA

Over 90% of the population relies on fuel wood and charcoal for their energy sources. In the last 15 years, the country has lost more than 1.3million hectares of natural forest cover i.e. a loss rate of 86,700ha of forest cover per year (FAO, 2007) as a result of human encroachment. At this rate, it is projected that all the natural forest cover would be whipped out in the next 20-25 years, if nothing is done to abate this forest loss. This calls for alternative energy sources and a development paradigm that includes taking advantage of the vertical spaces when developing human settlements; intensive and more efficient farming systems; and use of alternative renewable energy sources.

Uganda has an installed capacity of about 540MW electricity i.e. 390MW hydropower (MEMD, 2004) (Plate 1) and 150MW thermal (Plate 2). The electricity demand is currently projected at 450MW and growing at a rate of 8% per year (PPA, 2007). While the installed electricity capacity seems to be more than sufficient to meet current demand, today's power supply is riddled with frequent power-cuts and load-shedding that is due to insufficient water supply leading to reduced hydropower generation (40% of installed capacity), high technical and commercial losses of 38% to 41% generated electricity; and delayed development of alternative additional power supply. This is further compounded by the low coverage of the national grid system that serves less than 10% of the total households in the country. The apparent mindset in government on centralized grid systems is jeopardizing distribution and access to electricity to the wider public, especially those in remote areas not readily served by the national grid system.

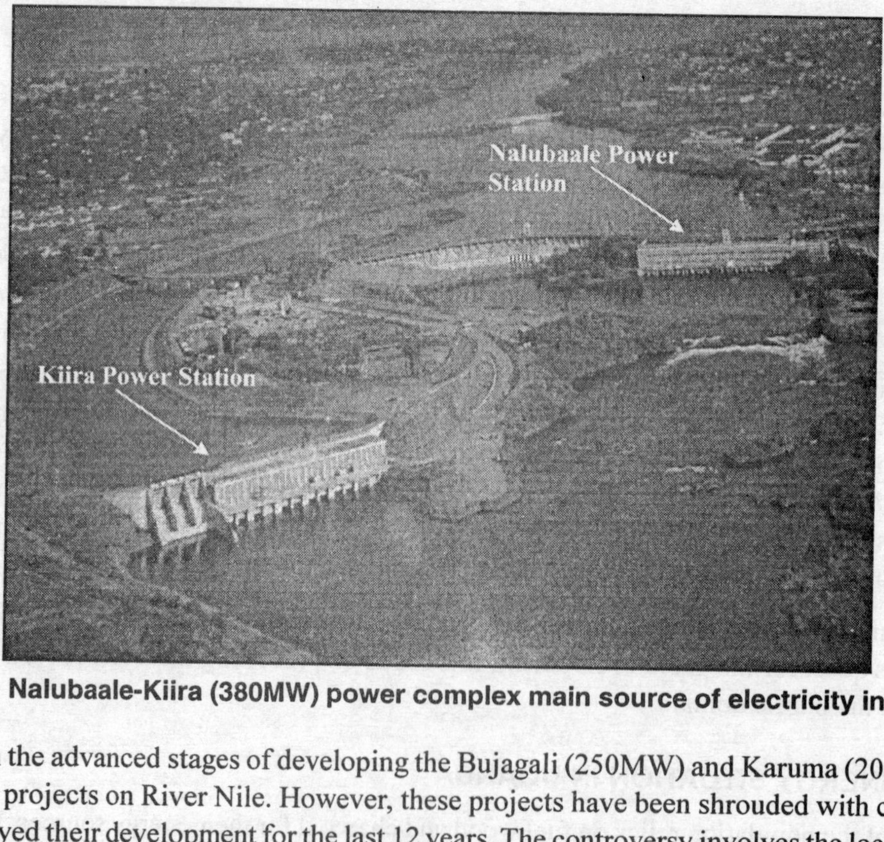

PLATE 1: Nalubaale-Kiira (380MW) power complex main source of electricity in Uganda

Uganda is in the advanced stages of developing the Bujagali (250MW) and Karuma (200-300MW) hydropower projects on River Nile. However, these projects have been shrouded with controversy that has delayed their development for the last 12 years. The controversy involves the location of the projects, hydrological risks, climate change, costs, and ability to deliver designed capacity, politics, and affordability of the power from the projects, among others. Often rural households require energy for cooking, lighting, milling and refrigerating their farm produce. This energy requirement can easily be met by developing small-scale independent electricity grids based on solar, wind, small-medium hydro- and geothermal power. There are arguments that solar power is very expensive; there is limited wind capacity, and the available geothermal potential can only produce a maximum of 45MW (PPA, 2007).

PLATE 2: Aggreko Thermal (50MW) power plant in Jinja, Uganda

However, earlier studies have revealed a geothermal potential of 450MW and shown evidence of investments that depend on solar and wind. There is, therefore, no justification for not supporting the development of solar, wind and geothermal based systems in the country. In addition, these energy sources would actually serve as alternative and additional energy, especially for rural-urban households and would free substantial amounts of hydro- or thermal- based power in the national grid for industry and commercial purposes.

The country is in advanced stages of commercially exploiting its fossil fuel resources in the Albertine Graben. Since 1998, over 50 licenses for exploration have been issued to different oil companies, more than 10 Production Sharing Agreements (PSAs) signed, more than 30 Environmental impact assessment (EIAs) for oil exploration and exploitation carried out and a new Oil and Gas Policy 2008 put in place. The dominant companies in the region are Tullow Oil Uganda Operations PTY Limited, Heritage Oil & Gas, Dominion and Neptune. Significant oil discoveries have been made at several sites in the Albertine Graben that are estimated to have a yield potential of 100,000–150,000bpd (barrels per day) of crude oil (MEMD, 2008). Tullow and Heritage report even greater potential.

An Early oil Production Scheme (EPS), in partnership with Tulllow, is being pursued by government. It is intended to utilise 4000bpd of crude oil to generate electricity (85MW) and produce 500bpd of diesel and 500bpd of Kerosene. This is intended to alleviate the current energy crisis in the country. While this seems to be ingenious, the manner in which oil exploitation is being pursued, leaves a lot to be desired.

- The development is shrouded with a lot of secrecy between government, the oil companies and individuals in the public and private sector involved in the deals. The development is also characterized with political influence peddling and with little or no opportunity for public participation in the development process. The few cases where the public has participated, they were not given sufficient time to respond as mandated by the law and were consulted in a vacuum of information concerning the oil projects. This is compounded by the fact that not many people in the public can comprehend the usually highly technical issues that characterize the oil industry. In addition, there hasn't been any sector wide public awareness and education programme to accustom the public on matters concerning the industry e.g. on such as licensing, PSAs, EIAs, consultations, and how the development will affect the wellbeing of ordinary citizens.
- Environmental Impact Assessments on oil are currently being carried out hurriedly; consequently, they are riddled with inadequacies and gaps and do not thoroughly elucidate the likely project specific and cumulative impacts. The EPS is expected to link-up with other developments such as the Electricity Transmission lines from the Kaiso-Tonya area to Nkenda, access and maintenance road networks and associated industries and businesses, but the EIAs, if they were done, have not been made public and do not form an integral part of the EIAs prepared for oil exploitation. Also, the EIAs are not supported by Sectoral Strategic Environmental Assessments (SSEA).
- The recently developed Oil and Gas Policy (2007) is not supported by legal and institutional framework to implement it. It is also characterized with gaps and is incomprehensive. Therefore, the oil industry is developing in an environment that is devoid of explicit regulatory mechanisms. Yet, it is a well known fact that faulty oil processes are the single major causes of conflicts and the so called "oil curse" in the oil producing countries such as Nigeria, Chad, Algeria, Iraq, Angola, Sudan and others.

3.0 THE QUEST FOR BIO-FUELS IN UGANDA

Bio-fuel is defined as solid, liquid or gaseous fuel derived from relatively recently dead biological material. Unlike fossil fuels which are derived from long dead biological material, bio-fuels can be produced from any biological carbon source, although the most common sources are photosynthetic plants, which produce agro-fuels. Several plants (crops) are potential sources of biofuels such as soybean, cassava, sweet potatoes, sunflower, switch-grass, corn, wheat, sorghum, sugar-beet, groundnuts, sugarcane, palm oil, *Jatropha*, Rapeseed, *Pongamia pinnata*, Hemp, etc. Wood and its by-products can be converted into biofuels such as wood-gas, methanol and ethanol fuel. It is also possible to produce cellulosic ethanol from non-edible plants (Somerville, 2007), but this is economically unviable. Recent technology can now convert pollution into renewable biofuel, which countries should venture into with a view of mitigating pollution problems. For instance, there is potential to produce biofuel from waste processes such as landfill and sewage off-gassing or recycled vegetable oil, which is distinct from agro-fuel production. There is significant potential for biodegradable waste-based biofuel production in Uganda, but it has received little or no attention over the decades, yet it would contribute to waste management in the country.

Biofuel industries are expanding in Europe, Asia and the Americas where the fuel is used to power vehicles, heating homes and cooking stoves. There is also a growing quest to develop biofuel production capacity in Uganda. This quest for biofuels is being driven by several factors, including the declining stocks in fossil fuel in traditional production areas of the Americas, Europe, Asia and the Middle-east; the desire to reduce dependency on fossil fuels among the developed countries; and the need to reduce the carbon footprint. In developing countries, like Uganda, the quest for bio-fuels is essentially being driven by the desire of countries to free themselves from the bondage of imported fossil fuels and to make money for the starving economies. In Uganda, there are already bio-fuel trials in Masindi and Luwero Districts based on sunflower and Jatropha oils producing biodiesel (Ndawula and Turyahabwe, 2008). Also, sugar manufacturing factories in the country are producing small amounts of ethanol from molasses (Ndawula & Turyahabwe, 2008). Muzizi Tea factory in Kibaale district and Nyabyeya Forest College in Masindi are using electricity based on wood gasification (Ndawula, 2008).

While the quest for bio-fuels in Uganda seems to be ingenious, the manner in which it is being pursued leaves a lot to be desired. It is being pursued in an environment devoid of sufficient information on the social, economic and environmental impacts. Government of Uganda is spearheading the expansion of sugarcane, sunflower and palm oil plantations into gazetted and private natural forest estates for mainly bio-fuel production, in total disregard of the environmental and social consequences. The argument is that plantations encroached on natural forests long before bio-fuels became important and therefore bio-fuel production would not cause any worse impact than has already been caused.

In addition, bio-fuels are being development in an environment devoid of comprehensive policy, legal and institutional frameworks to govern their exploitation. It is argued that bio-fuels are cheaper and have a lesser carbon footprint than fossil fuels (Thurmond, 2007). While this may be true in specific terms, Uganda seems to be pursuing the bio-fuel venture in narrow perspective. There is no evidence that opportunity-cost and cost-benefit analyses of utilizing the available arable land for agro-fuels visa vie food production have been done. Also, no sectoral strategic environmental assessment has been conducted for the country to form basis for the development of bio-fuels. Bio-fuels have been criticized for their potential of taking away land critical for food production, diverting food from the animal and human food chain and increasing food prices (UN bio-fuels Report), thus negatively affecting livelihoods of the poor.

It should be recalled that most of the land cover is situated in the northern hemisphere (Europe, Russia, Britain, North America, China, etc) and less south of the Equator (i.e. Africa, South America, Australia, & Antarctica) (Plate 3). The South is dominated by least developed countries that are highly dependent on agriculture for survival and forests for their energy sources and it inhabits majority of the worlds poorest.

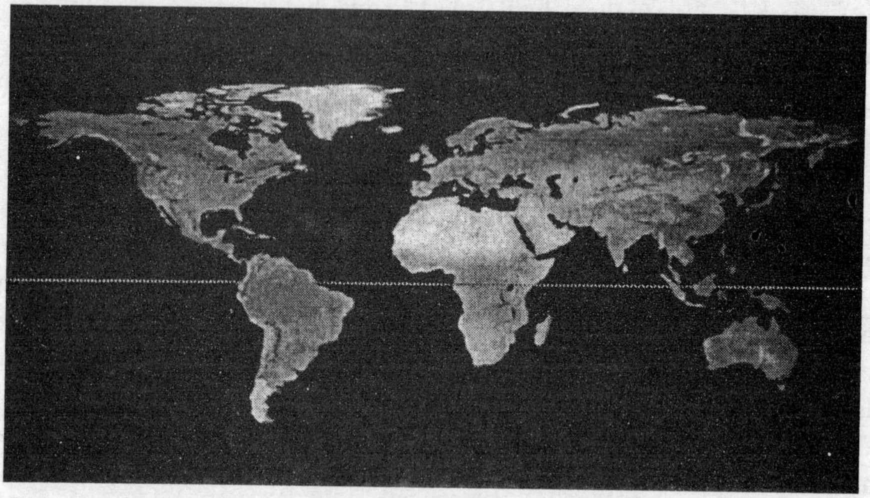

PLATE 3: World Map showing the Northern and Southern hemisphere across the Equator (Source: Al Gore (2005))

In social terms, bio-fuels would be most suitable to southern countries, since their communities are more dependent on biomass for energy. In addition, the southern countries usually have conducive weather (climate) for the growth of bio-fuel producing crops year-round compared to the north that has a greater part of the year in winter. However, in production terms, the north would be most suitable for bio-fuel production, since it has most of the land. Therefore, taking away the most valuable arable land for bio-fuel production would deny the greater population in the south cheap and ready access to food.

Bio-fuel production in the north is heavily subsidised, which would make it difficult for unsubsidised bio-fuels from the south to compete in the World Market. There is currently a conflict between Europe and America over America's subsidised bio-fuels that is undercutting market prices for bio-fuels produced in Europe. It is argued that bio-fuels from America to Europe receive double subsidisation: i.e. first, when they are being produced in America and second, through tax cuts as they enter Europe. Consequently, American biodiesel reaches Europe at a cost of US$300 per ton compared to US$450 price-tag for one tonne of bio-fuel produced in Europe (BBC, 2009). It is, therefore, important for the south, Uganda in particular, to assess its comparative advantage of producing bio-fuels against this framework.

Supporters of bio-fuels claim that a more viable solution to the increasing controversy over bio-fuels is to increase political and industrial support for and the production of bio-fuels from non-food crops and biodegradable waste materials, including cellulose (Somerville, 2007). However, although cellulosic material is abundant, it does not divert food away from the animal and human food chain. It is possible to produce cellulosic bio-fuels, however, it is technically difficult and commercially unviable to exploit. This challenge is a disincentive for bio-fuel producers to strictly adhere to cellulosic sources.

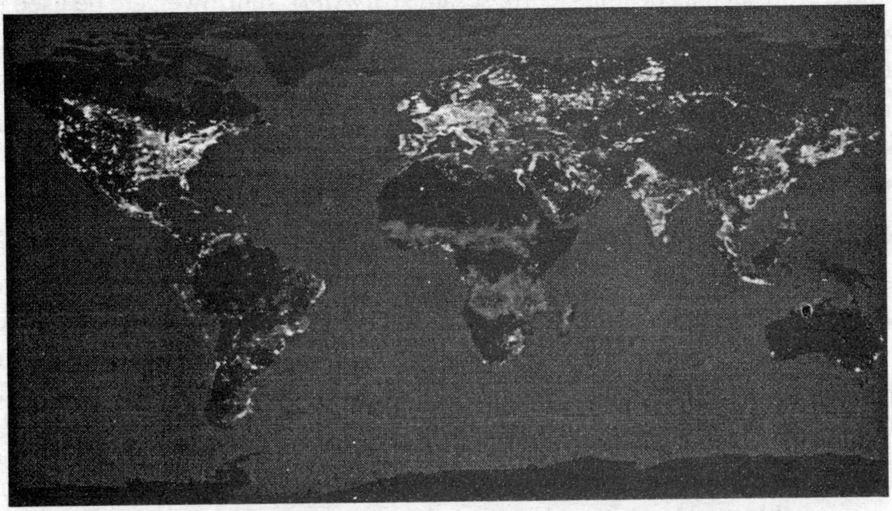

PLATE 4: The red and yellow colours show areas of significant biomass and fossil gas burning emissions (Source: Al Gore (2005))

It is also argued that bio-fuels emit less greenhouse gases when burnt than fossil fuels (EEA, 2006; Marshal, 2007; Evans, 2008). Despite the arguments, bio-fuels will still contribute to the carbon footprint in one way or another. Usually, the commonly used landfills and natural decomposition of organic matter produce a lot of methane with more global warming potential than Carbon-Dioxide (IPCC, 2007; EPA, 2007), which if not harvested/harnessed, would significantly contribute to greenhouse gases and climate change over time. Also, when the bio-fuels (biomass) are burnt, they generate significant amounts of CO_2, SO_2, NO_x dioxins, Chlorophenols and polycyclic Aromatic Hydrocarbons (Cedric, *et al.*, 2008) that have impact on global warming (Plate 4).

The electricity and greenhouse gas potential of landfills and other biodegradable wastes in Uganda are not thoroughly known, yet they could be burnt for heat or to generate electricity for public consumption. The advantage of many bio-fuels over other fuel types is that they are biodegradable and therefore are relatively harmless to the environment, if spilled. Hemp has been found to yield 4 times as much biomass as a forest can yield. An acre of hemp yields 10tons of biomass in 4 months, enough to produce 3,300litres of methanol fuel and 660kg of oil from the seed about the same as soybean. Five to seven (5-7kg) kilogram's of Jatropha seed yields 1-1.5 litres of biodiesel in Uganda (Ndawula and Turyahabwe, 2008), implying that 1.0 ton of Jatropha seed would on average produce 208.3litres of biodiesel. Use of this biodiesel fuel has demonstrated a 20% reduction in the requirement for conventional diesel fuel at 50% less cost. This obviously seems attractive. However, caution needs to be taken to ensure that a thorough understanding of the social, economic and environmental advantages and disadvantages is attained before large-scale ventures are initiated.

Therefore, in the quest of bio-fuels, there are important questions that demand answers. These include:

- Should a crop such as peanuts, corn, sesame (*Simsim*), sunflower, etc be used to make fuel or would villagers be better off eating the food? Should the oil seeds be pressed to produce oil for cooking or selling or for fuel and feed the high-protein seedcake to livestock, which in turn, people can either eat or sell, while using the livestock wastes (and crop wastes) to make compost to renew the soil, or to generate biogas for cooking and heating or both? Or should they grow a different crop altogether for its energy potential?
- Should grain crops such as wheat, oats, corn, sorghum, millet, etc. be processed to make ethanol fuel or should the villagers just eat the grain? If they use the grain for livestock feed, it can be used for ethanol and still feed the livestock i.e. the fermentation process converts the carbohydrates in grain into ethanol, leaving behind a high protein residue, which is high-quality feed for livestock that can be supplemented with forage. This could mean improved use of the available resources.
- In the light of the growing debate on climate change, is bio-fuel the best option, or should other renewable energy sources be sought?

Answers to these questions should be approached in multi-stakeholder consultative and participatory approaches, including the villagers themselves likely to be affected by these developments.

4.0 CONCLUSIONS

Continued dependence on biomass (firewood, charcoal and crop residues) energy by a greater proportion of the population is not sustainable. It is a threat to the environment and could stimulate climate change. If nothing is done the current natural forest cover will be whipped-out in the next 20-25 years.

Centralised grid hydropower sources alone are not enough to solve the energy crisis in Uganda.

The country is renowned to have very good policies, but often the challenge is the lack of comprehensive and adequate legal and institutional provisions and the lack of political commitment and will to implement and enforce the policies and legal provisions. It is also challenged by an inadequate institutional capacity to implement/enforce the provisions. Often, there is duplication of roles and responsibilities within and across the different government ministries and departments, which complicates the implementation and enforcement of the policy and legal provisions.

Uganda's Renewable Energy Policy (2007) sets targets of substituting 7,000m^3 and 10,000m^3 of fossil fuels with bio-fuels by 2012 and 2017, respectively. Similarly, it targets at increasing biogas digesters from 1000 today to 100,000 by 2017. Whether this is attainable remains to be seen.

The debate and controversy pertaining to balancing of food and bio-fuel production still continues to reign. Unless it is adequately resolved, bio-fuel production, especially in the southern countries, shall continue to be contested.

Usually it takes a long time (months) to produce the raw materials for bio-fuel production, yet it takes a very short time (a few hours) to burn the product.

5.0 RECOMMENDATIONS

There is need to utilise the existing biomass resources efficiently, expand on their acreage through replanting of indigenous tree species and establishing energy plantations. There is also need to develop new approaches of exploiting traditional biomass resources such as gasification, fermentation and decomposition to produce biodiesel, bio-ethanol and biogas. Commercial and domestic biodegradable wastes that are frequently disposed in landfills should be exploited as alternative sources of energy.

There is need for an energy mix that includes hydropower, solar, wind, geothermal, bio-fuels and petroleum-based thermal power. Government should focus on developing independent electricity grid systems. Future sustainable energy consumption requires cleaner energies, greater energy efficiency and decentralization of energy supply, along with the use of all ready-to-use renewable energy technologies in combination as local circumstance require. There is also need to exploit the benefits of nanoscience and nanotechnology in the energy sector. In the quest to develop modern energy sources, caution should be taken to avoid excessive greenhouse gas emissions and other pollutants to minimise the carbon footprint.

There is need to continuously test and review the policies, legal and institutional frameworks with a view of identifying gaps and challenges that should be purged. There is need for political commitment and will to implement and enforce the policy and legal provisions and streamline and build the necessary institutional capacity and avoid duplication of roles.

There is need to conduct social, economic and environmental assessment for bio-fuel development in Uganda that includes explicit opportunity-cost and cost-benefit analyses. Assessments of the economic and financial viability, capacity and type of feed-stocks for bio-fuel production need to be conducted. Any change of land-use should be based on sound science and empirical data. Also, the decision to invest and develop bio-fuels in Uganda should be reached in a multi-stakeholder participatory manner. There is need for guarantees and guidelines to ensure that the quest for bio-fuels does not infringe on the human and animal food chain.

While it is good to set impressive policy targets, it is important that the targets are realistic and achievable.

The southern countries, especially those within the tropics and along the Equator, need to take advantage of solar, wind and geothermal energy in the region. Hydropower development should desist from large infrastructure, because of the unreliability of water (hydrology & climate change), the prevalent poor performance of large hydropower dams, and the significant financial, social, economic and environmental burden they frequently impose on developing countries.

Ugandans should be very cautious in the manner in which they adopt the emerging bio-fuel-based economies, lest be plunged into anarchy and a dependency on the north for food and their very survival. The south in general and Uganda in particular should assess their comparative and competitive advantage of producing bio-fuels against those produced from highly subsidised economies like America and Europe. Also, they should put in place mechanisms of controlling or preventing price hikes of food (food insecurity) as a result of introduction of bio-fuel production. Government needs to adopt and inculcate the culture of transparency and accountability to the public. The current secrecy is the centre of doubt and lack of confidence in what government does.

REFERENCES

BBC (2009) Radio Interview with European Oil Officials broadcast on the 25th February, 2009.

Briens, C. Piskorz, J. and Berruti, F. (2008). Biomass Valorization for Fuel and Chemicals Production - A Review, International Journal of Chemical Reactor Engineering, 6, R2

Environmental Protection Agency (2007). Non CO_2 Gases Economic Analysis and

European Environment Agency (2006). How much bio-energy can Europe produce without harming the environment? EEA Report no. 7

Evans, J. (2008). "Biofuels aim higher". *Biofuels, Bioproducts and Biorefining (BioFPR)*. http://www.biofpr.com/details/feature/102347/Biofuels_aim_higher.html.

Globeco biodegradable bio-diesel www.globeco.co.uk; Friends of Ethanol.com Biodegradable ethanol www.friendsofethanol.com/fact.html;

International Panel on Climate Change (IPCC) (2007). Third Assessment Report 2007. www.grida.no/climate/ipcc_tar.

Inventory: Global Warming Potentials and Atmospheric Lifetimes,U.S. Environmental Protection Agency, accessed February 12th 2009. www.epa.gov/nonco2/

Marshall, A.T. (2007). Bioenergy from Waste: A Growing Source of Power.

Low Cost Algae production System Introduced

www.energyarizona.org/Archive/200708280001_low_cost_algae_production_introducted.php

Hemp Biomass for Energy www.fuelandfiber.com/Hemp4NRG/Hemp4NRG.htm

Ndawula, G. (2008). The status of bio-fuel production in Uganda. Ministry of Energy and Mineral

Development (MEMD). Uganda. PPT presentation.

Ndawula, G. and Turyahabwe, E. (2008). MFP in Luwero and Masindi District and Co-generation at Kinyara Sugar Works. Ministry of Energy and Mineral Development (MEMD), Uganda. 8p.

Power Planning Associates, PPA (2007). Economic and financial Analysis of the Bujagali II Hydropower Project, Uganda

Somerville, C. (2007). Development of Cellulosic Biofuels (PDF). U.S. Department of Agriculture.www.usda.gov/oce/forum/2007%20Speeches/PDF%20PPT/CSomerville.pdf.

Retrieved on 15-02-2009.

Thurmond, W. (2007). www.prleap.com/pr/80099

UN bio-fuels Report - www.esa.un.org/un-energy/pdf/susdev.Biofuels.FAO.pdf.

www.ciafactbook.org

www.fao.org

12

Brief Report on the Current Status of Biofuels Industry in Zambia

Christopher Hakachite
National Institute for Scientific and Industrial Research,
Sustainable Use of Underutilized Genetic Resources Programme,
Kitwe, Zambia
E-mail: hakachris@yahoo.com

Zambia is self-sufficient in all its energy sources with the exception of petroleum, which is wholly imported. A considerable amount of foreign exchange is used to import this commodity. The sharp increase in global demand for oil versus the non renewable deposits means that a global oil crisis is not far fetched. For this reason Zambia has decided to explore alternatives to oil and one such alternative being considered is biofuels, which include ethanol and biodiesel. In 2007, the Zambian Government adopted a new National Energy Policy. The policy adequately covers issues to deal with biofuels. The government through its appropriate institutions has developed both standards for Ethanol and Biodiesel to be produced in Zambia and a draft National Biofuels Strategy. The feedstocks which can be used to produce liquid biofuels have been assessed and are known. The cultivation of Jatropha, one of the preferred feedstock has been demonstrated both at small scale and commercial level. While basic research in biofuels has been done, a lot more remain to be done. For instance, research in the agronomy, processing, application as well as environment and biodiversity of the biofuels crops, utilisation of wastelands for production of non-edible crops such as Jatropha need to given attention.

KEY WORDS: *Biofuels, Feedstocks, Jatropha, Ethanol, Biodiesel, Regulatory framework*

1.0 INTRODUCTION

Biofuels are renewable fuels that are predominantly produced from locally produced biomass feedstocks or by products from industrial processing of agricultural or food products. They contain no petroleum, but can be blended with petroleum to create a biofuel blend. Crops such as sugar-cane, sweet sorghum, cassava; can be used to produce ethanol and jatropha can also be processed to produce biodiesel. Both ethanol and biodiesel belong to the class of fuels called biofuels. The utilisation of biofuels has been recognised as an effective way of fulfilling the country's energy requirements.

It is encouraging that the Zambian Government has made impressive advances in instituting the new energy policy and regulatory frameworks. However, certain aspects of regulation require further attention. Secondly, assessments of the preferred feedstocks for the biofuels industry have been done. From these evaluations, sweet sorghum and jatropha were ranked first for ethanol and biodiesel production, respectively.

Jatropha has existed in Zambia for more than 300 years. Currently the crop is cultivated in a number of different ways. Both small scale and commercial farmers are already engaged in jatropha production. The seed yields obtained in Zambia range from 0.25 to 0.75 kg per tree per year.

Basic research in biofuels, specifically, the production of Jatropha for the then Refined Oil Products was initiated by National Institute for Scientific and Industrial Research in 1990/91. The focus then was to produce oil for use in soap making and oil for lighting for the local communities. At present there is insufficient information from the Zambian R & D institutions on both biofuels and biofuels crops.

There are several other challenges to the biofuels industry in Zambia. They include issues relating to food security; land use; environment, biodiversity and conservation; socio-economic impact; wasteland utilization, biotechnology; processing plants even human and labour rights relating to biofuels cultivation. In spite of these challenges, there are some few projects which are being undertaken, mainly by the private sector and none implemented by government. What is exciting is that experts, key stakeholders, major players and government through its various line ministries, are charting the way forward for Zambia to realize the identified benefits from its biofuels industry. Thematic issues in the energy sector are being addressed at various levels. However, modalities have to be quickly addressed by government for Zambia to be one of the largest producers of biofuels in Africa.

2.0 BACKGROUND

Zambia is self-sufficient in all its energy sources with the exception of petroleum, which is wholly imported. A considerable amount of foreign exchange is used to import this commodity. All petroleum products are currently imported in the form of spiked crude oil transported through the 1760 Km TAZAMA pipeline and refined at INDENI Petroleum Refinery (a company in which the Zambian Government owns 50% shares), with a capacity of 1,100,000 metric tonnes per year, from Oil Marketing Companies in the country, uplift the refined products from this refinery for sale to consumers.

The sharp increase in global demand for oil versus the non renewable deposits means that a global oil crisis is not far fetched. This will mean the domestic oil demand will soon outstrip the current supply. Therefore, Zambia has decided to explore alternatives to oil and one such alternative being considered is biofuels. Ethanol offers an opportunity for blending with petrol or being used as neat, thereby reducing the expense on oil importations. On the other hand biodiesel offer an opportunity for blending with conventional diesel or to be used as a stand-alone fuel.

Zambia has a great potential to benefit from the use of its natural resource endowment base and in particular to harness alternative sources of energy. The country is well positioned to take advantage of its vast water and land resources coupled with the favourable climatic conditions to become a leader in the exploitation and use of alternative sources in the region and beyond.

While Zambia is gifted with renewable energy resources that can be sustainably exploited to supplement fossil fuels, these energies have remained unexploited. Several barriers such as lack of research into the sustainable approaches to biofuels industry development and economic and financial considerations, environmental as well as the socio – political barriers have inhibited the enhanced utilization of the renewable energy sources.

3.0 POLICY AND REGULATORY FRAMEWORK

In 2007 the Zambian Government adopted a new National Energy Policy. This was published by the Ministry of Energy and Water Development. The aim of the new policy is to create conditions that will ensure the availability of adequate supply of energy from various sources, which are dependable, at the lowest economic, financial, social and environmental cost consistent with national development goals. This new policy adequately covers issues to deal with biofuels. Further, the Ministry of Energy and Water Development jointly with Ministry of Commerce Trade and Industry are working on modalities which will declare biofuels a priority sector under the Zambia Development Agency.

Government has further issued a Statutory Instrument under the Energy Regulations Act incorporating Biofuels in the definition of energy. This has given the Energy Regulation Board (ERB) the mandate to regulate the production and use of biofuels in Zambia.

In July 2008, the Ministry of Science, Technology and Vocational Training (MSTVT) through its major players and stakeholders in biofuels prepared a position paper on Jatropha in Zambia in which the various concerns in growing Jatropha have been highlighted including the proposed way forward. Good and well formulated biofuel strategies would be necessary to address the critical issues of the biofuels industry in Zambia.

3.1 The National Energy Policy

The first National Energy Policy was formulated in 1994. A review of this policy was necessitated by the significant changes in the social, political, environmental and economic situation that Zambia had gone through. The new policy therefore recognizes the cross cutting nature of energy. Of interest for this paper is the Biomass component of the policy.

Biomass energy is the form of energy from organic matter such as woodfuel (firewood and charcoal), agricultural waste, industrial/municipal organic wastes, energy crops and products and animal waste. In Zambia, biomass energy, woodfuel in particular, significantly contributes to total energy consumption. The other energy resources that are being promoted are biofuels, gel fuel and biogas.

The current energy situation indicates that wood fuel (firewood and charcoal) would continue to dominate Zambia's energy consumption. In 2004, it accounted for over 70% of the total national energy consumption. Households accounted for about 88% of woodfuel (National Energy Policy, 2007).

Utilization of biofuels has also been recognized as an effective way of fulfilling the country's energy requirements considering the disruptions in petroleum supply which are being experienced locally and even on the international market.

Other components of the biomass energy include gel fuel, briquettes and biogas.

In the biomass energy sector in Zambia, the current legal and institutional frameworks are:-

- The Forest act
- Environmental Protection and Pollution Control Act
- and the Energy Regulation Act.

The New National Energy Policy is anticipated to yield desirable outcomes for Zambia. These include:—

- Security of supply of energy in the country;
- Efficient production and supply of energy;
- Minimization the negative environmental and health effects from the energy production, transportation and use;
- Reduction dependence on imported petroleum to locally available energy supplies;
- Promotion of the use of renewable energy.

The vision of the new National Energy Policy is:-

'To provide well developed, managed, reliable and sustainable energy resources for the improvement of the quality of life of all Zambians'

The new policies in the biomass sub-section are highlighted below.

The new policy therefore seeks to ensure environmentally sustainable exploitation of the biomass resources by ensuring efficiency through better management and introduction of new technologies i.e. biofuels and gel fuel.

The new policy measures on biomass include:

(i) To apply appropriate Financial and Fiscal Instruments for stimulating the production and use of biomass

(ii) To raise public awareness on the benefits and opportunities of other biomass energy and develop capacity for their implementation

(iii) To develop the regulatory framework of biomass production and utilization

The following are the new policy measures on woodfuel.

(i) To ensure better management of woodlands and forests as a sustainable source of woodfuel

(ii) To improve the technology of charcoal burning

(iii) To promote appropriate alternatives to woodfuel and reduce consumption

(iv) To encourage utilisation of agro, forest and sawmill residues for combustion and gasification

(v) To improve government revenue collection from the woodfuel industry

(vi) To establish community based woodfuel resource management systems

The new policy measures on biofuels are:-

(i) Using biofuels in the fuel mix

(ii) Ensure secure supply and stabilisation of prices of fuels by promoting the utilisation of biofuels for transport as an alternative to petroleum

(iii) To ensure availability of data and information on market demand , resource assessment and applicability of biofuels

(iv) Provide legal and Institutional framework for the biofuels sub-sector

(v) Support investment in biofuels industry through appropriate incentives, standards, research and implementation strategy.

The policy measure for the other forms of biomass energy is to promote other forms of energy as an alternative to woodfuel use which has negative impact in the environment.

3.2 Biofuels Standards

The government through its appropriate institutions which include the Zambia Bureau of Standards developed, in December 2008, both standards for Denatured Fuel Ethanol for Blending with Gasoline, for use as Automotive Spark-Ignition Engine Fuel and Automotive Biodiesel to be produced in Zambia. The Energy Regulation Board is working on all the Licenses that would be applicable in the biofuels industry.

3.3 National Biofuels Strategy

The Government through the Ministry of Energy and Water Development had done a number of studies with regards to the biofuels strategy. To this effect a draft National Biofuels Strategy has been developed and it is expected to be finalised by mid year 2009. The strategy will give a direction as to how the Biofuels programme in Zambia will be implemented.

3.4 Biofuels Association of Zambia

One of the efforts made by Government in establishing the Jatropha industry in Zambia was its facilitation in the formation of the Biofuels Association of Zambia (BAZ). It is an association of firms, companies and individuals involved and /or planning to be involved in the production of biofuel crops and/or biofuels in Zambia. Currently BAZ has more than 60 small-scale ordinary members and more than 40 corporate undertakings (Lubozhya, 2007).

4.0 ASSESMENT OF THE PREFFERED BIOFUELS FEEDSTOCKS IN ZAMBIA

There are a number of feedstocks that are known which can be used to produce liquid biofuels. For biodiesel, possible sources of feedstocks include algae, animal fat, cotton seed, jatropha seed, rape seed, sunflower seed, soy bean, peanuts, oil palm seed, avocado and castor seed. For bioethanol, possible sources include maize, sugar cane, sweet sorghum, cassava, sweet potato, wheat and switch grass (Sinkala, 2008). A new generation of 'ligno-cellulosic' bioethanol also includes a range of forest products such as rotational coppices and energy grasses. The choice of feedstocks depends on the climate and the food/geo/techno politics prevailing in a country or region. For bioethanol the main issues to consider are ethanol content, octane number, fractional distillation temperature/flash point and calorific value. For biodiesel feedstocks the most important criteria is assurance of high feedstock supply as described by the iodine number (IN), net calorific value, cold filter plugging point (CFPP), cetane number and the distillation curve (Yamba, 2007).

In assessing feedstocks for biofuels development in Zambia, the various feedstock sources were subjected to a set of criteria in matrix form comprising 12 parameters. These parameters include Scope of ownership of the biofuels wealth at national and individual levels, production technology, job creation, resilience against external influence, diversity of products from a feedstock, minimum size of investment required, market scope for the products, land requirements, water requirements, food security/soil fertility, geographical coverage of ownership in the country and effect on environment (Sinkala, 2008).

For feedstocks to produce ethanol, the highest ranking in terms of the above criteria became sweet sorghum followed by sugar cane. Sweet sorghum, among others, is drought resistant and can therefore be grown throughout Zambia. Sugar cane industry, although restricted to adequate water resources, already exists in the country and only requires a little bit more additional investment.

Jatropha curcas followed by palm became the highest ranking for the production of oil/biodiesel. Jatropha, like sweet sorghum, is drought resistant and can be grown country-wide. Most people in the biofuels industry within Zambia are therefore engaged in Jatropha. Oil Palm is high-yielding; however, it can only be grown in selected high-rainfall areas of Zambia. Promotion activities in the production of oil palm in the Luapula and Northern provinces of Zambia by the Ministry of Agriculture and Cooperatives are on-going and fairly good progress is being registered. The private sector is also expressing interest in the same crop. There is already a substantial investment by Zampalm in the production of oil palm. However, most people in biofuels industry are engaged in Jatropha.

5.0 GERMPLASM COLLECTIONS OF BIOFUELS CROPS

The Ministry of Agriculture through the Zambia Agricultural Research Institute (ZARI) undertook a germplasm collection of Jatropha in 1999 in the Eastern Province of Zambia (Position paper on Jatropha in Zambia, 2008). Attempts to plant the collection proved futile due to lack of funding. The National Institute for Scientific and Industrial Research (NISIR) introduced Jatropha in 1991 with the aim of producing oil for Refined Oil Products (ROP) for use in the processing of soap and other cosmetics. NISIR imported this seed from India for use in the trials and also established research plots of Jatropha which later were the source of seed for the cultivation promotion activities in selected communities in the Copperbelt Province. The Golden Valley Agricultural Research Trust (GART), University of Zambia (UNZA) and the Zambia Agricultural Research Institute (ZARI) have on-going research programmes in sweet sorghum and so have a good germplasm collection of this crop. However, there is urgent need to make collections and characterisation of the jatropha growing in Zambia. The Forestry Department under the Ministry of Tourism, Environment and Natural Resources and the Zambia Agricultural Institute under the Ministry of Agriculture and cooperatives are making efforts to attend to the germplasm concerns of mainly Jatropha as well as the other biofuels crops.

6.0 CULTIVATION OF JATROPHA

In Zambia, Jatropha has existed for more than 300 years. The plant is cultivated in a number of different ways depending on the purpose. In general, the plant is cultivated as hedges (Photograph 1) around homes and crop fields, as supports (Photograph 2) for bath shelters, as ornaments (Photograph 3) forming part of home garden plants, as support for other crops such as vanilla, and as commercial plantations (Photograph 4) for production of e.g. biodiesel (Sinkala, 2008).

Jatropha can be planted using seeds directly, cuttings, or seedlings. Cuttings are reported to have a lifespan of 10 – 15 years while plants grown from seeds/seedlings give a life span of over 50 years. Jatropha is relatively easy to establish, but like any other crop requires proper land preparation. Tilling land, which will improve land aeration and water infiltration, is especially important. What preparation is made is however site specific as different soils have different properties in relation to plant growth, and if such land preparation is done for intercropping.

Some of the activities that are involved in establishing a Jatropha plantation are land clearing, tilling/pitting, manure application, and irrigating (if rain-fed plantation management is not the option). Once land is cleared, either tillage, pitting, or both are carried out. In sloppy areas, it is much better to use pits than tillage to minimize erosion, especially for non-intercrop plantations. Pits for Jatropha planting are prepared at selected intervals. The inter-row and intra-row spacing selected depends on the region and climatic conditions (e.g. low rainfall / desert conditions), purpose (e.g. only Jatropha or intercropped with other plants), and soil conditions. In general, spacing should increase with

Photograph 1: Jatropha as a fence (Kapala-Katongo Village, Mpika District, Zambia)

Photograph 2: Jatropha as a bath shelter support (Kapepa Village, Fisenge Area, Luanshya, Zambia)

Photograph 3: Jatropha as a garden ornament (Tayali Village, Chief Chibesakunda, Chinsali, Zambia)

Photograph 4: Jatropha as a commercial plantation (Thomro Biofuels Farm, Chief Mungule, Cibombo District, Zambia)

increase in rainfall, because of the anticipated large size of Jatropha plants with increase in rainfall. Some spacings used are 2m x 2m (a plant density of 2500 plants per hectare), 2.5m x 2.5 m (a plant density of 1600 plants per hectare), 3m x 3m (a plant density of 1111 plants per hectare), 4m x 4m (a plant density of 625 plants per hectare; when intercrop is to be used in early stages of plantations (Sinkala, 2008; Mwamba, 2002). When Jatropha is grown as a hedge, a fence or for soil conservation, spacing as small as 0.15m between plants can be used depending on the purpose. Therefore, the actual spacing is determined on the basis of end-use.

Fertilizers containing small amounts of calcium, magnesium, and sulphur are used. At Thomro Biofuels Farm, chicken and treated sewage manure appears to give good results. Other commonly applied sources of nutrients include farm yard manure, nitrogen fertilizers, phosphorus fertilizers, compost and green manure, and potassium and sulphur based fertilizers. If irrigation is not to be used, transplanting is done at the on-set of the rains.

The five main activities in rain-fed Jatropha plantation management are weeding – which generally is done when necessary; pruning – to ensure that plants have a particular canopy type for easy harvesting; pest control – though some growers have indicated that jatropha pests in Zambia are not a major problem; seed harvesting and clearing for fire guard. Weeding and seed harvesting are probably the most labour intensive jobs in a Jatropha plantation.

Jatropha seed yields vary considerably. This variation can be attributed to a variety of factors, such as climate, level of management, site quality, and variety/genetics. Seed yields range from 0.25 to 0.75 Kg per plant per year (Mwamba, 2002).

7.0 BENEFITS OF THE BIOFUELS FOR ZAMBIA

Mbolela (2007) categorised the benefits that Zambia would get from a developed Biofuels industry into economic, social and environmental benefits.

On the economic front he outlined the following as benefits of the biofuels for Zambia;

- Enhance National Energy Security
- Reduce the burden on the imports of energy
- Increase investment in the priority sectors of the economy
- Positive effects on the balance of payments
- Improved sectoral productivity, growth and linkages leading to higher contributions to GDP
- Increase share in contribution of renewable energy to the energy supply mix at macro level
- Wealth and job creation for local communities
- Potential export

On the social front the following are the outlined benefits of biofuels for Zambia are;

- Contribution to poverty reduction
- Contribution to more equitable distribution of the resources (reduction of wealth disparities)
- Increase access to energy
- Capacity building

On the environmental front the following are the benefits of the biofuels for Zambia;

- Reduction of emissions to the environment
- Reduction of Green House Gas emissions (CO_2, CH_4, N_2O) aimed at enhancing global environmental integrity
- Reduction of emission impacting on air, water, land and biodiversity

- Provision of healthier, safer working environments, particularly in underground mining operations.

8.0 BIOFUELS RESEARCH IN ZAMBIA

The National Institute for Scientific and Industrial Research has been involved in research on biofuels at laboratory level as well as field level; specifically the agronomic studies in the production of Jatropha since 1990/91 agricultural season. The University of Zambia under the School of Engineering and School of Agricultural Sciences has been involved in research on biofuels. The preliminary results from this on-going research on the production of Feedstock from Sweet Sorghum Stalks for Ethanol Extraction are very encouraging. The Gorden Valley Agriculture Research Trust and Ministry of Agriculture and Cooperatives are researching on production of ethanol from sweet sorghum using out-growers.

9.0 CHALLENGES TO THE BIOFUELS INDUSTRY IN ZAMBIA

The Ministry of Science, Technology and Vocational Training (2008) committee composed of major players and stakeholders in biofuels prepared a position paper on Jatropha in Zambia. This committee observed a number of concerns on Jatropha production as well as the development of the biofuels industry in Zambia. Some of these concerns have been raised both locally and internationally. These include the invasiveness, the socio-economic impact, environmental impact, land use and food security. These issues have created anxiety and uncertainties in the development of the biofuels industry and the suitability of Jatropha as an energy crop in particular. The following paragraphs give a summary of these concerns raised by the committee.

9.1 Food security

The question of whether Jatropha growing for biofuel production is going to impair food security can not be answered for lack of facts based on actual experiences. A major criticism often leveled against biomass in general, particularly against large-scale bio fuel production, is that it could divert agricultural production away from food crops, especially in developing countries. The basic argument is that energy-crop programmes compete with food crops in a number of ways (agricultural, rural investment, infrastructure, water, fertilizers, skilled labour etc.) and thus cause food shortages and price increases. The argument should be analysed against the background of the world's (or an individual country's or region's) real food supply and demand situation (ever-increasing food surpluses in most industrialized and a number of developing countries), the use of food as animal feed, the under-utilized agricultural production potential, the increased potential for agricultural productivity, and the advantages and disadvantages of producing biofuels.

In Zambia, empirical observations regarding the cultivation of Jatropha as a species at non-commercial level has so far not shown any problems and that it has not been reported to exhibit any harmful characteristics to disqualify it as an energy crop. There is however need for establishing facts based on actual experience of its threats to food security.

9.2 Land use

In Zambia 42 million ha is arable land of which nearly 6 million ha or 14% is under cultivation with a further 294,000 ha being irrigated. This implies that nearly 36 million ha of arable land is not being utilized and can be made available for developmental activities.

Many parts of Zambia have excess land but there are also areas with serious deficits. The agricultural population is not a homogenous group and there are large differences in farming practices, size of farms and living conditions of households within the sector.

Since the real hectarage of Jatropha that is currently on the ground is not known, the expansion of farmland for biofuel production may exacerbate the problem of landlessness by dislocating farmers without clear land titles.

9.3 Utilisation of Zambia's Wastelands for Jatropha Production

There is a general myth that jatropha can be grown anywhere including marginal land. However, this has not been studied in Zambia. Fortunately there is reasonable chunk of wasteland, degraded forest land; and unutilized public land and fallow land. Wasteland categories may include the following; gullied and/or ravenous land, under-utilized degraded forest land, shifting cultivation areas, degraded land under plantation crops, mining/industrial waste land and steep sloping areas.

9.4 Environment, Biodiversity and Conservation

In Zambia, there exists no legislation or policy that deals with invasive plants except for the 1953 noxious weed act which however lacks institutional enforcement.

Promotion of Jatropha may increase the risk of damaging biodiversity and creating more green house gas emissions than would be prevented.

The ability for Jatropha to become an invasive alien species is not ascertained. Considering the plant profile, it is evident that *Jatropha curcas* has some characteristics of becoming an invasive alien species because *Jatropha curcas,* the preferred feedstock of bio-diesel in Zambia, has traits which are commonly found in invasive species. These traits include rapid growth, high water use efficiency and long canopy duration. It's feared that should such a crop be introduced, they could become invasive and displace indigenous species and result in a decrease in biodiversity. There is also great concern of Jatropha plantations becoming unmanageable due to the increased global carbon dioxide levels which the plants would easily convert and utilize for photosynthesis, hence their rapid growth.

9.5 Socio-economic impacts

The rising demand for agricultural biofuels is translating into higher market prices for some woody materials and agricultural products. The Agricultural Outlook forecasts that between 2007 and 2016 the rapid growth of the biofuel industry is likely to keep food prices high and increasing throughout the next decade at least. More generally, increased commodity prices can have serious consequence for food-importing developing countries with implications for agricultural production and food security.

On the positive side, the need for greater amounts of feedstock could also potentially create employment opportunities, and thereby increase rural incomes as the harvesting of biomass tends to be a labour intensive process.

As the production of biofuels tends to favour large-scale and industrial agricultural practices, farmers utilizing traditional agricultural methods may be effectively excluded from the production of biofuel feedstock.

The production of biofuel feedstock also has implications for indigenous and local communities. Several reports have drawn attention to the issue of populations being excluded from plantations without their prior and informed consent and an apparent lack of consideration of traditional land tenure and rights.

9.6 Biotechnology

The progress and accomplishment through the biotechnology have, at the same time, evoked tremendous concerns among researchers themselves as well as the public at large.

In biofuels industry, the criticism is that biotechnology could be a legalized entry of genetically modified crops through the assertion that it has to be grown faster, resist pests and create seeds less resistant to crushing or to press.

9.7 Human and Labour Rights Relating to Jatropha Cultivation

It is unverified that Jatropha oil production requires minimum amounts of labour inputs. This is because in order to prepare the land, set-up nurseries, plant irrigate, fertilise, prune, harvest and process the seeds for oil production, labour availability and labour costs are elements to be seriously accounted for especially in the first year of establishment of the crop. Labour needed for crop maintenance and harvest will increase to substantial levels in the subsequent years. This may lead to a shift from use of human labour to mechanisation resulting in loss of employment opportunities for the communities.

9.8 Incentives for Biofuels

There are a number of calls for Government to facilitate availability of funds to enable participation by citizens in the biofuels industry.

9.9 Application of Jatropha oil

Since Zambia does not manufacture the diesel engines and has no control over their designs, questions arise as to how compatible our current fleet of vehicles will be with the Jatropha oil; what ratios of blending Jatropha oil and diesel will need to be blended; what effects the various components in the Jatropha oil might have with regard to wear and breakdown of engines etc.

9.10 Country Biofuel Requirements Vs Land Requirements

There are intentions of the country to get into biofuel production but there are no estimates in terms of targets for the fuel to be produced as it is from such information that feedstock crops to be grown and the areas to be grown would have to be related.

9.11 Processing Plants

Currently there is no refinery with known capacity that would be taking in the feedstock so that the focus is not on exporting raw materials except for one pilot plant by Oval Biofuels.

9.12 Jatropha Varieties

There is no officially registered released variety of Jatropha. The Jatropha being established in different parts of the country is from different sources, both local and imported, and there is no information on the varieties that are being grown in Zambia. People have been indiscriminately bringing in varieties indicating for research purposes but have ended up multiplying the same.

10.0 CURRENT AND FUTURE BIOFUELS PROJECTS

Currently there are no projects on biofuels being implemented by the Zambian Government, however the private sector have taken a leading role with several projects of growing energy crops for the production of biofuels.

One company, Oval Biofuels involved in growing energy crops has set up a pilot plant for the production of biodiesel with the production capacity of 2000 liters per day. There is no other known refinery that would take the feedstocks locally so that the focus is not on exporting raw materials but processing them locally into biodiesel or ethanol. There are also plans for large scale production of biofuels by sugar producing firms. For example, the Zambia Sugar PLC has intentions of venturing into ethanol production by 2011.

Some investors in the agrofuel production in Zambia have chosen to grow crops such as Jatropha through huge numbers of out-growers using contracts that will last many years, even up to thirty years. These contracts serve to transfer control over production from the farmer to the company, through a system of loans, extra charges, service payments and prices determined by the company.

The government, through the University of Zambia and research institutions such as the National

Institute for Scientific and Industrial Research (NISIR), Golden Valley Agricultural Research Trust (GART) and Zambia Agricultural Research Institute are researching into the production of ethanol from locally available feedstocks, mainly sorghum.

11.0 THE WAY FORWARD

The committee of biofuels experts, constituted by the Ministry of Science, Technology and Vocational Training in 2008; key stakeholders; major players and individual experts in the biofuels industry have proposed attempts that would help this industry to succeed. These propositions are outlined below.

11.1 Science Related Research and Development

In order to ensure viability and sustainability of the biofuels industry, there would be need for research and development in the following areas:

11.1.1 Agriculture

(i) Seed – yield, adaptability and oil content

(ii) Ecological adaptability of the energy crops

(iii) Plant management/agronomy

(iv) Pest and Disease control

(v) Water use efficiency

(vi) Soil improvement

(vii) Mycorrhiza technology applications (Bhojvaid, 2006)

(viii) Tissue culture technology application (Bhojvaid, 2006)

(ix) Jatropha in Agro-forestry (Bhojvaid, 2006)

(x) Utilisation of Zambia's Wastelands for Jatropha Production

11.1.2 Processing

(i) Pre-processing

1. drying
2. harvesting and de-husking improvement
3. heating
4. seed soaking in oil
5. filtration

(ii) Extraction

1. Reclamation of chemicals
2. Minimize cost of reagents
3. Development of affordable local technologies
4. Use of by-products

11.1.3 *Application*

(i) Primary

1. bio-diesel
2. electricity generation
3. straight oil
4. organic fertilizer
5. charcoal
6. biogas
7. glycerine
8. lighting
9. soap

(ii) Secondary

1. medicine
2. cosmetic industry
3. mosquito repellent

11.1.4 *Environment and Biodiversity*

(i) Invasiveness

(ii) Environmental Education

(iii) Effect on Water Table

(iv) Effect on Soil

(v) Monitoring of Existing Plantations

(vi) Carbon sequestration potential

11.2 Other Issues

In addition to the above, the country needs to look into related other areas as highlighted below:

11.2.1 Legal and Institutional Arrangements

The Committee of Experts on Biofuel used the lack of legislation as part of evidence to determine that *Jatropha* species commonly cultivated in Zambia has not been classified as an invasive species. Stakeholders need to be engaged to undertake a national risk assessment or a strategic environmental assessment and continue conducting research in order to widen the scope of understanding the plant. In the absence of single legislation, developers need to be encouraged to put in place sound management systems of Jatropha plantations and early detection measures should they see the plant growing in wild or other than where it was planted, until such legislation is put in place.

11.2.2 Seed improvement and other propagation methods

Since the industry is new, there are no formal seed improvement programmes being undertaken. The only place that could have served as a source of seed in the country is not very well looked after. This therefore is a challenge to the National Institute for Scientific and Industrial Research (NISIR), Zambia Agricultural Research Institute, School of Agriculture Sciences at the University of Zambia, Gordern Valley Agricultural Research Trust and the Forestry Department.

11.2.3 Application of Jatropha oil

Since Zambia relies on imported engines and has no control over their designs, there is need to start looking at issues of how compatible our current fleet of vehicles will be with the biodiesel and bioethanol.

11.2.4 Processing Plants

There is need to develop a refinery with known capacity that would be taking in the feedstock so that the focus is not on exporting raw materials.

11.2.5 Jatropha Varieties

There is need to develop a data bank on registered and released varieties of Jatropha. The Jatropha being established in different parts of the country is from different sources, both local and imported, and there is no information on the varieties that are being grown in Zambia.

11.2.6 Food Security

There would be need to sensitize farmers on the biofuels issues so that they can make informed decisions in order to reduce the risk of food security and income stability.

11.2.7 Land Use

Where there are organizations that are recruiting farmers to grow Jatropha on their land, there is need to ensure that all mutual agreements between large scale energy crop growers with communities and individuals in customary land jurisdiction are clearly defined, documented and legally established.

There is also need to establish the real hectarage of Jatropha (and other biofuels crops) that is currently on the ground through a country wide survey.

11.2.8 Country Biofuel Requirements Vs Land Requirements

Since there are intentions of the country to get involved in biofuel production, there is need to have estimates in terms of targets for the fuel to be produced as it is from such information that feedstock crops to be grown and the areas to be grown would have to be related.

11.2.9 Government Frameworks on Biofuels

These should be put in place urgently to build confidence and energize players.

11.2.10 Other Sources of Biofuels

There is need to undertake economic studies on other non-food crops to be used as feedstock in the production of bio-diesel along with Jatropha.

12.0 SUPPORT TO THE BIOFUELS SECTOR

The Firth National Development Plan (FNDP) for Zambia has recognized the importance of the biofuels sector to the development of the economy. A provision of a substantial amount has been made which will be funded by the Government with the help of the cooperating partners.

13.0 THEMATIC ISSUES IN ENERGY UNDER THE SECTOR STRATEGY FOR RESTRUCTURING THE SUGAR INDUSTRY IN ZAMBIA

The Department of Energy Experts in the Ministry of Energy and Water Development recognized (in their January 2009 report on the status of the biofuels industry in Zambia) the following thematic issues in energy under the sector strategy for restructuring the sugar industry in Zambia.

13.1 Diversification strategy: Ethanol production for fuel blending

Ethanol can be produced using the sucrose part of sugar cane and or molasses. With the current EU Reforms on Sugar Industry, Sugar manufacturing companies can diversify sugar cane production to ethanol. Ethanol can be blended with petrol at a ratio of up to 10% without vehicle modifications. The blended fuel will emit less GHG emissions compared to pure petrol. This would also afford the country an opportunity to import less crude oil feed stock finished petroleum products. This would ensure improved energy security, improved balance of trade, reduction in GHG emissions and would improve the agriculture sector.

A number of studies have been undertaken to show the viability of ethanol production for fuel blending. But no tangible pilot project is in place to demonstrate how ethanol blending can be done. As for biodiesel, Oval Biofuels has a pilot plant where they are producing 2000ltrs/day of biodiesel from soya beans and utilizing the fuel in their fleet and mining machinery at their Nickel mine. The experience of Oval Biofuels has at least afforded the proponents of biofuels to have a hand on experience on handing, storage and transportation of biodiesel.

However, there has been of no such initiatives in ethanol production for blending. The Zambia Sugar Plc is in the process of installing a distillery for ethanol production in 2011. The company will be able to produce about 18 million litres of ethanol per year and this will meet the Country's current fuel demand at a blending ratio of 10%.

There is need to conduct a Frontend package. Front-end refers to the early stages of product development (i.e. opportunity identification and exploration). Commonly the modifier "fuzzy" is also used, addressing the difficulties of managing and planning this process. This planning will provide the proponents of ethanol production for fuel blending the opportunity to come up with the project drawing, determine cost of storage, handling, transportation and who is responsible for what a particular time. This study may culminate into a project.

All project promoters will be required to their own EIA since they are project specific. However, an environmental framework study can be done.

The estimated cost of this study is not well known however we are awaiting feedback from our stakeholders to give us the exact estimates for this study.

13.2 Co-generation of electricity

The Zambia Sugar is already undertaking a cogeneration of electricity using baggase. However, with the expansion programme, there would be surplus electricity which the company would need to sell to the national grid. Therefore, a study would be appropriate to look at the cost of production, economic tariffs and what policy and regulatory framework should be put in place with regards to selling of the surplus electricity to the potential buyers.

13.3 Diversification strategy - Ethanol production for Gel fuel

Gel fuel is a compound of ethyl alcohol (ethanol) and organic pulp (cellulose). The two are gelatinized with addition of water forming a clear and transparent compound with gel like consistency.

Gel fuel is used for domestic use and can be a good alternative to use of charcoal for cooking. A number of households in Lusaka are using gel fuel for cooking due to load shedding. These stoves and fuel are procured from retail shops like Spur and Game which import the gel fuels from South

Africa. The production of ethanol for gel fuel locally would be used locally and the excess exported. A programme on ethanol production would deal with both fuel blending and gel fuel. Therefore, a study that has been recommended for ethanol for fuel blending can also incorporate possibilities of gel fuel production.

14.0 CONCLUSION

Zambia has a great potential for producing biofuels. With the vast land, abundant water, good climatic conditions and good biofuels policies, Zambia is destined to be one of the largest producers of biofuels in Africa. However there would be need for Government to come up with modalities that would ensure that there are a number of projects in the sector to ensure that the production and use of biofuels in Zambia is a reality.

ACKNOWLEDGEMENTS

I am grateful to the Executive Director of National Institute for Scientific and Industrial Research, Dr. Lewanika Mwananyanda, and Mrs Jane Chinkusu, Chief Science and Technology Officer in the Ministry of Science, Technology and Vocational Training of the Government of the Republic of Zambia, for the nomination to participate in this important International Roundtable on 'Sustainable Utilization of Energy and Biodiversity Resources for Wealth Creation and Development in Kampala, Uganda (10-13 March 2009). I am equally grateful to the Manager- Research and Development, Dr Ray Handema and Mr Muchelen'ganga C., both of NISIR for the guidance and encouragement they provided. I wish to acknowledge the assistance given by Mr Mbolela N. of the Department of Energy in the Ministry of Energy and Water Development for the literature provided. May I also thank the many other colleagues for the encouragement given. Lastly but not the least, I acknowledge the efforts that all experts and key stakeholders have made and will continue to make for the advancement of the biofuels industry in Zambia.

REFERENCES

Bhojvaid P.P., 2006. Biofuels towards a greener and secure energy future., The Energy Research Institute, TERI. NewDelhi, India.

Lubozhya B., Biofuels Experiences in Zambia.
http://www.un.org/esa/sustdev/sdissues/energ/op/biofuels/biofuel

Mbolela N., 2007., Overview on Alternative Energy Sources with Specific Focus on Biofuels. In: The Report on Proceedings of the Colloquium on Alternative Energy Sources in Zambia; National Science and Technology Council; Lusaka; Zambia.

Ministry of Energy and Water Development. 2007; The National Energy Policy; Government of the Republic of Zambia.

Ministry of Energy and Water Development. 2009; Brief Report on the Current Status of Biofuels Industry in Zambia; Government of the Republic of Zambia.

Ministry of Science, Technology and Vocational Training. 2008; Draft Position Paper on Jatropha in Zambia; Government of the Republic of Zambia.

Mwamba C. K., 2002. Photo-documentation on the promotion of soap and fuel production from *Jatropha curcus* oil; National Institute for Scientific and Industrial Research; Lusaka; Zambia.

Pio P., 2007., Evaluation of Zambia's Capacity to process and Utilize Biofuels as a Source of Alternative Energy; A Perspective from Industry In: The Report on Proceedings of the Colloquim on Alternative Energy Sources in Zambia; National Science and Technology Council; Lusaka; Zambia.

Sinkala T., 2008. Jatropha and its Potential for Alleviation of Poverty and Environmental Degradtion in Zambia.

Yamba F.D., 2007., Technology Options for Sustainable Biofuel Development in Zambia; Challenges and Opportunities In: The Report on Proceedings of the Colloquium on Alternative Energy Sources in Zambia; National Science and Technology Council; Lusaka; Zambia.

Renewable Energy

13

Adoption of Energy Efficiency Measures by Households in Shared-meters Apartments

Michael Kiza
Ministry of Energy and Mineral Development, Uganda

In June and July, 2007, a survey was carried out to determine difference in adoption of energy efficiency measures between households in shared-meters apartments and those in self-metered apartments. The study area was Kampala, Entebbe, Mukono and Jinja. The survey revealed that 74.8% of those living in rented apartments share electricity meter and the monthly electricity bills are paid through the landlords, either as part of rent or as separate fixed monthly charge. The study also revealed that households with shared meters are not conscious of energy bills and very reluctant to adopt energy efficient technology and practices. From this study, it recommended that law should be enacted to compel landlords to provide self-meter for each household.

1.0 INTRODUCTION

1.1. Introduction

Uganda has a rapidly growing population currently estimated to be about 30 million with a growth rate of 3.57%. Agriculture still dominates the countries economy, accounting for about 30% of the total GDP in 2006 and employing over 80 % of the workforce. Since 1999/2000, the number of households has been rising. From 4.2 million in 1999/2000, they increased to 5.2 million in 2005/2006. The Eastern region had the largest average household size of 5.6 compared to Central region with 4.8. The survey showed that the youth (18-30 years) migrated more than any other group of the population.

The UNHS (2005/2006) categorizes the dwelling type (i.e. the type of building or a group of buildings in which the household lives) into huts, single detached houses, and apartments (tenements) and several one-room apartments among others. The report showed that over 65 % of the urban population leaves in apartments.

As shown in Table 1, most of the people in Ugandan towns and cities (73.6% in Kampala) live in one-room apartments. Some of these apartments share electricity meters and the electricity bills are included in the monthly rent, i.e., a fixed monthly charge for electricity is added on top of the agreed rent. Because tenants in these apartments do not choose how much energy to use after the monthly rent has been determined, they have no price incentive to conserve energy, and therefore may be reluctant to adopt energy efficient technologies. This in turn leads them to using more electricity than tenants in otherwise similar situations, but having individually metered apartments. Moreover, the cost of the extra energy use, if added to tenants' monthly rent, may be more than tenants would be willing to pay for that energy separately.

TABLE 1: Distribution of Households by Number of rooms used for Sleeping (Source: UNHS, 2005/6)

Number of Rooms	2005/2006					
	Kampala	Central	Eastern	Northern	Western	Uganda
One	73.6	50.9	58.1	80.1	36.5	**56.3**
Two	13.0	24.4	20.7	13.2	30.1	**22.0**
More than two	13.5	24.7	21.2	6.8	33.4	**21.8**

1.2. Residential Electricity Access and Utilization

Access to electricity still remains very low. According to UNHS (2005/2006), the proportion of households using electricity in rural areas was very small (4%). At national level 11% of households used electricity as the main source of lighting. Even in urban areas, less than half of households used electricity as the main source of lighting. Table 2 shows the distribution of various lighting fuel use in Uganda.

TABLE 2: Distribution of Lighting Fuel by Residence and Region (%) 2005/6 (*Source: UNHS, 2005/6*)

Residence Rural/Urban	Tadooba	Lantern	Electricity	Other*	Total
Rural	79.1	12.3	4.0	4.7	**100.0**
Urban	31.2	23.4	41.2	4.2	**100.0**
Region					
Kampala	13.1	20.5	60.6	5.7	**100.0**
Central	64.6	17.6	15.1	2.8	**100.0**

Eastern	81.2	12.3	5.0	1.6	**100.0**
Northern	79.9	7.6	1.4	11.1	**100.0**
Western	76.0	16.1	4.2	3.7	**100.0**
Uganda	**70.7**	**14.2**	**10.5**	**4.6**	**100.0**

**Includes firewood, biogas*

The UNHS (2005/6) report also reveals that a very small proportion of households (less than 1%) used electricity as the main source of energy for cooking, 78% of the households depended on firewood for cooking and 18% on charcoal. Overall, 96 % of the households depended on biomass for cooking purposes sustainability. Variations in residence show that charcoal was mainly used in urban areas (66%) while firewood was more prominent in rural areas (89%).

2.0 ADOPTION OF ENERGY EFFICIENCY MEASURES

2.1 Energy Efficiency Surveys

Since 2003, the Ministry of Energy and Mineral Development has been promoting efficient utilization of electricity as one of the means to reduce power demand. In particular, the ministry has been promoting replacement of incandescent bulbs with compact fluorescent lamps, switching-off lights when not necessary, as well as many other energy saving measures.

In May/June 2006, a survey was carried out on residential consumers lighting system in Kampala in order to gain insight on the level of adoption of energy efficient technologies, focusing on lighting as an indicator. The key objectives of the survey were to establish:

1. The types and nature of residential lighting presently used,
2. The normal lighting practice,
3. The extent of consumer awareness of energy savers that is now available in the market.

The survey was conducted in a total of the 527 households, of which 19% were from Rubaga, 20.1% were from Makindye, 19.2% were from Kawempe, 19.0% were from Nakawa and 22.8% were from Central division.

The result of the survey showed that 58.6% of the lighting fixtures found in the different rooms in the household had incandescent lamps. Although the survey indicated that 52.2% of the respondents were living in rented houses, it did not establish who was directly responsible for paying the electricity bill. This is a very important question which was overlooked by the first survey. The survey also did not consider other energy consuming appliances like refrigerators, iron, fans, etc., as well as energy conservation measures like switching lights off during the day.

The resistance to change is one of the most important factors that hinder adoption of new, efficient technologies. Most especially when the people involved are not directly affected by the consequences of their activities. This is true when it comes to utility bills. If the householder does not pay directly for the service, or if the services are paid for communally, there is a tendency of loss of responsibility and hence mismanagement.

It is therefore important to study the behavioural differences between tenants who pay their electricity bills directly and those whose bills are included in the rent. This study will thus help the government to come up with policy/regulatory measures to improve energy use in rented apartments.

The existence of these electricity-included rent contracts raises three questions that needed to be addressed:

1. What is the rate of adoption of energy efficient technologies among the electricity-included rent as compared to self-metered apartments?
2. How much extra energy is used by tenants in these apartments?, and
3. Why is electricity bill always included in the rents?

A second survey was conducted in June-July 2007 to answer the above questions.
The study was carried out in Jinja district in areas like Mbiko, Nile zone, Mukono in area like Seeta, Butembe, Kigunga, Kawuga and Kampala in areas like Kireka, Mengo, Kamwokya and Namuwongo, Entebbe. A total of 250 households were sampled.

2.3 Questionnaire Design

A comprehensive questionnaire framed by us and approved by the authorities in the ministry was used in the study. The questionnaire is attached in the annex.

2.4 Scope of the Study

Since the Government has not defined energy efficiency standards for home appliances, except in the areas of lighting whereby CFLs are promoted, the study focused on CFLs. Awareness on, and adoption of, CFL was considered to reflect the awareness on. and adoption of, energy efficient appliances. The study also considered the behavioural changes in electricity usage.

Bearing in mind that the Government was already distributing free the CFLs in these districts, the respondents were asked to give the number of CFLs and other energy efficient lighting fixtures) they had before and after the CFL programme, as well as the number of free CFLs received. The number of CFLs, the respondents had before the distribution campaign was used to determine the willingness to adopt energy efficiency measures.

3.0 RESULTS AND OBSERVATIONS

Out of the households surveyed, only 36% were living in their own houses, the rest lives in rented facilities (either detached houses or apartments). This is consistent with the UNHS (2005/6) report that revealed that 33.9% of households in Kampala and 73.3% in central Uganda live in own houses.

As the majority of the residents in Kampala City, Entebbe, Jinja, and Mukono towns live in rented apartments, the major objective of the study was to find out how electricity bills are paid by the tenants and how the mode of payment affect the adoption of energy efficient applies and conservation practices. The adoption of energy efficient lighting fixtures was considered to represent the adoption of energy efficiency measures in general. Energy efficient lighting fixtures are considered to include CFLs and fluorescent tubes. Those using incandescent lamps were considered inefficient.

The survey revealed that 74.8% of those living in rented apartments share electricity meter and the monthly electricity bills are paid through the landlords, either as part of rent or as separate fixed monthly charge. It was also revealed that most landlords charged between USh 10,000 to 20,000 per month for electricity supply to 1-2 room apartments. These charges are fixed, irrespective of what kinds and numbers of appliances and hours the household operated. It was surprising to note that, even households with only one bulb and a radio pay the same amount of money as those with multiple high-energy consuming equipment appliances.

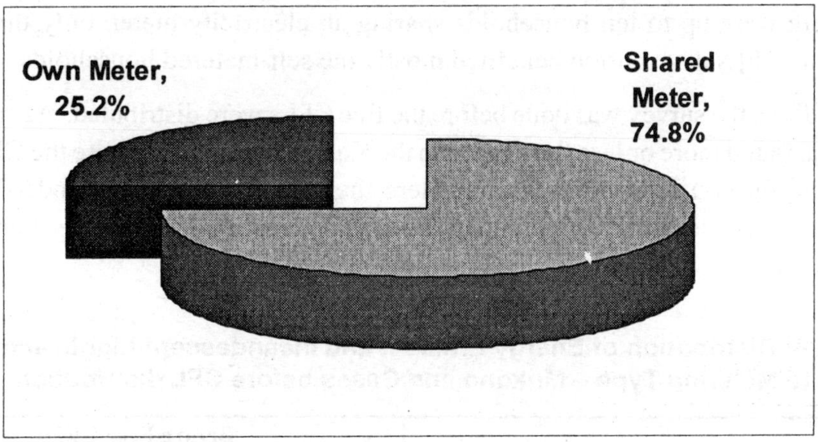

FIGURE 1: Distribution of Electricity Metering Type

With this kind of anomalies, can there be any incentive to adopt energy efficient measures? By the time of this study, the CFL distribution had already covered the whole of Kampala and most arts of Entebbe. All the areas surveyed in Kampala and Entebbe had received.

TABLE 3: Distribution of Energy Efficient and Incandescent Lights according to Metering Type – Kampala and Entebbe Cases

Metering Type	Before Free CFLs		After Free CFLs	
	Incandescent	Energy Efficient Lights	Incandescent	Energy Efficient Lights
Shared Meters	71.4%	28.6%	67.3%	32.7%
Self Meters	46.7%	53.3%	31.6%	68.4%

Table 3 shows the distribution of energy efficient and incandescent lightings among the households with shared meters or self meters for Kampala and Entebbe areas. Out of those with shared meters, 71.4% of the lighting fixtures used before the free distribution of CFLs were found to be incandescent lamps. In contrast, 53.3% of the lighting fixtures used by self-metered households before the free CFLs distribution were of energy efficient types.

After the distribution of the free CFLs in Kampala and Entebbe areas, the share of CFLs in self-metered households shot up to 68.4% from 53.3%. On the contrary, among the shared meters households, the share of CFLs rose only from 28.6% to 32.7%. This can be attributed to the CFLs distribution procedure, whereby three free CFls were given per connected meter. In some cases, even where there were up to ten households sharing an electricity meter, only three CFLs were given. Hence, the CFLs distribution benefited mostly the self-metered households.

In Mukono and Jinja, the survey was done before the free CFLs were distributed. As can be expected, the situation was found more or less the similar to the Kampala situation before the CFL distribution campaign. Among the household with shared meters, the share of CFLs was found to be only 24.4%, the balance being incandescent. While among the self-metered apartments, the share of CFLs was found to be 50.2%

TABLE 4: Distribution of Energy Efficient and Incandescent Lights according to Metering Type – Mukono and Cases before CFL distribution

Metering Type	Before Free CFLs	
	Incandescent	Energy Efficient Lights
Shared Meters	75.6%	24.4%
Self Meters	49.8%	50.2%

It can also be seen that the share of CFLs in Kampala was found to be higher than that in Jinja and Mukono. This may be attributed to the fact that much of the CFL promotional campaigns over the past few years have been concentrated in Kampala. It was not clear whether the CFLs were cheaper in Kampala than in Jinja.

During the survey, the energy conservational behaviour like switching off lights during the day-time, when artificial lights are not needed, were discretely observed. The observations are summarized in Table 5.

TABLE 5: Energy Conservation Behaviour

Metering Type	Lights Off	Lights On
Shared Meters	85.6%	14.4%
Self Meters	94.9%	5.1%

Households with shared meters are not conscious of energy bills. Hence, they bother less to switch off lights during the day. The practice of switching off lights during the day can save a lot of power from the grid and reduce the net bill paid.

4.0 CONCLUSIONS AND RECOMMENDATIONS

A significant portion of the generated electricity in Uganda is consumed by households. Hence, energy efficiency improvement in households is very critical in the overall national energy efficiency. The conclusion that can be drawn from this paper is that when tenants share electricity meters and/ or the electricity bills are included in the rent, there is less incentive to adopt energy efficiency and conservation measures. This study showed a clear difference adoption of energy efficient lighting between the self-metered households and those with shared meters.

In this era of electricity constraints, landlords should never include electricity bills in rents. The apartments with shared meters should not be encouraged. Yet this study showed that 74.8% of tenants share electricity meters. These are mainly households in the lower income category.

This study found that tenants in shared-meters apartments are less conscious of energy bills. If tenants are aware of the electricity tariff and do not want high electricity bills, they would be willing to adopt energy efficiency measures. However, the study also found that most of the tenants in shared meters apartments do not even know about the electricity tariff. So, while we cannot rule out the presence of shared-meters apartments, some of the explanation for its persistence may be installation costs of these meters. In fact, every new meter installed is considered by Umeme as new connection and hence attracts connection fee, even if the meter is added to an already connected house. This discourages segregation of meters.

One approach the landlords can adopt to improve energy efficiency in this shared-meters scenario would be to incorporate energy efficiency in house construction and installation of fixed appliances. However, since these are generally low cost apartments, such landlords have little incentive to invest in energy-efficiency.

This study has also shown that households in self-metered apartments tend to be relatively more energy efficient. These households are more sensitive to the energy costs since they directly receive and pay monthly bills.

From this study, it suffices to recommend that law should be enacted to compel landlords to provide self-meter for each household.

14

Need and Challenges in Developing Renewable Energy Resources: A Networked Approach of CSIR

Sandeep Kumar Tiwari
R&D Planning Division,
Council of Scientific & Industrial Research (CSIR), New Delhi, India
E-mail: sktiwari@csir.res.in, drtiwarisk@yahoo.co.in

Indiscriminate use of fossil fuels has led the world to a situation wherein the business-as-usual scenario would increase concentrations of Green House Gases (GHGs) that could subsequently result in high temperatures and irreversible climate change. Such a scenario is not only unsustainable but would be catastrophic. The energy sector in world over urgently needs reformation through policy interventions. There is need to enhance low carbon and renewable-energy share in the total energy mix to contain and subsequently reverse the damage to the environment. This paper presents various projected scenarios and Indian initiatives to reduce the CO_2 emissions. It also discusses the challenges in renewable energy development and enhancing energy efficiencies, and Council of Scientific and Industrial Research's approach in dealing with developmental challenges in the area of energy.

KEY WORDS: *Climate change, Renewable energy, Biomass, Biodiesel*

1.0 INTRODUCTION

Energy affects all aspects of development such as social, economic, scientific and environmental– including livelihoods, agricultural productivity, health, population levels and education.

India ranks sixth in the world in total energy consumption. The country has very small hydrocarbon reserves (0.4% of the world's reserve), though rich in coal and abundantly endowed with renewable energy potential in the form of solar, wind, hydro and bio-energy. Like many other developing countries, India is a net importer of energy, much of her primary energy needs being met through imports mainly in the form of crude oil and natural gas.

World Energy Outlook (2007) predicted that world consumption of petroleum and other liquid fuels will grow from 83 million barrels oil equivalent per day in 2004 to 97 million in 2015 and 118 million in 2030. Much of the overall increase in liquids consumption is projected for the nations of non-OECD Asia, where strong economic growth is expected. To meet the increase in liquids consumption in the International Energy Agency (IEA, 2007), liquids production is projected to increase by 14 million barrels per day from 2004 to 2015 and by an additional 20 million barrels per day from 2015 to 2030.

According to World Energy Outlook (WEO) *2008 Fact Sheet,* the impact of the credit crisis on world economic growth prospects, higher energy prices and some notable new policy initiatives have left their mark. Energy use grew more slowly to 2030 than projected in 2007, but the overall trends are broadly unchanged: persistent dominance of fossil fuels – oil, gas and coal – in the energy mix; a rising share of emerging economies in global energy consumption; an increase in the consuming countries' reliance on imports of oil and gas; and an inexorable rise in global CO_2 emissions.

The report of World Energy Outlook (2007) indicated that India and China will account for maximum increase in energy demand between 2005 and 2030. Globally, there will be 55% increase in energy demand. India and China will account for 45% of increase in energy demand. Fossil fuels will remain dominant source of energy accounting for 84% of overall energy demand.

According to World Energy Outlook (2008), the rising global consumption of fossil fuels would increase the greenhouse-gas emissions and global temperatures, resulting in potentially catastrophic and irreversible climate change. The projected rise in emissions in the Reference Scenario (Business As Usual; BAU), in which no change in government policies is assumed, would double the concentration of those gases in the atmosphere to around 1000 ppm of CO_2-equivalent by the end of this century. In the emerging scenario, there is urgent need for all countries to devise and implement policies for CO_2 stabilization and subsequent reduction.

India's overriding priority is for economic and social development while preserving environment and maintaining the ecological balance. India is faced with the challenge of sustaining its economic growth while dealing with the issue of climate change. Recognizing it to be a global challenge, India is engaged in multilateral negotiations in the UN Framework Convention on Climate Change. While discussing various policy frameworks on reduction of CO_2, this paper discusses the possible energy options and challenges for India to achieve the global objective of reducing GHG and at the same time achieving its developmental goals.

2.0 GLOBAL POLICY FRAMEWORKS FOR REDUCING ATMOSPHERIC CO_2

World Energy Outlook has observed that without new policies in place global primary energy demand will increase by 45% in 2030. This would put global energy-related CO_2 emissions at 40.6 Gt in 2030, with an annual growth rate of 1.6%. Such a trajectory puts the world on track for a long-term concentration of greenhouse gases of over 750 ppm of CO_2 in the atmosphere. This scenario is not sustainable- environmentally, socially and economically.

Various policy interventions are proposed by International Energy Organization, Intergovernmental

Panel on Climate Change (IPCC) and others. These are basically to reduce the fossil fuel uses in energy mix and enhance the share of nuclear and renewable energy on the supply side and increasing the energy efficiency on the demand side.

For ensuring lower emissions scenarios by 2030, the WEO (2008) suggested the following Climate-policy frameworks:

550 Scenario – This scenario corresponds to a 3°C global temperature rise where greenhouse gases would stabilize at 550 ppm of CO_2 equivalent by 2030. It takes into account the increase in energy demand but would present a different fuel mix.

450 Scenario – This corresponds to a with a 2°C rise in global temperature wherein GHG (Green House Gases) would stabilize at 450 ppm of CO_2 equivalent. This scenario is consistent with reducing global GHG emissions by 50% from current levels by 2050.

The share of low-carbon energy – hydropower, nuclear, biomass, other renewables and fossil-fuel power plants equipped with Carbon Capture and Storage (CCS) – in the world primary energy mix would need to expand from 19% in 2006 to 26% in 2030 (550 ppm scenario). This calls for US$ 4.1 trillion more investment in energy-related infrastructure and equipment than in the Reference Scenario – equal to 0.2% of annual world GDP. While improved energy efficiency would deliver fuel-cost savings of over US$ 7 trillion.

Energy Technology Perspectives (IEA, 2008) proposed a range of **ACT** Scenarios showing how CO_2 emissions could be brought back to current (Reference) levels by 2050, and a set of **BLUE** Scenarios outlining how they could be reduced to 50% below current (Reference) levels.

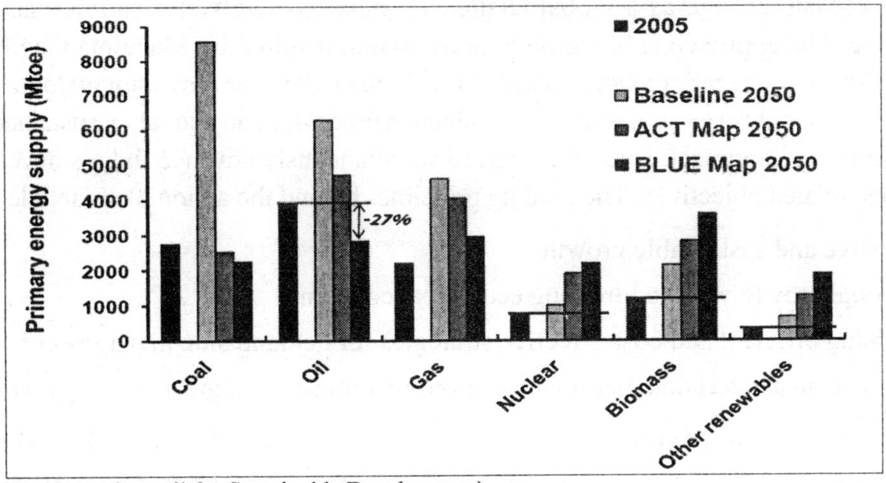

(Source: World Business Council for Sustainable Development)

FIGURE 1: Projected ACT and BLUE scenarios

These low GHG emission scenarios would require reformation of energy sector. The policy interventions would therefore be required to ensure the following:

- efficient clean coal technologies with Carbon Capture and Storage (CCS)
- change to the low carbon/ renewable energy mix
- energy efficiency on demand side

Improved energy efficiency makes the biggest contribution to lowering emissions in both scenarios, which according to WEO is 54% in the 450 ppm scenario. The transport sector is also an important component in reducing GHG by enhancing fuel efficiency and policy interventions for encouraging mass transport.

Reducing the emissions worldwide to 50% by 2050 would be an extremely challenging target. Beyond 2030, it will be difficult to achieve such a target simply with the technologies that are available today. Energy Technology Perspective (ETP, 2008) indicated that in all sectors new technologies will be needed to bring the costs of the low-carbon transition down especially the transport sector which would require low carbon technologies such as electric and fuel cell vehicles. It is true that without policy interventions, oil use by transport sector in many countries will increase significantly by 2050.

In its Fourth Assessment Report the Intergovernmental Panel on Climate Change (IPCC, 2007) suggested key sectoral mitigation technologies, policies and measures. After assessing suitability and requirements of policy interventions, India has formulated national action plan to adapt and mitigate climate change.

3.0 NATIONAL (INDIA) ACTION PLAN ON CLIMATE CHANGE

Recognizing climate change as a global challenge, India has evolved a national action plan on climate change. The approach is based on a global vision inspired by Mahatma Gandhi's dictum- *The earth has enough resources to meet people's needs, but will never have enough to satisfy people's greed.* Thus, India emphasizes on sustainable production processes and equitable/ sustainable lifestyles across the globe. The approach is to take steps to simultaneously advance India's development and climate change related objectives. The guiding principles behind the action plan include:

- Inclusive and sustainable growth
- National growth while addressing ecological concerns
- Devising efficient and cost-effective strategies for demand side management
- Deployment and technologies for adaptation and mitigation of green house gas emissions, etc.

In consonance with the principles and approach to deal with the challenge of climate change, eight national missions are identified representing multi-pronged, long term integrated strategies for achieving its goals. The missions are:

a) <u>**National Solar Mission**</u>: It is to significantly increase the share of solar energy in the total energy mix while recognizing the need to enhance other non-fossil options such as nuclear, wind and biomass energy.

b) **Enhanced Energy Efficiency**: While anticipating a saving of 10,000 MW to be achieved by the end of XI Five Year Plan (XI FYP) (2012) through various scheme and programmes of Bureau of Energy Efficiency (BEE) and designated state agencies, four new initiatives would be launched to improve energy efficiency in energy-intensive large industries/facilities, accelerating shift to energy efficient appliances by making these affordable and fiscal incentives for demand side management and to promote energy efficiency.

c) **Sustainable Habitat**: The mission is meant to improve energy efficiency in buildings, management of solid waste and modal shift to public transport. It includes design and energy conservation in building recycling of materials, bio-chemical conversion of urban waste, waste water use, mass transport systems etc.

d) **National Water Mission**: The mission would ensure integrated water resource management to recycle, conserve water and enhance water use efficiency by 20%.

e) **Sustaining the Himalayas Ecosystem**: The mission would evolve measures for sustaining and safe guarding the Himalayan glaciers and mountain eco-system.

f) **Green India**: The mission is proposed to enhance forest cover for preserving ecological balance and maintenance of biodiversity. The target is to bring 33% area under forest cover from 23% at present. Forests are regarded as effective CO_2 sinks.

TABLE 1: Projections for Electricity generation

	Electricity Generation (TWh)				Growth (% p.a.)		Change vs. RS		Share %	
	1990	**2005**	**2015**	**2030**	**2005-15**	**2005-30**	**2015**	**2030**	**2015**	**2030**
Total generation	**11802**	**18197**	**24654**	**31240**	**3.1**	**2.2**	**-3.5**	**-11.7**	**100%**	**100%**
Coal	4429	7334	9755	10716	2.9	1.5	-12	-32.2	40%	34%
Oil	1324	1186	1118	844	-0.6	-1.4	-4.1	-9.2	5%	3%
Gas	1725	3585	5150	6602	3.7	2.5	-3.4	-18.2	21%	21%
Nuclear	2013	2771	3262	4144	1.6	1.6	5.8	26.5	13%	13%
Hydro	**2145**	**2922**	**4089**	**5403**	**3.4**	**2.5**	**7.6**	**11.6**	17%	17%
Renewables (excluding hydro)	**166**	**398**	**1276**	**3532**	**12.3**	**9.1**	**16.7**	**42.8**	5%	11%
Biomass and waste	124	231	511	1166	8.2	6.7	25.6	38.7	2%	4%
Wind	**4**	**111**	**623**	**1800**	**18.8**	**11.8**	**13.5**	**39.9**	3%	6%
Geothermal	36	52	100	190	6.8	5.3	1.2	9.9	0%	1%
Solar	1	6	44	352	29.7	20.6	19.3	11.5	0%	1%
Tide and wave	1	1	2	24	12.3	18.2	24	108.1	0%	0%

(Source: WEO 2007)

g) **Sustainable Agriculture:** The mission is meant to devise strategies to make agricultural plants more resistant to climate change.

h) **Strategic Knowledge for climate change**: It envisages establishing mechanisms for enlisting collaborator of global community to identify challenges and responses to climate change and would ensure funding of high quality and focused research into various aspects of climate change.

It is envisaged that these missions would help in adapting and mitigating the climate changes in a holistic manner.

4.0 ANALYSIS OF CO_2 STABILIZATION SCENARIOS VIS-À-VIS INDIA'S INITIATIVES

The United Nations Framework Convention on Climate Change calls for *stabilization of green house gas concentrations in the atmosphere at a level that would prevent dangerous anthropogenic interference with the climate system.* As against business-as-usual (BAU) the targets of achieving 450 ppm and 550 ppm levels of atmospheric CO_2 includes energy mix on the supply side while achieving energy efficiencies on the demand side. Table 1 indicates the technology options for achieving the target GHG reductions.

The future technology scenario on the supply side is indicated in Table 2. For India, coal will be the dominant component in its energy mix because of the fact that India is richly endowed with coal while having high dependence on imports of oil and gas. Its share is expected to remain high in the energy mix and therefore, there is need to install efficient power plants with carbon capture technologies. The present state of development and commercial viability of CCS technologies is not very attractive to India. However, in future energy mix, once the Coal plus CCS technologies becomes competitive; these will have significant contribution in Indian Energy mix.

TABLE 2: Key Technology Options

Supply side	Demand side
• CCS power generation	• Energy efficiency in buildings
• Coal-IGCC	• Energy efficient motor systems
• Nuclear	• Efficient ICE's
• Solar-PV	• Heat pumps
• Solar-CSP	• Plug-ins and electric vehicles
• Wind	• Fuel cell vehicles
• Biomass-IGCC & co-combustion	• Industrial CCS
• 2nd generation biofuels	• Solar heating

(Source: Energy Technology Perspectives 2008 (IEA)).

As projected in Table 1, the share of coal and gas in world energy mix would decrease whereas that of gas, nuclear and hydro would remain the same. However, there would be a significant increase in share of non-hydro renewables which include biomass, wind and solar. The growth of power generation through wind, biomass and solar would be very high from 2005-2030, expecting growth factors of 6, 16 and 60 times. With entry of private players, such as Moserbaer and Indosolar, a large capacity to produce SPVs is created in India with the aim of generating 1000 MW by 2017 as against 125 MW today.

Nuclear energy is low carbon energy resource available to India. India is poorly endowed with Uranium. Its available ores are low grade (~0.1% Uranium) compared to 12-14% uranium in ores available in other countries. This makes Indian nuclear fuel 2-3 times costlier than international prices. However, India possesses large reserves of thorium which can be converted into fissile material. To utilize this vast reserve for future power generation, India has envisaged three stage nuclear power programme which involves setting up of Pressurized Heavy Water Reactors (PHWRs) in the first stage, follow by Fast Breeder Reactors (FBRs) and finally the reactors based on Uranium 233 – Thorium 232 cycle in the third stage. In the first stage, PHWR would efficiently utilize natural uranium for yielding plutonium fuel required for the second stage FBR programme. While utilizing plutonium, FBRs will recycle spent uranium from PHWR to breed more plutonium fuel. Thorium as blanket material in FBRs would produce uranium 233 to fire this third stage. India has more than 20 years experience in developing FBR technology. It is critical link to developing India's nuclear power programme. Without this, it would be difficult for India to go beyond 10,000 MW nuclear capacity based on indigenous uranium resources. India has now entered into US-India nuclear deal, which has also been ratified by NSG (Nuclear Supplier Group) and IAEA. These steps will allow India to enhance the share of Nuclear energy to over 5% in the energy mix.

India has a vast potential of renewable energy. The potential and implementation status of these is shown in Table 3. The total installed energy capacity is approximately 10,000 MW which is about 6% of total installed capacity with 2% contribution to electricity mix (Murthy, 2008). The potential for generation of renewable energy lies at about 172,000 MW. Huge untapped potential of ~ 160,000 MW and starving grid (about 12% peak demand shortage), present tremendous scope and opportunity in Renewable energy sector. India is second largest in the world in wind energy installations and one of the largest producers of the renewable energy in the world. A number of grid connected power plants have come up through private entrepreneurs in the range of 6-15 MW in Andhra Pradesh and Chhattisgarh states in India. There are encouraging signs wherein sugarcane based co-generation and power supply to the national grid got a boost when large number of sugar mill owners opted for enhanced co-generation and supply surplus power to the grid.

The Government of India is committed to generate 10% of its power from renewable energy by 2012. The regulatory framework has already been developed for regulatory purchase standards and tariffs for renewable energy. Many states have also permitted adequate wheeling and banking facilities for such projects (Ernst and Young Research).

TABLE 3: Estimated cost of power generation from renewables

S.No.	Source	Capital cost	Estimated cost of generation	Total Installed Capacity	Estimated Potential
		Rs. Cr./MW	Rs./kwh	MW*	MW
1.	Small Hydro Power	5.00-6.00	1.50-2.50	2181	15,000
2.	Wind Power	4.00-5.00	2.00-3.00	8757	45,000
3.	Biomass Power	4.00	2.50-3.50	1586	19,500
4.	Bagasse cogeneration	3.50	2.50-3.00	379.00	
5.	Biomass Gasifier	1.94	2.50-3.50	60.20	
6.	Solar Photovoltaic	26.50	15.00-20.00	6.72	1700
7.	Energy from Waste	2.50-10.00	2.50-7.50	46.5	

(Source: MNRE, GoI)

India has also put in place many provisions and mechanisms to encourage growth of renewable energy. For example, provisions of Electricity Act 2003 necessitate all State Electricity Regulatory Commissions (SERCs) for fixing minimum percentage of procurement from renewable energy. MNRE (Ministry of New & Renewable Energy), GoI has announced number of financial incentives for renewable energy based power generation which included interest and capital subsidy, 80% accelerated depreciation tax holiday under section 80IA. In addition, soft loans are provided through Indian Renewable Energy Development Agency (IREDA) and also some nationalized banks and other financial institutions for identified technologies/systems.

The fiscal incentives offered to renewable energy sector include accelerated depreciation in the first year of installation, exemption/reduction in excise duty concessions on import of components, equipments and materials for generation based on renewable energy.

On the supply side, the projections of world's energy consumption scenario (Figure 2), without policy interventions (BAU scenario), world over the major energy consumption would be in buildings where as transport and industry would also have significant share. The focus therefore would have to be on development of energy efficient technologies for buildings, transport sector and industry.

Figure 3 shows the Energy intensity which is a measure of energy efficiency of a nation's economy of various countries. The energy intensity of India is better than US and China but there is scope for further improvement. While having very low per capita consumption of energy, the energy efficiency in Indian industry has increased steadily in recent years. According to Bureau of Energy Efficiency,

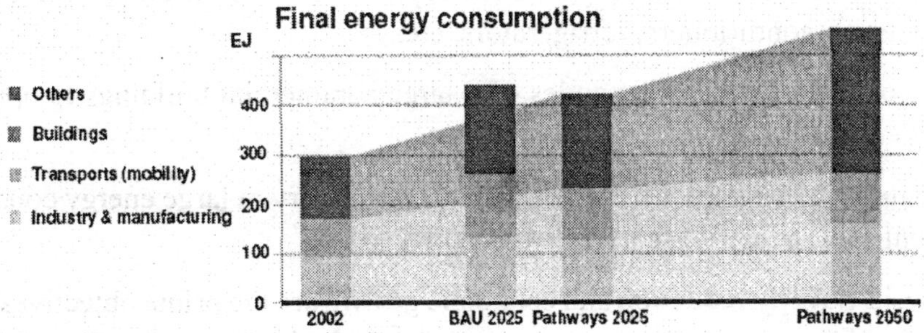

(Source: World Business Council for Sustainable Development)

FIGURE 2: World's energy consumption scenario

India has policies to promote energy efficiency which include:

a) All government buildings should meet 10% of their energy needs through renewable energy by 2010, progressively increasing to 20% by 2020 and 30% by 2030.

b) Enhancing efficiency of power plants - new plants to adopt more efficient and clean coal technologies.

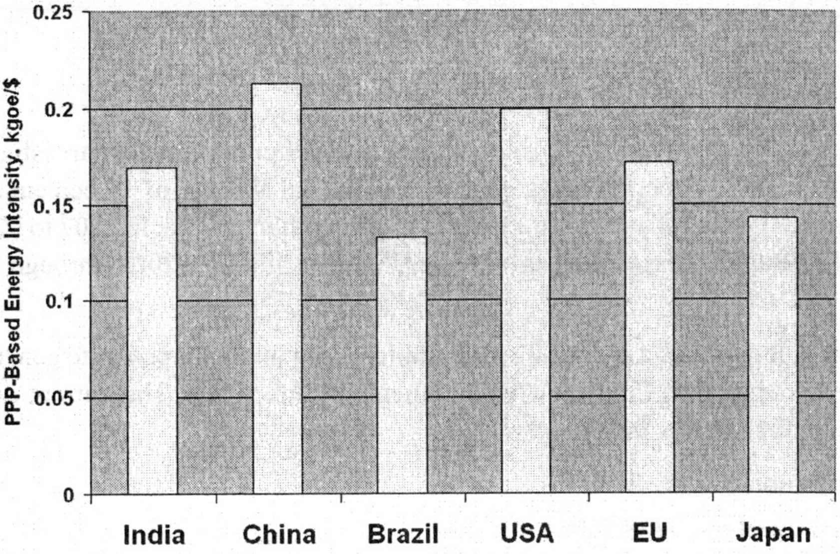

(Source: Bureau of Energy Efficiency, Ministry of Power, Govt. of India)

FIGURE 3: Comparison of Energy Intensity of Developed/Developing Economies

c) Introduction of labelling programme for appliances - star based labelling for fluorescent tube lights, air conditioners, refrigerators, etc.

d) Energy conservation building codes - for large commercial buildings to optimize building energy demand.

e) Energy audits of large industrial consumers- mandatory in large energy consuming units in nine identified industrial sectors.

In the final analysis, poverty reduction and economic growth are the prime objectives of the national policy of India. Quality energy and energy security would be important for sustained development and poverty alleviation so as to climb up in Human Development Index. India has registered a growth of 9% per annum which has been achieved with an energy growth of less than 4% per annum. In spite of development issues, India is making all efforts to contribute to the cause of climate change.

5.0 THE R&D CHALLENGES/ OPPORTUNITIES FOR INDIA: A REVIEW

Once the importance of renewable energies are identified, to achieve twin objective of development and potential energy resource, having low-carbon footprints, it becomes imperative to develop these resources to their optimal potentials. In India, these resources are yet to be taped to their fullest which requires development of technologies with high conversion efficiencies. A Strength, Weaknesses, Opportunity and Threats (SWOT) analysis of these resources in context of India is given in Table 4. Only areas where CSIR has research capabilities are reviewed for interventions in succeeding sections.

5.1 Solar Energy

In context of India, Photo Voltaics (PV) presents tremendous opportunity for harvesting solar energy. In the World over, the PV market had seen the growth at an average of 15 percent annually from 1990-2000. Global PV installations are expected to grow from 3.9 GW in 2007 to 52 GW by 2012 and the global investments expected to grow from USD 27.5 billion in 2004 (through 148.4 in 2007) to 600 billion in 2020.

Photovoltaic devices provide a promising platform for using that solar power to generate electricity and additionally for production of new environmentally safer fuels. The current distributions of photovoltaic devices in use are:

(a) polycrystalline Si (52.3 %)

(b) single crystalline Si (38.3%)

(c) a-Si (4.7%), Si ribbon (2.7%)

(d) CdTe (1.6%)

(e) copper-indium-diselenide (0.2%).

TABLE 4: SWOT of renewable sources – Indian perspective

Resource	Strength	Weakness	Opportunity/Challenges	Threats
Small Hydro	- Eco-friendly, minimal civil works - Large potential (15020MW) - Considerable indigenous manufacturing capability of turbines, generators and controllers	- Lack of grid connectivity - Fluctuations in input energy - Only for captive generation	- Hybrid systems with oil/gas - Storage of energy into batteries and transporta- tions to places of require ment - Hydrogen generation through water electrolysis.	- Problems of ecology & environment - Social problem of resettlement - Dependence on rivers. The water availability may be affected by erratic precipitation.
Wind	- 7000 MW grid connected installed capacity - Large potential (45000 MW)	- Low wind power density - High installation cost - Off-shore potential locali- zed to few regions - Unreliable / highly variable - Uneven & discontinuous - Technical sophistications, cost, issues of control, store and distribution.	- Turbines for low wind power density winds - Storage in batteries and distribution	- Social concerns related to noise, visual, electromagnetic interference.
Biomass	- Large potential for bioethanol, biodiesel, biogas. - Large availability of biomass as waste, non- edible oil, algae , etc.	- Scattered and uneven availability - Requires large land area	- Technology gaps in *Biomethanation of waste * Bio ethanol from lingoce llulosic material * Development of elite varieties of Jatropha * Gainful utilization of glycerol /Jatropha deoiled cake	- Unresohred issues related to Jatropha toxicity - Higher returns on bioethanol for other uses (consumption, industrial, etc) - Food vs fuel
Solar	- Huge potential	- Large area requirement - High installation cost - Low efficiency of available Tech. - Remained confined to small/ domestic uses - Non availability during night	- Developing High efficiency conversion tech. - Vastly untapped resource - Developing low cost SPVs	- Fluctuations in availability leading to uncertainty in quality & quantity of power.

However, the maximum efficiency is still well below 50%. The researchers believe that reduction of solar cell manufacturing cost and increase in efficiency can be achieved with thin-film technology and/or new device architectures. Therefore, the opportunity exists for investigating in new materials, materials integration and/or hybrid device architectures with the ultimate goal of increasing efficiency above 50%.

The other aspect is availability of raw material to fulfill huge demand in photovoltaic manufacture. The extremely high growth rates of the photovoltaics industry during the last years led to the current temporary shortage in silicon feedstock as the silicon production capacity did not grow at the same pace. Like in any other industry, raw material availability and the corresponding prices, will influence demand and supply, and will be a key driver for new developments and innovation within the industry. Significant expansions of production capacities are under way, which increased the 94 MW thin-film shipments in 2005 by a factor of five by 2008 and could even reach 1000 MW by 2010. In the present scenario of high cost and raw material availability problems, thin-film solar cells would gain importance. A comparison of choice of material for thin-film solar cells is given below:

Gallium-arsenide (GaAs) – It has efficiencies above 30% but is very expensive, mainly suited for space application. Arsenic is toxic.

Cadmium teluride (CdTe) - Cadmium being toxic, its use in materials covering large surface area is not desirable.

Copper-indium-gallium-arsenide (CIGS) – Low availability of Indium, 0.08 ppm, increased costs in case of higher demand.

Amorphous Si (a-Si) - Low-cost but has low conversion efficiency.

Crystalline Si (c-Si) - Abundant raw-material (227000 ppm), high efficiency, thin-film technology provides potential to reduce the cost of wafer based cells, very attractive option to explore

In general, material availability is not expected to be a major constraint in the development of PV technologies. This is particularly true for silicon based technologies. Problems of material availability can possibly occur in the case of thin-film technologies involving scarce elements, such as indium and selenium (for example, $CuInSe_2$ material) and tellurium (CdTe material), particularly if PV growth and usage rates reach very high levels.

5.2 Wind Energy

Renewable energy sources such as wind energy are indigenous and can help reduce dependence on fossil fuels. With abundance of wind energy resources in many parts of the country, especially along the long coastline, electricity generation through wind energy provides a viable and environmentally friendly option. With an installed capacity of 3,595 MW, India is now recognized as a leading country in the world in wind energy. The reassessed gross wind power potential of the country stands at 45,000 MW. However, the technical potential, assuming 20% grid penetration, works out to be 13,000 MW. In Indian wind conditions, two areas of research could be identified:

- development of indigenous low cost turbines
- development of cost effective light weight rotor blades

5.3 Biomass

Biomass can potentially play a role in an effort to reduce net emissions of greenhouse gases. In fact, forests and agriculture have substantial potential for offsetting emissions by serving as a sink augmenting the GHG absorption, particularly CO_2, through photosynthesis. Forests and agriculture can offset GHG emissions by increasing production of biomass commodities, which can serve either as feedstock for electricity generating power plants or as blend/substitute for fossil fuel based gasoline. Biofuels mitigate GHG emissions because their usage spares fossil fuel use. The net emission savings from biomass amount to approximately 95 percent of the emissions from extraction and combustion of an equivalent amount of fossil fuels.

Biomass energy and biomass fuel are becoming more and more viable options for a sustainable future. Biomass is abundantly available in India in the form of animal waste, domestic waste, industrial organic waste, agriculture/farm waste, algae, slaughter house waste, etc. Some of the challenges are discussed as below:

5.4 Bio-ethanol

While cellulosic materials are utilized as value added products which fetch higher returns, the use of these materials for production of fuels raises the issue of 'food vs fuel'. Therefore, the lignocellulosic biomass is better option to be utilized for production of biofuels. The polysaccharides in lignocellulosic materials (crop residues, house wastes, leftovers from forestry industry) can be used for bioethanol production. The cellulose present cannot be enzymatically hydrolyzed to glucose without a physical and chemical pre-treatment to remove lignin layer. The physical and chemical protection of cellulose must be removed before subsequent hydrolysis and fermentation steps for bioethanol production.

The challenges in bioethanol production include:

- improvement of fermentative bacterial strains by genetic modifications
- development of reactor for simultaneous fermentation and ethanol recovery at elevated temperature.

5.5 Biodiesel

India has conceived a plan to launch 'National Biodiesel Mission'. The basic premise on which the mission was based was to identify and develop plant species having non-edible oil bearing seeds which could be cultivated in forests and waste lands, and have little or no water/irrigation requirements. '*Jatropha curcas*' is identified for following characteristics:

- Jatropha is foremost among 100 most productive oil bearing plants.
- It can be grown in low rainfall, low fertile and degraded, and waste land areas.
- Plants can be grown without much effort as animals do not eat it.
- Seeds can be collected after three years from plantation. It is claimed to yield about 10 tons of seeds from 5 year old plants.
- This corresponds to 1200 to 2000 liter oil/year/hectare.
- Unskilled workers and low level technique required.

However, agro climatic variability exists and therefore elite varieties of Jatropha are being identified/ developed to address the following challenges:

- Virus resistance
- Stress resistance
- High oil bearing seed yield per plant
- High oil content in oil bearing seeds
- Toxic deoiled seed cake causes disposal problem

Downstream processes include value addition to by-products like glycerine and de-oiled seed cake.

5.6 Algae

Algae are regarded as next generation biofuels as it not only provides solution to the fuel problem but also helps in fixing atmospheric CO_2 through photosynthesis. The advantages of algae biomass are as follow:

- Oil content in algae is reported between 10 to 50%
- Oil yield per year per hectare reported is 100,000 liter
- After extracting oil, biomass can be used for producing ethanol and biogas.

There are many issues indicated below which need to be resolved before tapping the full potential of this resource:

- Requires the high oil containing algae to be produced on large scale
- Requires harvesting and extracting oil quantitatively from algae
- Cultivation of algae on large scale needs to be developed
- Requires skilled persons and lot of engineering inputs for optimal growth of algae

- The factors that most influence cost are biological i.e. need for highly productive organisms capable of high levels of conversion of sunlight to biomass.

5.7 Energy Storage batteries/ batteries for vehicles

Storage of energy from renewable resources in itself is an important area of development. It would serve as a bridge between generation and utilization of renewable energy. The technology of storing renewable energy still suffers from a number of drawbacks. For example, batteries in PV systems are subject to performance losses that stem from five basic sources: 1) limited time and energy available to recharge the battery, 2) hardware failure, 3) inadequate battery maintenance, 4) improper system design, and 5) improper battery design.

The intermittent nature of the PV (as well as wind) energy source and its consequent effect on power system reliability are the issues which need interventions.

With increase in GDP and per capita income in India, there is an increase in sales of small cars as well as on transportation of consumer goods. As a result, concerns about air pollution, environmental degradation, and petroleum consumption are prompting policy makers to seek advanced transportation alternatives including hybrid cars/Electric vehicles. Automobile manufacturers are developing vehicles which either replaces the internal combustion engine with an electric motor or which provide a hybrid configuration with a combination of an IC engine (ICE) and an electric motor. In some concepts, fuel cells replace the IC engine in a hybrid vehicle configuration. Appropriate batteries having high energy capacity and power density with longer cycle life are the challenges for development. Lithium-ion batteries show promise for following reasons:

— Lithium-ion battery have higher volumetric and gravimetric energy than competing battery technologies, enabling smaller and lighter battery packs

— Lithium-ion also promises reduced cost, increase performance and longer life

The challenges before researchers in developing Li-ion batteries are:

— high cost of Li-ion batteries (2-4 times).

— scope of improvement in calendar life (about 15 years).

— intolerant to overcharge, crush and high temperature.

5.7.1 Hydrogen

Electricity generated from renewable resources can be converted and stored as hydrogen energy. Hydrogen has an excellent energy density by weight, which leads to it being used for spaceships like the space shuttle. It is believed that fuel cells, using hydrogen as a fuel, will be able to replace most internal combustion engines and will be able to solve most grid load balancing needs in the future.

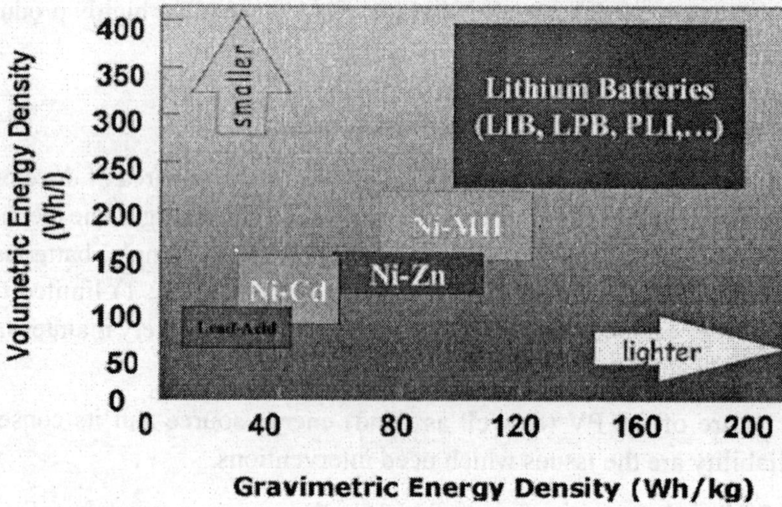

(Source: US Department of Energy)

FIGURE 4: Weight vs energy density of various batteries

In future, it is envisioned that the hydrogen economy would use non-fossil energy sources to produce hydrogen gas as a stored energy source for use in various sectors of the economy. If energy were in place to produce the hydrogen, then hydrogen could be used as an intermediate energy storage form to replace greenhouse-gas emitting alternatives like gas and oil.

Mainly due to cost reasons and partly also due to the not yet sufficiently advanced state of infrastructural development, hydrogen has not yet found its way into wider spread energy applications. The issues involved in hydrogen economy to take shape are many and there is lot be done before it could become a reality. The issues which need to be resolved cover all the areas of hydrogen economy, viz., production, storage, transportation and usage. The challenges which remain to be addressed are:

- Development of alternative (less costly) polymeric membranes
- Reduction or replacement of noble metals, primarily platinum as electrocatalyst
- Development and cost effective production of bipolar plates
- Development of storing materials with high hydrogen absorption capacity and easy release under thermal/chemical interferences.

5.8 Energy Efficient Devices/ Systems

The energy efficiency and demand side management could never be more emphasized than now in light of projected energy scenarios (Figure 5) and rising costs of energy. There exists ample scope of enhanced energy efficiency at all stages of energy extraction, conversion, transportation and consumption. While the efficiencies of technologies for extraction are not covered in this paper, the energy conservation and efficiency are important aspects.

5.8.1 Transport (Mobility)

Transport sector is identified as big contributor to GHG emissions. Many metros in India have switched to CNG fuel based transport system which is a cleaner fuel option. Also, many states in India have opted for 5% ethanol mix in petrol. Planning Commission of India has suggested the following measures to bring about the efficiency in transport sector:

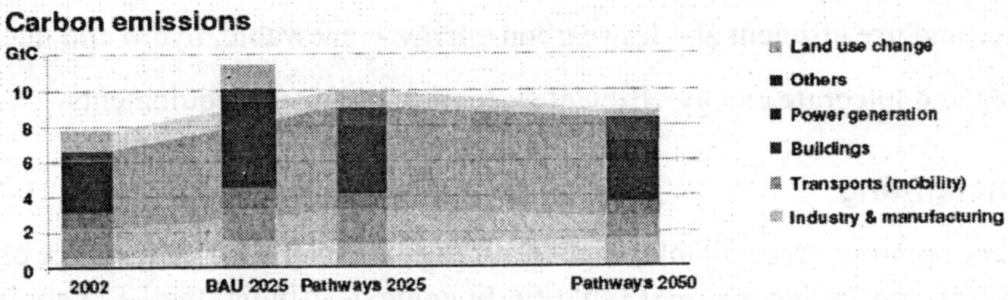

(Source: World Business Council for Sustainable Development)

FIGURE 5: Projected CO_2 emission scenarios

- promotion of mass transport system in metros in India
- promotion of fuel efficient vehicles
- diverting freight transport from road to railways
- increasing uses of biofuels in vehicles
- battery/fuel cell operated transport vehicles

The challenges would be development of non-fossil, low carbon liquid fuels, batteries and fuel cells along with vehicle engine modification to suit their use.

5.8.2 Energy efficient buildings

The global contribution from buildings towards energy consumption, both residential and commercial, has steadily increased, reaching figures between 20 and 40% in developed countries, and has exceeded the other major sectors, industrial and transportation. In 1990, the residential, commercial and institutional sector consumed 31% of global energy and emitted 1900 mega tones of carbon and by 2050 its share would rise to 38% and 3,800 mega tonnes, respectively. By having energy efficient buildings, it is said that the global emissions can be reduced by 40% by 2050.

According to World Business Council for Sustainable Development (Figure 1), the buildings would consume about one-third of the world's energy. If current trends continue, by 2050, buildings will consume almost as much energy as industry and transport combined. Worldwide energy consumption for buildings is expected to grow by 45% from in 2025.

Growth in population, increasing demand for building services and comfort levels, together with the rise in time spent inside buildings, assure the upward trend in energy demand will continue in the future. For this reason, energy efficiency in buildings is today a prime objective for energy policy at regional, national and international levels. The actions required to bring about the higher energy efficiency in buildings include:

- reduction in energy demand - by life style change, insulation and proper ventilation of building.
- produce and use efficient and low carbon energy - renewable, hydro and nuclear.
- upgrade and integrate energy efficient systems - appliances, equipments.

5.8.3 Efficient lighting

While a number of options are available, Light Emitting Diodes (LEDs) as a source of light appears to have promise in conserving energy (Table 5). In context of India, the LED application would bring relief to the burgeoning energy requirement for road lighting, industry, office, home lighting and lighting of places for vigilance purposes. It would also pave the way for rural electrification through solar energy storage using PV systems. The energy savings are indicated in Table 6.

TABLE 5: Efficiencies of various lighting devices

Light Source	Efficacy (lumens/Watt)
Incandescence	18-20
Sodium vapour	60-70
Mercury vapour	40-120
Metal Halide	50-60
CFLs	80-125
LED	20-60 (but high efficiency 85-90% as the energy is not lost as heat; R&D to enhance lumens per watt output)

TABLE 6: Energy savings by various lighting devices

Light Source	Annual Saving	Investment	Pay Back Period
Tube Light	Nil	Nil	Nil
CFLs	47450	27,570	6 months
LED	1,01,105	1,13,000	14 months

(Source: N.B. Soni. of Theoretical & Applied Information Tech., p.1083, 2008)

The commercial and societal advantages of developing LEDs/OLEDs could be:

- Substantial savings in electrical energy consumption;
- Reduction in carbon dioxide pollution; and
- The creation of a new lighting industry, with many new, high quality jobs.

5.8.4 Coatings

Coating is an emerging area which has potential to significantly reduce power consumption on account of lighting and cooling during day time in hot weather condition that prevail in India. Two types of coatings are being developed.

a. For glass panels in building for high light throughput

b. For walls and roofs for reflection of heat radiations to keep building cool during summers

The glass panels are increasing used in large commercial buildings like malls for aesthetics and high sunlight throughput. However, these allow heat radiations in sunlight to transmit which in turns heats up the environment inside. In order to keep the light throughput high and reflect the heat radiations, smart coatings are developed. Also, the walls and roofs of the buildings are heated by sun rays during summers which raises the inside temperature as well as causes heat island effect. Suitable coating reflecting infrared radiations are being seriously looked into to reflect infrared radiations and keep the inner building cool by 5-7°C.

6.0 CSIR: SOLUTIONS THROUGH NETWORK APPROACH

Council of Scientific and Industrial Research (CSIR) is one of the world's largest public funded organization having 37 laboratories across India with about 4500 scientists and technocrats working in diverse areas of science and technology. CSIR has strong capabilities in all core areas of sciences. Development of renewable energy (excluding hydro) happens at the periphery of chemical, biology and physical sciences. This requires multidisciplinary inputs and networking of resources. SWOT of CSIR's capabilities in development of renewable energy are shown in Table 7. In its futuristic programmes (XI FYP) CSIR has initiated programmes on both supply side and demand side energy management.

The other aspect of CSIR development programmes is providing total solution through its multi-dimensional capabilities. Two case studies are discussed below to emphasize the point:

Case Study 1: ***Developing End-to-end-Technologies for Biodiesel***

CSIR has been working on this alternative source of fuel for many years. The networked approach of CSIR (Table 8) is illustrated by the fact that capabilities of many laboratories are utilized in developing total technology solutions for production and utilization of biodiesel from jatropha

TABLE 7: Multidisciplinary Capabilities of CSIR in renewable energy

Resource	No. of Patents/ Publications*	Opportunity/ Challenges (S&T interventions)	Networks and Multidisciplinary capability in CSIR labs	
Biomass	Resource Personnel -53 Publication-31 Patents -9 (*source- Energy Research and CSIR 2008)	- Biomethanation of waste (Tannery, poultry, Municipal waste, etc) - Bioethanol from hemi-cellulosic waste viz. baggasse - Bioethanol from Algae - Elite plant variety development - Biodiesel from Jatropha , Pongamia, Algae - Liquefaction of biomass to liquid fuel / producer gas	CLRI NCL IIP CSMCRI NIIST NBRI CSMCRI IICT IIP NCL NIIST IIP CIMFR NCL	(Chemical/ Biological) (Chemical/ Biological) (Chemical/ Biological) (Chemical/ Biological) (Chemical/ Biological) (Biology) (Chemical/ Biological) (Chemical/ Biological) (Chemical/ Biological) (Chemical/ Biological) (Chemical/ Biological) (Chemical/Engineering) (Chemical/Engineering) (Chemical/Engineering)
Solar	Resource Personnel -91 Publication -60 Patents -2	- Alternatives to silicon based SPV - Solar-thermal technologies - Nano materials for SPVs - High efficiency/low cost SPVs - Storage batteries	NPL NCL NAL CSMCRI CECRI NPL	(Physical/Material) (Chemical/Material) (Engineering/Material) (Chemical/Material) (Chemical/Engineering) (Physical/Chemical)
Wind	Resource Personnel -27 Publication-17 Patents -5	- Light weight / high strength national for rotor blades - Low cost turbines - Storage batteries	NAL NAL NPL	(Engineering/ Materials) (Engineering) (Physical/ Chemical)

TABLE 8: CSIR networks in developing biodiesel technologies

S.N.	Activity	Capabilities in CSIR Labs						
		NBRI	NEIST	CSMCRI	IICT	MERADO	IIP	NCL
1.	Generation of Plantation Material	✓	✓	✓	.	.	.	.
2.	Storage and transportation	✓	.	.	✓	.	.	.
3.	Oil extraction from different plants	✓	.	✓	✓	✓	.	.
4.	Blending technologies	.	.	.	✓	.	✓	.
5.	Biodiesel production/ transesterification	.	.	✓	✓	.	✓	✓
6.	Quality Control	.	.	✓	.	.	✓	.
7.	Testing of blends of biodiesel/ diesel for emissions and engine performance	.	.	✓	.	.	✓	.
8.	Value addition/ processing of by-products (Glycerine/ deoiled cake)	.	.	✓	✓	.	✓	.

a. Generation of plantation material

- Inadequate seed availability is a major problem. *Jatropha curcas* was never grown in organized sector and there is very little information available on germplasm, if at all, besides the problem of composition and consistency. One germplasm, which grows in one part of India, may not fruit in other region. Also, it can have wide variations in seed yield and oil content.
- NBRI (National Botanical Research Institute), Lucknow and CSMCRI, Bhavnagar have made a concerted effort in collecting germplasm from various states, creating nursery trails of germplasm, nursery experiments and raising successful nurseries of selected materials.

b. Extraction of oil: This is an important area wherein the oil purity and yield would have an impact on the cost of biodiesel and the byproduct. NBRI, Lucknow has developed protocols for extraction of oil from Jatropha. IICT, Hyderabad has the experience and capability of developing efficient oil expellers. MERADO, Jalandhar has expertise in developing innovative indigenous oil expellers.

c. Storage and transportation: No authentic data is available on shelf life of Jatropha seeds, oil and biodiesel in India. NBRI, Lucknow has been engaged in generating data on storage of oil under various conditions. IICT, Hyderabad has expertise and experience in similar activities, which can be used to generate data on shelf life of seeds, oils and biodiesel.

Table 9: CSIR Laboratory networks in hydrogen economy programmes

S.N	Development Aspect	Activity	Capabilities in CSIR Labs								
			NCL	IICT	IIP	CECRI	NEERI	CMERI	CIMFR	CSMCRI	CGCRI
1	Generation	Ethanol/ Methanol/ Methane Reforming	✓	✓	✓	.	.	.	.	.	.
		Electrolysis of Water	✓	.	.	✓	.	.	.	.	.
		Photoelectrochemical conversion	.	.	.	.	✓	✓	.	.	.
		Coal gasification	.	.	✓	.	.	.	✓	.	.
		Biomass gasification	.	.	✓	.	.	.	✓	.	.
2	Storage	Carbon Nanotubes	✓	.	.	.	.	.	.	✓	.
		Metal-organic frameworks	✓	✓	.	.	✓	.	.	.	.
		Mesoporous materials	✓	.	.	.	✓	.	.	✓	.
3	Utilization	Polymer Electrolyte Fuel Cell	✓	✓	.	✓	.	.	.	.	.
		Solid Oxide Fuel Cell	.	.	.	✓	.	.	.	.	✓
		Polymer Electrolyte									
		Direct Methanol Fuel Cell	✓	✓	.	✓	.	.	.	.	.

d. **Biodiesel production/transesterification/blending:** CSIR laboratories have been doing R&D on transeterification and have successfully converted jatropha oil to biodiesel that met international specifications. Many CSIR laboratories have successfully developed catalysts for transesterification of oils even having high Free Fatty Acid (FFA) contents.

e. **Testing of blends:** Indian Institute of Petroleum (IIP), Dehradun and CSMCRI, Bhavnagar, have capabilities of testing blends of Diesel/Biodiesel for emissions and engine performance. While IIP is testing blends on engine performance vis-à-vis emissions, CSMCRI, has studied the engine performance of vehicle run over 10,000 km.

f. **Effective generation/ utilization of by products:** Value addition to the Jatropha oil cake, which is a non-edible matter, is another important area and will have impact on the cost of biodiesel. Many CSIR laboratories are working on deoiled cake for value addition. Glycerine, which is another by-product of transesterification, needs to be converted into valuable intermediate for chemical industry.

Case Study 2: ***Hydrogen Economy: Production, storage and utilization of hydrogen***

Various laboratories of CSIR are presently engaged in different aspects of hydrogen economy as indicated in Table 9. As indicated in the table, the competency/expertise developed within CSIR labs, in various emerging areas is being networked to provide total technology solutions for hydrogen economy. The networks are being used for:

a) **Development of hydrogen generation and storage systems**: This involves development of systems for more efficient production of hydrogen from water, biomass and fossil fuels. It also involves synthesis of wide range of carbon-based hybrid materials; carbon nanotubes based composites, novel metal hydride systems for hydrogen storage.

b) **Development of fuel cell technology**: Work on various technologies of fuel cells has been undertaken by CSIR Laboratories. These technologies include:

   i. ***Polymer Electrolyte Fuel Cells***- involves development of new cost effective ion exchange polymers as well as catalysts

   ii. ***Direct Methanol Fuel Cells***- involves development of micro-fluidic based DMFCs. Development of prototypes with indigenous bipolar plates, carbon clothes and other peripherals.

   iii. ***Solid Oxide Fuel Cells*** - to develop new interconnected ceramic materials with improved conductivity and ability to withstand higher thermal and mechanical stresses and to address fabrication aspects of SOFC.

   iv. ***Direct Borohydride Fuel Cells***- involves development of unique electrode-electrolyte combinations and polymer composite electrolyte with high protonic conductivity.

7.0 CONCLUSIONS

To avoid "abrupt and irreversible" climate change, the world needs a major decarbonisation of energy system. Limiting temperature rise to 2°C will require significant emission reductions in all regions, as well as technological breakthroughs. Mitigating climate change will have significant benefits such as improved energy security and reduced local air pollution.

India needs to develop technologies that aim to enhance access to clean and efficient energy, improve energy security, and promote environmental protection. While searching for alternate technologies and energy resources, India should not lose the responsibility of social development. This means meeting the needs that are essential for economic growth and fighting poverty while at the same time leaving a smaller environmental footprint.

CSIR is engaged in developing R&D solution by pooling in expertise and resources within CSIR system. The networking of expertise available within CSIR on different aspects of energy provides unique advantage in conducting R&D in Energy related areas.

8.0 ACKNOWLEDGEMENTS

The author is thankful to Prof. S. K. Brahmachari, DG CSIR and Secretary, DSIR, Govt. of India for giving the opportunity to attend and contribute to the conference. The author also acknowledges the constant encouragement and useful suggestion by Dr. Naresh Kumar, Head R&D Planning Division in preparing this paper. Heartfelt thanks are also goes to NAM S&T Centre, New Delhi for providing all necessary information and help to attend the Conference.

REFERENCES

Annual Reports. Ministry of New & Renewable Energy (MNES), Govt. of India. www.mnes.nic.in

Bureau of Energy Efficiency, Ministry of Power, Govt. of India. www.bee-india.nic.in

David Howell. Energy Efficiency and Renewable Energy, US Department of Energy, June 2007

Energy Research and CSIR. CSIR, New Delhi, (2008)

Ernst & Young. Background Paper on Indian Power Sector. India Energy Summit 2007.

Fourth Assessment Report: Intergovernmental Panel on Climate Change (IPCC-2007). www.ipcc.ch

Kirit S. Parikh. Integrated Energy Policy, Planning Commission, Govt. of India. (2006). www.planningcommission.nic.in

S.S. Murthy. Indian J. Power & River Valley Development (Annual Review), 91-97, (2008)

Narendra B Soni. J. Theoretical & Applied Information Technology, 1083-1087, (2005-2008)

National Action Plan on Climate Change (NAPCC), Govt. of India (2008)

World Business Council for Sustainable Development, Switzerland. www.wbcsd.org

World Energy Outlook 2007- China and India Insights (IEA). www.iea.org

World Energy Outlook 2008 (IEA). www.iea.org

ABBREVIATIONS

IEA	International Energy Agency
WEO	World Energy Outlook
ETP	Energy Technology Perspective
IPCC	Intergovernmental Panel on Climate Change
GHG	Green House Gases
BEE	Bureau of Energy Efficiency
MNRE	Ministry of New & Renewable Energy
WBCSD	World Business Council for Sustainable Development
NAPCC	National Action Plan on Climate Change
XI FYP	Eleventh Five Year Plan
SWOT	Strength, Weaknesses, Opportunity & Threats
NSG	Nuclear Supplier Group
CSIR	Council of Scientific & Industrial Research
CECRI	Central Electrochemical Research Institute
CGCRI	Central Glass and Ceramic Research Institute
CIMFR	Central Institute of Mining and Fuel Research
CLRI	Central Leather Research Institute
CMERI	Central Mechanical Engineering Research Institute
CSMCRI	Central Salt & Marine Chemicals Research Institute
IICT	Indian Institute of Chemical Technology
IIP	Indian Institute of Petroleum
NBRI	National Botanical Research Institute
NCL	National Chemical Laboratory
NEERI	National Environmental Engineering Research Institute
NIIST	National Institute of Interdisciplinary Science & Technology
NPL	National Physical Laboratory

15

Status and Potential of Renewable Energy in India

Arvind Kumar* **and Rajiv Sharma****
Department of Science and Technology, New Delhi 110016, India
*E-mail: arvindkumar25@rediffmail.com
**E-mail: rajivdst@nic.in

The increasing population and fast-growing economy of India have necessitated a greater demand for energy which is mostly met by fossil fuels. The excessive use of fossil fuel is causing considerable environmental and resource degradation. Most of the developing countries including India are heavily dependent on imported energy particularly oil and petroleum products. India has had a negative energy balance for decades, which has forced the purchase of energy from outside the country. This dependence on fossil fuels, which are imported, exacerbates its foreign exchange debt burden. For India, new and renewable energy development and deployment is of great importance from the point of view of long term energy supply security; decentralisation of energy supply, particularly for the benefit of the rural population; environmental benefits and sustainability. Renewable energy has been appropriately given the central place in India's National Action Plan on Climate Change. Intergovernmental Panel on Climate Change (IPCC) in its Fourth Assessment Report on 'Mitigation of Climate Change' identified renewable energy technologies as one of the key mitigation technologies for energy supply, transport, buildings, agriculture and waste management. It was also observed that renewable technologies are available for mitigating the climate change, however, this require appropriate policy and financial support. These factors, along with the country's large endowment of renewable resources, suggest that the development of renewable energy will go a long way in meeting the challenge of providing clean power in India. It would also be a great opportunity for the country to increase the share of renewable energy in the overall energy mix. Moreover, to achieve the goal mission 2012 'Power for All' and to achieve GDP growth rate of 8 per cent, and realize the dream of "India shining", it is necessary that the contribution of renewable energy should be enhanced. This paper discussed the present energy scenario in India, status of renewable energy programmes, research and technology development, government policies and issues related to renewable energy financing, institutional strengthening and infrastructure development.

KEY WORDS: *Renewable energy, Energy security, Global warming, GHS, Climate change, Solar energy, Wind energy, Biomass energy, SHP*

1.0 INTRODUCTION

Energy is a basic requirement and important input for economic development. Every sector of Indian economy including agriculture, industry, transport, commercial, and domestic need inputs of energy. India's increasing population and fast-growing economy have necessitated a greater energy demand. As a result, consumption of energy in all forms has been steadily rising. This growing demand of energy resulted in increasingly dependence on fossil fuels such as coal and, oil and gas. Rising prices of oil and gas, and potential shortages in future led to concerns about the security of energy supply needed to sustain our economic growth. Increased use of fossil fuels also causes environmental problems both locally and globally. This has serious economic, social and environmental implications. After the Rio Summit in 1992, a lot of concerns have been expressed about environmental degradation, global warming, emission of green house gases, ozone layer depletion and other global issues related to the environment. Consequently, lots of interests have grown for developing, promoting and adopting clean and environmentally friendly technologies. The renewable energy are generally environmentally friendly and are likely to play an increasing role in near future because of their capabilities to aid economic growth in underdeveloped region beyond the reach of centralized power generation and distribution system, and their ability to help combat both global warming and other environmental problem.

Since conventional energy sources such as coal and crude oil in the country are limited, there is an urgent need to focus attention on development of renewable energy sources and use of energy efficient technologies. The exploitation and development of various forms of energy and making energy available at affordable rates are the need of the hour. Against this background, the country urgently needs to develop a sustainable path of energy development. Promotion of energy conservation and increased use of renewable energy sources are the twin planks of a sustainable energy supply. Fortunately, India is blessed with a variety of renewable energy sources, the main ones being biomass, biogas, the sun, wind, and small hydro power. Municipal and industrial wastes can also be useful sources of energy, but are basically different forms of biomass. Today, India is one of the few leading countries in the development and utilization of renewable energy. The country is blessed with various sources of non-conventional energy and its sustainable utilization would benefit people in the far-flung regions of the country.

India has made sustained efforts during the past quarter century and as a result, a number of technologies and devices have been developed and have become commercially available. These include biogas plants, improved wood stoves, solar water heaters, solar cookers, solar lanterns, street lights, pumps, wind electric generators, water-pumping wind mills, biomass gasifiers and small hydro-electric generators. Grid interactive renewable power development in the country is taking place at greater pace. Energy technologies for the future such as hydrogen, fuel cells, and bio-fuels are being actively developed.

Notwithstanding the above, due to its vast market potential for renewable energy projects, and a relatively well-developed industrial, financing and business infrastructure, India is perceived as an excellent country for developing Clean Development Mechanism (CDM) projects. As such, India has emerged as one of the most favoured destination for CDM projects globally, with renewable energy projects having the major share. National renewable energy plans offer ample opportunity for CDM projects and technological innovation. CDM and the emerging carbon market in general have potential to contribute to the financial viability of renewable energy projects.

2.0 ENERGY SCENARIO IN INDIA

The total primary energy supply in India was 565.82 Mtoe (million tonnes of oil equivalent) in year 2006 and accounted for 4.8% of the world's primary energy supply (IEA, 2008). While commercial primary energy grew at 5.3% over the period, non-commercial energy grew at only 1.6%, which is a reflection of industrialisation. As a result, the share of commercial energy grew from 28 per cent in 1950 to around 65% in 2008 with an associated decline of non-commercial energy. Per capita energy consumption was just 28.3% of the world average at slightly over 510 kg oil equivalent. India accounted for 4.9 per cent (557.97 TWh) of the world's electricity consumption in 2006 (IEA, 2008) with an installed capacity of 147,402.81 MW as of December 2008 (Table 1). Of this, 93,392.64 MW is accounted for by thermal power plants, 36,647.76 MW by large hydro plants, 4,120 MW by nuclear, and the balance from renewable sources (Table 1).

TABLE1: Installed Capacity for different sources of energy

Sources of power	Capacity MW	%
Total Thermal	93,392.64	64.6
Coal	77,458.88	53.3
Gas	14,734.01	10.5
Oil	1,199.75	0.9
Hydro Power	36,647.76	24.7
Nuclear	4,120.00	2.9
RES**	13242.41	7.7
Total	147,402.81	

As on 31.12.2008, source: CEA.
** Renewable Energy Sources (RES) include SHP, BG, U&I and Wind Energy.

Per capita electricity consumption was 503 kWh as against the world average of 2659 kWh (IEA, 2008). Despite the significant growth in electricity generation, shortage of power continues to exist primarily due to the growth in power demand outstripping the growth in generation and generating capacity addition. The country experienced an estimated 10 to 12% energy shortage during normal and 15-18% shortage of peaking power. Moreover, the 2001 census recorded 12.5% of urban households and 56.5% of rural households as still un-electrified. There is also considerable environmental and resource degradation because of a higher dependence on fossil fuels. This dependence on fossil fuels, which are imported, exacerbates its foreign exchange debt burden. These factors, along with the country's large endowment of renewable resources, suggest that the development of new renewable energy (NRE) will go a long way in meeting the challenge of providing clean power in India. When considering NRE power options, both grid-connected power and distributed power generation are important, especially for India, which has a high population living in rural areas. The status of different renewable technologies and related issues are discussed below.

3.0 STATUS OF RENEWABLE ENERGY RESOURCES AND TECHNOLOGY IN INDIA

Renewable energy technologies have in the last two decades not only matured enough as supplements to conventional power, they also provide viable solutions in the off-grid mode to meet the electricity needs of remote villages in the country. Wind, small hydro and biomass are the major sources of grid interactive renewable power. The grid interactive and offgrid renewable power installed capacity as at September 2008 was about 13,451 and 279 MW, respectively (MNRE, 2008). Grid interactive renewable power projects are essentially private investment driven and almost all the renewable power capacity addition is coming through this route. Since renewable power would need to compete with conventional electricity, the focus is towards reducing the capital cost of projects and increasing their capacity factors, with the eventual aim of reducing the unit cost of renewable power generation.

3.1 Grid Interactive Renewable Power

Wind Energy

India is the fourth largest producer of wind power in the world after Germany, Spain and USA with a wind power generation achievement of 9,587 MW, most of which come through commercial projects (MNRE, 2008). This sector is growing at over 35 % over last four years.

Resource assessment and potential: The wind speeds in India are in the low wind regime with average wind speeds between 17 and 24 km/h. The potential for wind power generation for grid interaction has been estimated at about 48,500 MW taking sites having wind power density greater than 250 Wm^{-2} at 50 m hub-height with 3% land availability in potential areas for setting up wind farms at 12 ha/MW. For wind resource assessment in the country, 1050 wind monitoring and wind mapping stations were established covering 25 States and Union Territories. So far, 216 potential sites have been identified in the country with wind power density in excess of 200 Wm^{-2} at 50 m height. Wind has the highest potential in the country and is expected to contribute 60% of the power generation from renewable sources.

Demonstration and commercial projects: An aggregate demonstration wind power capacity of 71 MW has been established at 33 locations in 9 states (Table 2). Demonstration projects were implemented through the State Governments, State Nodal Agencies or State Electricity Boards. Demonstration projects were taken up only in those states where commercial wind power projects have not yet been initiated/taken off. A total capacity of 9587 MW have been established up to November, 2008, mainly in Tamil Nadu, Gujarat, Maharashtra, Andhra Pradesh, Karnataka and Rajasthan.

TABLE 2: State-wise demonstration wind power projects

State	Demonstration Projects (MW)
Andhra Pradesh	5.4
Gujarat	17.3
Karnataka	7.1
Kerala	2.0
Madhya Pradesh	0.6
Maharashtra	8.4
Rajasthan	6.4
Tamil Nadu	19.4
West Bengal	1.1
Others	3.3
Total	**71.0**

Wind electric generators of unit sizes between 225 kW and 1.65 MW have been deployed across the country. A cumulative total of over 54 billions units of electricity have been fed to the State Electricity Grids.

Technology development and manufacturing base: Wind electric generators are being manufactured in the country by a dozen manufacturers, through joint ventures under licensed production, subsidiaries of foreign companies under licensed production and Indian companies with their own technology. An indigenization level up to 70% has been achieved in machines of unit sizes up to 500 kW. The import component is somewhat more in higher capacity machines. The current annual production capacity of domestic wind turbines is about 2,500 MW. Two types of wind turbines namely stall regulated and pitch regulated are being deployed in the country and abroad for grid-interactive power. The stall regulated wind turbines have fixed rotor blades whereas pitch regulated wind turbines have adjustable rotor blades that change the angle of attach depending upon wind speed. Wind turbines are also available with lattice, steel tubular and concrete tubular towers. India has established a good manufacturing base with more than 12 manufacturers of wind turbines and allied equipment. A new concept of mega wind farms owned by the private sector is being tested in India to increase the penetration of wind power, and invite greater participation from the private sector. The advantage of such approach will be reduced capital cost. Mega wind farms can also negotiate a better power purchase agreement with the utilities.

Guidelines for setting up projects: The MNRE has been issuing guidelines for wind power development since July 1995 in order to bring about balanced growth of the sector. These guidelines relate to preparation of detailed project reports, micro-siting, selection of wind turbine equipment, operation & maintenance, performance evaluation, etc.

Promotional policy: A package of fiscal and financial incentives is available which include concessions such as 80% accelerated depreciation, concessional custom duty on specified items, excise duty exemption, sales tax exemption, income tax exemption for 10 years, etc. In addition, State Electricity Regulatory Commissions (SERCs) are determining preferential tariffs. Indian Renewable Energy Development Agency (IREDA) provides loan for setting up wind power projects.

Small Hydro Power

The hydropower is a renewable, non-polluting and environmentally friendly source of energy. Hydro power projects are generally categorized in two segments i.e. small and large hydro. In India, hydro projects up to 25 MW capacities have been categorized as Small Hydro Power (SHP) projects. While Ministry of Power, Government of India is responsible for large hydro projects, the mandate for the subject small hydro power (up to 25 MW) is given to Ministry of New and Renewable Energy(MNRE). Small hydro power projects are further classified as, micro (< 100 kW), mini (101-2000 kW) and small hydro (2001-25000 kW) power project. Out of the total power generation installed capacity of 147,402.81 MW (December, 2008) in the country, hydro power contributes 36,647.76 MW *i.e.* about 25 per cent excluding small hydro power project.

Small Hydro Power (SHP) Programme is one of the thrust areas of power generation from renewable in the country. It has been recognized that small hydropower projects can play a critical role in improving the over all energy scenario of the country and in particular for remote and inaccessible areas. The government is encouraging development of small hydro projects both in the public as well as private sector. Equal attention is being paid to grid-interactive and decentralized projects. The aim is that out of the total grid interactive power generation capacity that is being installed, 2% should come from mini/small hydro. The focus of the SHP programme is to lower the cost of equipment, increase its reliability and set up projects in areas which give the maximum advantage in terms of capacity utilisation.

Potential: An estimated potential of about 15,000 MW of small hydro power projects exists in India. About 4,861 potential sites with an aggregate capacity of 12,841.81 MW for projects up to 25 MW capacity have been identified in 29 states (Table 3) (MNRE, 2008).

TABLE 3: State wise identified small hydel sites up to 25 MW capacity (as on 30.6.2007)

S.No.	Name of State	Identified number of sites	Total capacity (MW)
1	Andhra Pradesh	377	250.50
2	Arunachal Pradesh	452	1243.47
3	Assam	40	119.54
4	Bihar	74	149.35
5	Chhatisgarh	132	482.82
6	Goa	4	4.60
7	Gujarat	287	186.37
8	Haryana	23	36.55
9	Himachal Pradesh	457	2019.03
10	Jammu & Kashmir	208	1294.43
11	Jharkhand	89	170.05
12	Karnataka	468	1940.31
13	Kerala	207	455.53
14	Madhya Pradesh	85	336.33
15	Maharashtra	221	484.50
16	Manipur	99	91.75
17	Meghalaya	90	197.32
18	Mizoram	53	135.93
19	Nagaland	84	149.31
20	Orissa	206	217.99
21	Punjab	204	270.18
22	Rajasthan	55	27.82
23	Sikkim	70	214.33
24	Tamil Nadu	155	373.46
25	Tripura	10	30.85
26	Uttar Pradesh	211	267.06
27	Uttaranchal	354	1478.24
28	West Bengal	141	213.50
29	A&N Island	5	1.15
	Total	**4861**	**12,841.81**

The central government is giving financial support to state governments/agencies for identification of new potential SHP sites & preparation of state perspective plan.

Installed capacity: The total installed capacity of small hydro power projects as at April, 2007 was 1,975 MW from 602 projects and 219 projects with aggregate capacity of 650 MW are under construction (Table 4). The total installed capacity as at September 2008 was 2,220.99 MW (MNRE, 2008). The small hydro is also seen as a potential source along with PV for providing decentralized power in remote areas. A target of adding about 1,400 MW during the 11th Plan (2007-2012) has been fixed.

Further, a comprehensive resource assessment for all the renewable energy sources including small hydro and mapping of potential sites/locations on a GIS platform is receiving utmost attention. The aim is to map renewable energy potential in the country and bring it on a GIS platform with information necessary to take investment decisions to set up projects. Computer models have been developed that takes into account the regional flow duration curves, geological and seismological data, vegetation cover etc. for identification of potential sites. Software packages have been developed incorporating regional hydrological models to enable users to rapidly estimate the hydropower potential and other salient features of potential sites. Ramachandra et.al. (2005) have developed s spatial decision support system for **Wind energy potential assessment.**

Private sector participation: The major thrust is to achieve development of the small hydro projects through private sector participation. Beginning of the 21st century saw near commercialization in the small hydro sector. Private sector entrepreneurs found attractive business opportunities in small hydro and state governments also felt that the private participation may be necessary in tapping the full potential of rivers and canals for power generation. The private sector has been attracted by these projects due to their small adoptable capacity matching with their captive requirements or even as affordable investment opportunities. In line with Government of India policy, 18 states have announced their policy for inviting private sector to set up SHP projects. The Government of India announced the Electricity Act in 2003, Electricity Policy in 2005 and Tariff Policy in 2006 to create a conducive environment for investments in the power sector. Small hydropower projects are now governed by these policies and the tariff is decided by the State Electricity Regulatory Commissions (SERCs) as per the Tariff Policy.

Incentives: A package of incentives which includes fiscal concessions such as customs duty concessions, income tax exemptions on projects for power generation for 10 years etc. are available to small hydro power projects. Some states are also giving sales tax and electricity tax exemptions. The State Electricity Regulatory Commissions in Andhra Pradesh, Himachal Pradesh, Karnataka, Maharashtra, Chhattisgarh and Punjab have announced preferential tariff. New SHP projects both in the public and private sector are eligible for subsidy. Subsidy is also provided for renovation and modernization of existing as well as languishing SHP projects only in the public sector. A special incentive package has been developed for the promotion of the SHP programme in the north-eastern states, Sikkim, J&K, Himachal Pradesh and Uttarakhand. In case of private sector projects, subsidy

TABLE 4: State-wise details of the installed/under installation projects

S.No	State	Projects Installed		Projects under Implementation	
		No	Capacity (MW)	No	Capacity (MW)
1	Andhra Pradesh	57	178.850	11	17.50
2	Arunachal Pradesh	68	45.240	56	41.82
3	Assam	3	2.110	4	15.00
4	Bihar	7	50.400	9	7.60
5	Chattisgarh	5	18.050	1	1.00
6	Goa	1	0.050	-	-
7	Gujarat	2	7.000	-	-
8	Haryana	5	62.700	1	6.00
9	Himachal Pradesh	61	141.615	13	64.00
10	J&K	32	111.830	5	5.91
11	Jharkhand	6	4.050	8	34.85
12	Karnataka	69	416.500	17	103.50
13	Kerala	16	98.120	5	39.55
14	Madhya Pradesh	9	51.160	5	39.90
15	Maharashtra	29	209.330	3	13.50
16	Manipur	8	5.450	3	2.75
17	Meghalaya	3	30.710	3	1.70
18	Mizoram	16	17.470	3	15.50
19	Nagaland	9	20.670	5	12.20
20	Orissa	6	7.300	8	60.93
21	Punjab	29	123.900	-	-
22	Rajasthan	10	23.850	-	-
23	Sikkim	14	39.110	4	13.20
24	Tamil Nadu	14	89.700	4	13.00
25	Tripura	3	16.010	-	-
26	Uttar Pradesh	9	25.100		
27	Uttarakhand	87	75.670	35	61.25
28	West Bengal	23	98.400	16	79.25
29	A&N Islands	1	5.250	-	-
	Total	**602**	**1975.595**	**219**	**649.91**

is released after successful commissioning and commencement of commercial generation from the project, to the financial institution (FI) for offsetting it against the term loan provided to the developer. Current subsidy for the grid interactive small hydro power projects upto 25 MW capacity, both in public as well as private sector, are given as following:

	Special Category States (NE Region, Sikkim, J&K, HP & Uttaranchal)	Other States
Small Hydro Power projects	Rs 2.25 crores ×(C MW)^0.646	Rs 1.50 crores × (C MW)^0.646

Where C is the capacity of the project in MW. The subsidy is released after completion of the project. The project when commissioned is required to be tested for its performance by AHEC, IIT Roorkee. The subsidy is released after project attaining 80 per cent of the envisaged generation, as per DPR for three consecutive months.

Total 19 states have announced policies for setting up commercial SHP projects through private sector participation. The facilities available in the States include wheeling of power produced, banking, buy-back of power, facility for third party sale, etc. There is scheme for providing financial assistance for Renovation and Modernisation and capacity up rating of Small Hydro Power projects in government sector to extend the life of these stations with improved performance and reliability.

Standards, manufacturing base and consultancy services: In order to ensure project quality/ performance, emphasis has been given to adhere to IEC/International standards for equipment and civil works. The subsidy available is linked to use of equipment manufactured to IEC or other prescribed international standards. The equipment in the project is required to confirm to the IEC standards. Recently AHEC, IIT Roorkee has been given assignment to revisit the existing standards and come out with standards/manuals/guidelines for improving reliability and quality of small hydro power projects in the country. India has a wide base of manufacturers of equipment for small hydro power projects. State-of-the-art equipment is available indigenously. Thirteen manufacturers fabricate almost the entire range and type of SHP equipment. Manufacturers' capacity is estimated at about 250 MW per year. In addition, there are about 5 manufactures that are producing micro hydel and watermill equipment. Consultancy services in the field of small hydro projects are available from a number of Government/private consultancy organizations. The government is strengthening technical institutions to provide such services. For example AHEC and IIT Roorkee are providing full range of technical services in the field of small hydro including survey and investigation, DPR preparation, project design etc. On site testing facility has been created at AHEC to test SHP stations for their performance.

Biomass Power

Biomass power/cogeneration programme is being implemented with the objective to promote technologies for optimum use of country's biomass resources for power generation. Different biomass sources are available in different regions. In addition to agro-residues that are available, it is possible to have dedicated plantations on wasteland or degraded lands that are not normally used for agriculture. The main conversion process for power generation from biomass is thermo chemical or biochemical. The thermo-chemical processes involve combustion/cogeneration, gasification or pyrolysis. The technologies being promoted include combustion/cogeneration and gasification either for power in captive or grid connected modes or for heat applications.

Biomass Resource Potential: The current availability of biomass in India is estimated at about 540 million tonnes per year including residues from agriculture, agro-industrial residue, forestry and plantations. Principal agricultural residues include rice husk, rice straw, bagasse, sugar cane tops and leaves, trash, groundnut shells, cotton stalks and mustard stalks, etc. It has been estimated that about 70-75% of these residues are used as fodder and fuel for domestic cooking and for other economic purposes leaving behind 120–150 million tonnes of usable agro industrial and agricultural residues per year which could be made available for power generation. By using these surplus agricultural residues, about 16,000 MW of grid quality power can be generated with presently available technologies. In addition, about 5,000 MW of power can be produced, if all the 550 sugar mills in the country adopt modern techniques of co-generation. Thus, the country is having biomass power potential of about 21,000 MW.

Achievements: A total of 203 biomass power and cogeneration projects aggregating to 1677 MW capacity has been installed in the country for feeding power to the grid (Table 5). In addition, around 171 biomass power and cogeneration projects aggregating to 1850 MW of electricity are under various stages of implementation (MNRE, 2008). Cogeneration projects in sugar mills include 82 projects with installed capacity aggregating to 690 MW. Another 107 projects are under implementation aggregating to 1280 MW. States which have taken leadership position in implementation of cogeneration projects are Andhra Pradesh, Tamil Nadu, Karnataka and Uttar Pradesh, The leading States for biomass power projects are Andhra Pradesh, Karnataka, Chattisgarh, Maharashtra and Tamil Nadu.

Manufacturing base: Manufacturing capability exists in the country for the equipment/machinery required for setting up Biomass Projects. Except for some critical control equipment, and high efficiency turbines, most of the equipments can be procured from indigenous sources.

Boilers: A number of large manufacturers have established capabilities for manufacturing spreader stoker fired, travelling grate/dumping grate boilers; atmospheric pressure fluidized bed boilers and circulating fluidized bed boilers. Due to recent upsurge of interest in co-generation for surplus power, leading manufacturers are further upgrading their capabilities for high efficiency boilers.

Steam Turbines: Almost all combinations – condensing, single extraction/double extraction condensing, back pressure, etc. are now being offered in the country with full after sales services guarantees. The efficiencies of turbines now being offered are comparable to the best in the world.

Other Equipment: Apart from the main equipment, there is a well established capability and capacity

TABLE 5: State-wise/year-wise list of commissioned biomass power/co-generation Project (as on 30.11.2008) (in MW)

S.No.	State	Total
1	Andhra Pradesh	334.25
2	Chattisgarh	156.10
3	Gujarat	0.50
4	Haryana	6.00
5	Karnataka	274.28
6	Madhya Pradesh	1.00
7	Maharashtra	155.50
8	Punjab	28.00
9	Rajasthan	31.30
10	Tamil Nadu	308.70
11	Uttar Pradesh	372.50
	Total	1677.13

for manufacture of related equipment for use of biomass for energy including harvesters, balers, briquetting equipment, handling and firing equipment, pollution control systems, etc. Many multinational companies have set up manufacturing facilities in the country for such equipment.

Policies, Fiscal & Financial Incentives: Besides the Central Financial Assistance, fiscal incentives such as 80% accelerated depreciation, concessional import duty, excise duty, tax holiday for 10 years etc., are available for Biomass power projects. The benefit of concessional custom duty and excise duty exemption on equipments. In addition, State Electricity Regulatory Commissions have determined preferential tariffs and Renewable Purchase Standards (RPS). Indian Renewable Energy Development Agency (IREDA) provides loan for setting up wind power and bagasse cogeneration projects.

Solar power

India receives solar energy equivalent to over 5,000 trillion kWh per year. The daily average solar energy incident varies from 4 -7 kWh per square meter depending upon the location. The annual average global solar radiation on horizontal surface, incident over India is about 5.5 kWh per square meter per day. A new program has been initiated for development and demonstration of mega-watt capacity grid interactive solar power generation from January 2008, with a view to harnessing the vast solar energy potential in the country for power generation. Under this, grid interactive solar power generation projects up to a maximum capacity of 50 MW is supported by government. The registered companies, as project developers, are eligible to set up solar power projects on build, own and operate basis. Proposal from each project developer with a maximum aggregate capacity of 5 MW, either through a single project or multiple projects of a minimum capacity of 1 MW each, is considered. A maximum of 10 MW capacity solar photovoltaic power generation projects would be considered in a State. Preference would be given to the projects from the States where the State Electricity Regulatory Commissions (SERCs) have announced or are in the process of announcing tariff for solar power. In absence of tariff for solar power, the utility should provide the highest tariff offered for medium term power purchase or the maximum tariff fixed for power from any other renewable energy source, till the SERC announces a tariff for solar power, failing which project in that State would not be considered under the demonstration programme. The generation based incentive up to Rs.12 per kWh for solar photovoltaic power and Rs.10 per kWh for solar thermal power fed to the grid by the solar power project developers is provided, after taking in to account the tariff provided by the SERC or the utility. The concerned utility and the State Nodal Agency are involved in implementation and monitoring of the projects. Presently, about 2 MW is contributed from solar power in grid interactive renewable energy.

Waste-to-energy

The National Programme on Energy Recovery from Urban and Industrial Wastes in India aims at promoting efficient and proven technologies for the treatment, processing and disposal of wastes, not only as a means of improving the waste management practices in the country, but also for augmenting power generation. A wide range of waste material can be used to recover energy. Municipal solid waste comprises mainly domestic refuse, with some commercial waste. The total grid installed waste-to-energy generation capacity is about 56 MW.

3.2. Off-Grid Renewable Power

The unmet demand and supplementary power generation requirements in electrified and to-be electrified areas of the country are proposed to be met by developing distributed/decentralized renewable power facilities. Despite the fact that the unit cost of such generation is likely to be on the higher side due to reverse economies of scale and lower capacity factors, higher generating costs of such systems could be somewhat offset on account of the lower line losses. Distributed/decentralized renewable power generating systems are being set up using wind, biomass, hydro, solar or in any combination thereof in hybrid mode. The contribution from different sources is given in Table 6. In addition to this, to provide power and energy to far flung villages, several programme have been initiated such as biogas plants, solar thermal systems, photovoltaic devices, biomass gasifiers, etc. as well as the Remote Village Electrification, Village Energy Security Test Projects and Biogas Power Generation.

TABLE 6: Off-grid/Distributed Renewable Power (including Captive/CHP plants)

Renewable Sources	Capacity
Biomass Power / Cogen.(non-bagasse)	136.70 MW
Biomass Gasifier	102.21 MWeq
Waste-to- Energy	31.07 MWeq
Solar PV Power Plants and Street Lights	8.01 MWp
Aero-Generators/Hybrid Systems	0.72 MW
Total	278.71 MW

In addition to this, to provide power and energy to far flung villages, several programme have been initiated such as biogas plants, solar thermal systems, photovoltaic devices, biomass gasifiers, etc. as well as the Remote Village Electrification, Village Energy Security Test Projects and Biogas Power Generation. Under remote village electrification about 4,237 villages and 1,142 hamlets were provided with basic lighting/electricity facilities. Biogas plants of the approved models were installed across the country with financial support of the government, taking the cumulative installation to over 43.32 lakh biogas plants. About 4.35 lakh solar home lighting systems were installed and 6.97 lakh solar lanterns were distributed. A total of 7148 solar PV water-pumping systems have been installed across the country under solar photovoltaic water pumping programme. Under solar thermal energy demonstration programme total 6.37 lakh solar cookers were sold at subsidized rate. About 1342 wind pumps were installed for water lifting for irrigation and drinking purposes.

Under the programme for renewable energy for urban and industrial & commercial applications, a total of about 2.5 million sq.m. of collector area has been installed upto September, 2008. So far 269 Akshay Urja Shops and 516 energy parks were developed in 29 States/UTs. Systems in two States were sanctioned for installation of SPV devices and systems in urban areas. Under the programme for energy recovery from urban wastes the total capacity of 31.07 MWeq achieved.

4.0 RESEARCH, DESIGN AND DEVELOPMENT IN NEW AND RENEWABLE ENERGY

The aim of Research and Development (R&D) effort in the country is to make new and renewable energy industry competitive and to increase the share of indigenously designed, developed and manufactured new and renewable energy systems/devices consequently its eventual growth to a dominant position. A comprehensive policy in place to support R&D in new and renewable energy sector, including supporting R&D carried out by industry for market development. The broad framework provides guidelines for project identification, formulation appraisal, approval and financial support. Alternate Fuels (hydrogen, bio, synthetic and electric) to supplement and eventually substitute liquid hydrocarbons and development of vehicles for such fuels and their hybrids; Green Initiative for Power Generation (GIPS) based on Alternate Fuels for stationary and portable power generation applications to supplement and eventually substitute fossil-fuel systems. Country has made significant achievement in R&D area. Research & development activities have been taken up with national laboratories, universities, scientific and educational institutions and industry for improvements in the renewable energy. The industry is being encouraged to take up product development in collaboration with research and academic institutions. The research and technology development efforts in the areas of biogas technology for process improvement, enhanced methane production through microbial route, diversified utilization of biogas and slurry and purification and bottling of biogas. While there exists a manufacturing base for small hydro sector in the country, research and product development are required to improve performance parameters, meet international standards and reduce the cost of equipment through simplification and optimization of system design and engineering practices; reduction in the cost of all elements of SHP projects; improvement in equipment reliability and development of techniques for speedy implementation so as to reduce the gestation period. The Research and Development (R&D) efforts in the solar photovoltaic technology have been aimed at development of materials used in fabrication of solar cells and modules, different types of solar cell device structures, module designs, components, sub-systems and systems, with a view to reduce the cost and improve the overall efficiency at different stages. The government has been sponsoring Research and Development projects on different aspects of the PV technology in academic and research institutions, national laboratories, and industry, for development of new materials, processes, systems, production and testing equipment for solar cells and modules and electronics used in the PV systems. The country has developed a strong research base as well as indigenous production capabilities for manufacture of solar cells, modules and a wide range of SPV systems. The Indian SPV programme is one of the leading programmes in the world for decentralized applications of photovoltaic technology. Abroad based programme for developing new and emerging technologies like Hydrogen, Fuel Cells, Bio-fuels, Electric and Hybrid Electric Vehicles, Geothermal Energy and Tidal Energy is being implemented are some of the promising technologies for transport and power generation sectors. R&D for Renewable energy based hydrogen production techniques include, methods and materials for hydrogen storage and utilization of hydrogen energy as a fuel in internal combustion engines and fuel cells. The focus of research on fuel cell technologies has been on improvements in the fuel cell related materials, sub-systems and fuel cell stacks/systems. Indian Oil Corporation Ltd. is setting up hydrogen dispensing system in New Delhi with partial financial support from the government. The MNRE, which is the nodal Ministry for the development of bio-fuels in the country, has supported R&D projects to assess different sources for production of bio-diesel from non-edible oils and bio-ethanol. Magnetotelluric studies are conducted in the identified locations for assessing the geothermal power potential in the country. At the Solar Energy Centre, a research and technology evaluation facility was upgraded the existing sun simulator for testing of solar PV modules of dimensions up to 150 x 70 cm. A large size integrating sphere lumen meter for evaluating lamps, tube lights, CFLs and LEDs has been commissioned. Testing & evaluation facilities of other components & devices have also been completed. The Centre conducted testing of 8 evacuated collector based solar water heating systems and one solar cooker during the year. Evaluation of PV systems devices and sub-systems was continued during the year. The Centre for Wind Energy Technology (C-WET) carried out measurements of wind speeds at higher hub heights by deploying Sound Detection and Ranging instruments at Palghat Gap. A 45 kW wind-solar-biodiesel trigeneration system has been installed to partially meet the load demand of the Centre. Provisional Type Certification of one wind turbine model was completed recently. In addition renewal of Provisional Type Certification of one wind turbine model has been completed. Three draft Indian Standards on Wind Energy corresponding the International Standards have been modified and submitted to Bureau of Indian Standards. The establishment of Sardar Swaran Singh National Institute of Renewable Energy is in its final stages and the Institute is likely to be operationalised during the 11th Plan period.

5.0 INSTITUTIONAL CAPACITY BUILDING

Ministry of New and Renewable Energy (MNRE) is the nodal Ministry of the Government of India at the National level for all matters relating to new and renewable energy. The Ministry has been facilitating the implementation of broad spectrum programmes including harnessing renewable power, renewable energy to rural areas for lighting, cooking and motive power, use of renewable energy in urban, industrial and commercial applications and development of alternate fuels and applications. In addition, it supports research and development. The Solar Energy Centre (SEC), Centre for Wind Energy Technology (C-WET) and Sardar Swaran Singh National Institute of Renewable Energy (SSS-NIRE) are three technical institutions setup by the Ministry. While SSS-NIRE is in the final stages of establishment, SEC and C-WET are functioning. Indian Renewable Energy Development Agency Ltd. (IREDA), a 100% Government owned company is registered as a non-banking financial company with the Reserve Bank of India under the administrative control of this Ministry.

6.0 CONCLUSION

Development and application of renewable energy have brought about significant changes in the Indian energy scenario. In addition to grid interactive electricity generation, which has now started contributing significantly in the national electricity mix, the adoption of improved renewable technologies have benefited millions of rural people by meeting their cooking and other energy requirements in an environmentally friendly manner. The social and economic benefits include reduction in drudgery among rural women and girls engaged in the collection of fuel wood from long distances and cooking in smoky kitchens, minimization of the risks of contracting lung and eye ailments, reduction in deforestation, employment generation at village level and ultimately the improvement in the standard of living and creation of opportunity for economic activities at village level. Indian efforts for promoting renewable energy are in harmony with global concerns with strategies focus on working towards lowering the relative price of new and renewable power technologies through a continuous and focused research and development effort, and improved access to reliable, affordable, economically viable, socially acceptable and environmentally sound energy services and resources. The approach in India matches the global aim of ushering in a carbon-free economy; an economy based on a fuel mix mainly provided by the green or renewable energy technologies. For India, new and renewable energy development and deployment is of great importance from the point of view of long term energy supply security, decentralisation of energy supply particularly for the benefit of the rural population, environmental benefits and sustainability. In this context, the Indian renewable energy programme is a goal-oriented effort to meet the country's energy requirement in an environmentally sound way. However, to meet the energy requirement for over a billion of population which is still growing and in order to meet the projected GDP growth of eight to ten percent per cent per annum, demand for primary energy will increase many fold. This poses a formidable challenge but is perceived to be a great opportunity for the country to increase the share of renewable in the overall energy mix.

AKNOWLEDGEMENT

The author is thankful to NAM S&T Centre and Uganda National Academy of Science(UNAS) for giving an opportunity for discussion in the international roundtable on Sustainable Utilization of Energy and Biodiversity Resources for Wealth Creation and Development. The author is also thankful to DST for deputing him to the conference.

REFERENCES

CEA, 2008. Power Sector Report, Central Electricity Authority, Ministry of power, Government of India. http:\\ www.cea.nic.in.

IEA, 2008. Key World Energy Statistics. International Energy Agency, 9, rue de la Fédération 75739 Paris Cedex 15. www.iea.org

MNRE, 2008. Annual report 2007-08. Ministry of New and Renewable Energy, Government of India New Delhi.

http://www.mnes.nic\programmes\shemes(accessed 10 Feb. 2009)

http://www.poermin.nic\indian energy scenario(accessed 10 Feb. 2009)

Ramachandra, T. V., Rajeev, K. J., Vamsee Krishna,S., Shruthi,B. V., 2005. **Wind energy potential assessment spatial decision support system.** *Energy Education Science and Technology*, 14 (2): 61-80

Tyagi, A., 2008. Renewable Energy holds promising future in India. RenewableEnergyWorld.com

16

The Renewable Energy Policy for Uganda

Albert Rugumayo
Energy for Rural Transformation Programme,
Ministry of Energy and Mineral Development of Uganda
E-mail: rugumayo@energy.go.ug

In recent years, Uganda has experienced a number of challenges related to its energy supply. These includes a prolonged drought (leading to falling lake levels and limiting the generation capacity); the escalating international oil prices; the need to provide electricity to support rural transformation and meet the Millennium Development Goals and the need to support Governments' commitment to the Kyoto Protocol. There was also the need to reinforce the national energy policy; develop the renewable resource base and diversify the energy mix, currently based on traditional biomass. In this respect, government sought to develop a renewable energy policy, which addressed the above challenges and provided a framework for upscaling renewable energy development. In the process, a number of barriers to renewable energy development were identified. These include the high upfront investment costs, the inadequate legal and institutional frameworks, limited technical and institutional capacity, inadequate financing mechanisms, limited awareness of the benefits, unsustainable use of biomass, inadequate standards and quality assurance and insufficient information and data on availability. There were also a number of cross cutting issues, which include inadequate integrated resource planning, inadequate attention to training and research, limited participation of other sectors, limited attention to environmental sustainability and limited regional and international cooperation. These barriers are addressed in the renewable energy policy, together with the appropriate strategies, targets, institutional framework, programmes and the costs for interventions. This paper describes the process of developing the renewable energy policy and highlights the key aspects it addresses through the work of a task force, updating the resource database, creating a vision, goal, principles and stakeholder consultation.

KEY WORDS: *renewable energy policy, rural transformation, Kyoto Protocol, biomass, hydropower, solar energy*

1. BACKGROUND

In recent years, the Government has met number of challenges in meeting the energy needs of its people. These are explained below:

Firstly, is the unprecedented electricity supply deficit on the national grid due to the fall in Lake Victoria water levels as a result of a prolonged drought. This has necessitated the installation of 100 MW of emergency thermal diesel generation, to be followed by another 100 MW, to bridge the gap which was very expensive. Secondly, the escalating oil prices on the international market, which impose a heavy burden on the economy and constrains the individual consumers' budgets. Thirdly, the need to make electricity accessible to the rural population, through grid extension and mini-grids, considering that the level of electrification, is very low. This calls for alternative approaches to support rural transformation and meet the Millennium Development Goals. Fourthly, is the fulfilment of Government's commitment on greenhouse gas emissions reductions, under the Kyoto Protocol and to contribute to the global fight against climate change. In particular, Government would want to provide the necessary framework for private sector investors in renewable energy projects to benefit from the available facilities in emissions trading.

These were the challenges observed, while implementing the Energy Policy in general and the Power Sector Reform, as well as Government's commitment to poverty and gender responsive energy actions.

In view of the above, government sought to develop a policy framework that would respond to these challenges and encourage renewable energy investments. This was specifically because of the

(i) recognition that a number of renewable energy technologies have become commercially viable and therefore, need to be brought into the national energy supply mix; and

(ii) ever rising cost of fossil fuels make them too expensive for developing countries, iii) emissions from coal and fossil fuels are responsible for climate change.

2. INTRODUCTION

In order for the government to meet its commitment to promote the development and utilization of renewable energy, a number of barriers have to be addressed. The key areas to consider are categorized under i) renewable energy specific and ii) cross cutting issues:

2.1 Renewable Energy Specific Issues

i. Most renewable energy technologies have much higher upfront investment costs compared to other conventional energy options.

ii. Legal and institutional frameworks to support new renewable energy investments are still inadequate.

iii. There is limited technical and institutional capacity in the public and private sectors to implement and manage renewable energy investments.

iv. Financing mechanisms to support investments in renewable energy projects and to address the affordability of consumers are either inappropriate or inadequate.

v. There is limited awareness of the availability, benefits and opportunities of renewable energy within the public domain.

vi. Biomass energy resources are utilized inefficiently and therefore, unsustainably.

vii. There are inadequate standards and quality assurance for most RETs.

viii. There is insufficient information and data on renewable energy resources availability and technologies.

2.2 Cross-Cutting Issues

i. Inadequate integrated resource planning. This type of planning integrates supply planning considerations with demand side constraints and environmental planning considerations.

ii. Inadequate attention to training, research and development, and technological transfer in the energy sector.

iii. Inadequate integration of energy issues in the policies and regulatory instruments of non-energy sectors.

iv. Many energy activities, which include production, transportation and utilization, are carried out without paying sufficient attention to their implications on environmental sustainability.

v. Inadequate utilization of the instruments of regional and international cooperation, to support investments in renewable energy.

vi. Insufficient stakeholder participation in the planning and implementation of energy projects.

The overall objective therefore, was to formulate a policy framework that responded to the above challenges and would facilitate more rapid investments in renewable energy.

3. METHODOLOGY

The initial step was to institute a task force consisting of different stakeholders these included the National Environment Management Authority, the Uganda Electricity Transmission Company, the Rural Electrification Agency, the Electricity Regulatory Authority, the National Forestry Authority, the Meteorology Department and the Ministry of Energy and Mineral Development.

The next step was the updating of the renewable energy resource base. This was done by additional field studies, compiling the information and data from the various studies in the Ministry, other Government departments, consultants and other stakeholders. The vision, goal and principles on which the policy is based were then generated.

This was followed by the review renewable energy policies of other countries. These countries included; USA, European Union, Brazil, Kenya, Tanzania, Nigeria, Swaziland among others. Furthermore, reviews of various relevant studies and reports were also conducted. These ideas were incorporated into the several drafts of the policy.

Consultations were then done with a wide range of stakeholders. These included the private sector, consultants, project developers, Non-Governmental organizations, both locally and internationally. Ministries and Government institutions consulted in the policy formulation were; the Ministry of Water and Environment, the Ministry of Local Government, the Ministry of Justice and Constitutional Affairs, the Ministry of Lands and Urban Development, the Ministry of Finance, Planning and Economic Development, the Ministry of Agriculture, Animal Industry and Fisheries, the National Environment Management Authority, the National Forestry Authority and the Uganda National Bureau of Standards, the Electricity Regulatory Authority, the Uganda Electricity Transmission Company Limited, among others.

4. THE RENEWABLE ENERGY POLICY

4.1 The Policy, Vision, Goal and Principles

Government's Policy Vision for Renewable Energy is: *To make modern renewable energy a substantial part of the national energy consumption.*

The Overall Policy Goal is: *To increase the use of modern renewable energy, from the current 4% to 61% of the total energy consumption by the year 2017.*

The Key Principles on which this Policy is based are:

i. Energy is essential for poverty eradication, regional equity and socioeconomic development.
ii. Reliability, efficiency and sustainability are essential in the successful deployment of renewable energy technologies.
iii. Renewable energy enhances energy diversity, security and independence.
iv. Public-private partnerships will form the basic mechanism for renewable energy investments.
v. Energy pricing will be based on full economic, social and environmental costs, taking into account the affordability and social good.
vi. The avoided cost principle will be used for determining feed in tariffs.
vii. The gender dimension will be integrated in renewable energy planning and management.
viii. It is necessary to enhance stakeholder participation, including Government, the private sector and communities.

ix. It is necessary to enhance market competitiveness of renewable energy technologies.

x. It is necessary to ensure the sustainable supply and utilization of energy resources.

4.2 The Policy Objectives

In order to achieve the Policy Vision and Goal, the following supporting *Policy Objectives* will be pursued:

i. Maintain and improve the responsiveness of the legal and institutional framework to promote renewable energy investments.

ii. Establish an appropriate financing and fiscal policy framework for RET investments.

iii. Mainstream poverty eradication, equitable distribution and gender issues in renewable energy strategies.

iv. Acquire and disseminate information in order to raise public awareness and attract investments in renewable energy sources and technologies.

v. Promote research and development, international cooperation, technology transfer and adoption of standards in renewable energy technologies.

vi. Utilize biomass energy efficiently, so as to contribute to the management of the resource in a sustainable manner.

vii. Promote the sustainable production and utilization of biofuels.

viii. Promote the conversion of municipal and industrial waste to energy.

4.3 The Strategies

Various strategies have been elaborated to realize the policy objectives and are shown in Tables 1 and 2 provide the main targets.

TABLE 1: The Policy Objectives and the Strategies

Policy Objective	Strategies
1) Maintain and improve the responsiveness of the legal and institutional framework to promote renewable energy investments.	1. Publish a standardized Power-Purchase Agreement (PPA) with feed-in-tariffs. 2. Put in place legislation and regulations to promote appropriate use of RETs in other sectors. 3. Develop appropriate regulations for grid connections and wheeling of electricity generated from renewable energy. 4. Establish a National Energy Committee. 5. Establish a decentralized coordination framework to support the promotion of renewable energy investments at the lowest level.

Policy Objective	Strategies
	6. Create both a Renewable Energy Department and an Energy Efficiency and Conservation Department at the Ministry of Energy and Mineral Development , 7. Attract qualified personnel into the sector so as to support Renewable Energy Investments. 8. Integrate energy issues into non-energy sector policies and planning for sustainable energy service provision. 9. Introduce a Sector-Wide Approach (SWAP) in energy planning and implementation.
2) Establish an appropriate financing and fiscal policy framework for RET investments.	1. Implement, through public-private partnerships (PPP), innovative financing mechanisms, including targeted subsidies. 2. Introduce fiscal measures that favour renewable energy investments. 3. Implement innovative risk mitigation mechanisms and credit enhancement instruments. 4. Enhance social service provision through grant financing of renewable energy projects. 5. Develop financing schemes adapted to local needs, traditions, and experiences. 6. Take advantage of the Clean Development Mechanism, Emission Trading and Joint Implementation Programmes under the Kyoto Protocol. 7. Determine the feed-in-tariffs for renewable energy projects periodically.
3) Mainstream poverty eradication, equitable distribution, social services and gender issues in renewable energy strategies.	1. Study the linkages and mechanisms between poverty eradication, gender, and renewable energy. 2. Sensitize stakeholders on the linkages between gender, poverty and renewable energy. 3. Implement a comprehensive integrated renewable energy, gender sensitive, poverty alleviation plan. 4. Reinforce the gender related benefits of renewable energy in PEAP. 5. Mainstream HIV/AIDS issues in renewable energy plans, projects and activities.
4) Acquire and disseminate information in order to raise public awareness and attract investments in renewable energy sources and technologies.	1. Continuously acquire data on the renewable energy resource availability. 2. Develop the capacity to process this data 3. Develop and promote knowledge and exchange of information on renewable energy to all stakeholders. 4. Promote and stimulate renewable energy and energy efficiency markets through information dissemination.

Policy Objective	Strategies
	5. Incorporate renewable energy education into the curricula of educational institutions at all levels. 6. Develop and implement a comprehensive capacity building programme for the Renewable Energy Sub-sector.
5) Promote research and development, international cooperation technology transfer and adoption and standards in renewable energy technologies.	1. Promote appropriate R and D and local manufacturing capability in renewable energy technologies. 2. Allocate funds for R and D in Renewable Energy Technologies. 3. Set up a Research and Development Division under the Renewable Energy Department to liaise with other institutions on R and D in RETs. 4. Support the research initiatives in tertiary institutions and among other stakeholders. 5. Develop and adapt RET standards and certification processes. 6. Identify and enhance mechanisms to gain from technology skills transfer and from international experience.
6) Utilize biomass energy efficiently so as to contribute to the management of the resource in a sustainable manner.	1. Promote, in collaboration with NFA and MAAIF, the growing of energy crops. 2. Provide incentives for farmers to establish commercial *woodlots.* 3. Integrate biomass energy production and efficient utilization and its impacts on climate and health into the formal education system. 4. Licence charcoal production and transportation and encourage commercial production in an efficient and sustainable manner. 5. Promote the production and use of biogas at both household levels and large/industrial scale. Scale up household biogas units to 100,000 by 2017. 6. Scale-up the adoption of efficient charcoal fuel stoves from 20,000 currently to 2,500,000 households by 2017. 7. Increase the adoption of efficient fuelwood stoves from 170,000 currently to 4,000,000 by 2017. 8. Promote interfuel/intertechnology substitution in households, commercial buildings and industry. 9. Promote efficiency in intensive wood burning industries. 10. Promote biomass fired cogeneration in industries and institutions. 11. Offer training opportunities for *"Jua Kali"* artisans for manufacture, installation and maintenance of efficient cook stoves. 12. License encroached national forest reserves to investors.

Policy Objective	Strategies
7) Promote the sustainable production and utilization of biofuels	1. Develop appropriate legislation for the use of biofuels. 2. Adopt appropriate international standards for the manufacture and blending of biofuels with petroleum fuels. 3. License companies to blend up to 20% biofuels into gasoline and diesel. 4. Provide financial incentives for the production of biofuels. 5. Set up a biofuels standard testing facility at the UNBS for testing and monitoring purposes. 6. Monitor the standards of biofuels producers 7. Sensitize the public and stakeholders on the use of the biofuels. 8. Facilitate research on biofuels.
8) Promote the conversion of municipal and industrial waste to energy	1. Provide incentives for the conversion of wastes to energy. 2. Put in place fiscal measures that will discourage open burning or disposal of wastes without extracting their energy content.

4.4 Policy Actions

The strategies to achieve the policy objectives have been translated into policy actions in the form of *specific programmes* as indicated below:-

i. Power Generation Programme

This programme will support public and private sector investments in renewable energy generation and consists of two approaches; one for large hydropower schemes and one for small power schemes.

a) *Large Hydropower Schemes:* Sites will be tendered out according to the provisions of the Electricity Act, 1999 Sections 29 and 32. The developer will also arrange an appropriate financing package. Tariffs will be determined through negotiations, on a case by case basis.

b) *Small Power Schemes:* Basic studies of the various resources and sites will be carried out followed by promotion and tendering to the private sector, followed by their development. This will cover mini hydropower schemes, biomass cogeneration, wind power, peat, geothermal and solar thermal electric and limited to 20 MW installed capacity per plant. The feed-in- tariffs, for renewable electricity are presented in Annex 2. These will be reviewed periodically.

ii. Rural and Urban-poor Electricity Access Programme

Electricity access to rural populations and the urban poor require special packages to make connections and services affordable. The programme will enhance the on-going procedures for community schemes, where the cost of connection to the community is subsidized. It will also support the development of independent grids supplied by micro and pico hydros and biomass gasifiers to be managed by communities and solar PV systems in dispersed remote settlements. The programme will prioritize supporting electrification for productive uses and key social services.

iii. Modern Energy Services Programme

This programme will support renewable energy technologies such as improved wood fuel and charcoal, stoves, solar PV and solar water heaters. It will also incorporate the dissemination of biogas, liquefied petroleum gas (LPG) and kerosene for cooking as substitutes for wood energy.

iv. Biofuels Programme

This programme will support investments in the production and use of ethanol, biodiesel, methanol and biogas. Specifically, all dealers in petroleum products will be obligated to blend fossil fuels with biofuels up to 20%, as appropriate.

v. Energy Efficiency Programme

The programme seeks to implement the Energy Efficiency Strategy. The Government will promote efficient utilization of renewable energy resources, through the activities described in the *Energy Efficiency Strategy for Uganda.* The necessary legal instruments will also be put in place.

vi. Wastes for Energy Programme

This will cover the conversion of waste to energy through direct combustion, gasification or biological conversion to biogas.

4.5 Institutional Framework

The overall responsibility for this policy lies with the Ministry of Energy and Mineral Development. The Ministry will oversee and coordinate the implementation of this policy by various stakeholders and will ensure the effectiveness of these activities. Within the Ministry, a Renewable Energy Department is being created to specifically focus on the promotion of RE and RETs and an Energy Efficiency and Conservation Department is being created to spearhead the promotion of Energy Efficiency and Conservation.

The other main stakeholders include the Electricity Regulatory Authority (ERA), the Rural Electrification Agency (REA), which is the secretariat of the Rural Electrification Board (REB), the Uganda Electricity Transmission Company (UETCL), the Uganda Electricity Distribution Company

TABLE 2: The Targets

Programmes	Baseline	Cumulative Targets	
1) ***Power Generation***	**2007**	**2012**	**2017**
Hydropower plants (large) (MW installed)	380	830	1200
Hydropower plants (mini and micro) (MW installed)	17	50	85
Cogeneration (MW installed)	15	35	60
Geothermal (MW installed)	0	25	45
Municipal Waste (MW installed)	0	15	30
2) ***Rural Electrification and Urban Access***	**2007**	**2012**	**2017**
Electrified households through PREPS/ LIREPS and CIREPS	250,000	375,000	625,000
3) ***Modern Energy Services for Households***	**2007**	**2012**	**2017**
Improved woodstoves (No)	170,000	500,000	4,000,000
Improved charcoal stoves (No)	30,000	100,000	250,000
Institutional stoves (No)	450	1,500	5,000
Baking Ovens (No)	60	250	1,000
Kilns (lime, charcoal, brick...) (No)	10	30	100
Household Biogas (No)	500	30,000	100,000
Solar Home Systems (kWp)	200	400	700
Fruit driers (No)	3	1000	2000
4) ***Biofuels*** (Ethanol, Biodiesel) (m^3/a)	0	720,000	2,160,000
5) ***Energy Efficiency***	**2007**	**2012**	**2017**
Solar water heaters (m^2 installed)	2,000	6,000	30,000
Energy savers (No)	1,000,000	2,000,000	4,000,000
Industrial energy audits implemented (No)	20	70	300
Energy efficient equipment for industries implemented (No)	15	50	250

(UEDCL) and UMEME the concessionaire, the Uganda Electricity Generation Company (UEGCL), the Uganda National Bureau of Standards (UNBS), the National Environment Management Agency (NEMA) and the Directorate of Water Development (DWD).

Other stakeholders include; the Private Sector Foundation Uganda (PSFU), the Uganda Investment Authority (UIA), the Uganda Manufacturers Association (UMA), the Uganda Renewable Energy Association (UREA) and the Uganda Small Scale Industries Association (USSIA).

Other Government ministries participating are the Ministry of Finance, Planning and Economic Development, Ministry of Health, Ministry of Education and Sports, Ministry of Water and Environment, National Forestry Authority and Ministry of Agriculture, Animal Industries and Fisheries.

A special financial mechanism has been instituted to facilitate rural electrification and renewable energy investments. This is the Credit Support Facility (CSF) known as the *Uganda Energy Capitalisation Trust.* Participating Financial Institutions (PFIs) will include Commercial Banks, Development Banks and Microfinance Institutions.

4.6 The Financial Implications

The total financial resources required to implement the strategic interventions are of the order of UShs 6,500 billion or US$ 3.5 billion over the next ten years. It is estimated that 86% of these resources will come from direct private investment, while 14% have to be obtained from the Public Sector, either through Government resources or from Development Partners.

5. CONCLUSION

Government has prepared a comprehensive document that will support investments in renewable energy technologies in Uganda. It is hoped stakeholders will take keen interest in this area of national development. Future collaboration would be required to develop specific policies for particular areas like sustainable biomass use, energy from waste or biofuels.

REFERENCES

1. The Renewable Energy Policy for Uganda, Ministry of Energy and Mineral Development, November 2007, Kampala, Uganda.

2. The Energy Policy for Uganda, Ministry of Energy and Mineral Development, 2001, Kampala, Uganda.

3. The Poverty Eradication Action Plan PEAP 2002-2007, Ministry of Finance, Planning and Economic Development, 2002, Kampala, Uganda.

4. The Rural Electrification Strategy and Plan 2001-2010, Ministry of Energy and Mineral Development, Kampala, Uganda.

5. The Alternative Energy Resources Assessment and Utilization Studies, Ministry of Energy and Mineral Development, 2004, Kampala, Uganda.

6. A Guide to Good Regulation, Ministry of Finance, Planning and Economic Development, 2002, Kampala, Uganda.

7. The Indicative Rural Electrification Master Plan, Ministry of Energy and Mineral Development, 2008, Kampala, Uganda.

Biodiversity: Emerging Scenario & Applications

17

Status and Initiatives of Biodiversity and Energy Resources Utilization in Nepal

Dilip Subba
Nepal Academy of Science and Technology,
Khumaltar, Lalitpur, Kathmandu, Nepal
E-mail:dilipsubba2003@yahoo.com, dilipsubba@mail.com.np

Nepal is rich in biodiversity and energy resources and it carries huge potential of harnessing these resources in sustainable manner. The most potential renewable energy technologies are related to hydropower, biomass energy, solar energy and wind energy. Renewable energy technology such as hydropower, biogas, bio-briquette, solar water heating, improved cooking stoves, solar dryer, solar cooker and solar PV are already at the stage of commercial applications whereas other technologies are at developmental stage. Although different efforts and programs are being undertaken much is needed to be done to utilize these resources to fullest extent and create wealth and prosperity without further adverse effect on ecosystem and living system. There is need for consistent and coherent initiatives from government, public and private sectors, and honest commitment of political parties. The strength, status and initiatives being taken to utilize the various energy resources, socio-economic prospects, associated ecological and health implications, issues of challenge and government policy and plan are elucidated in this paper.

1.0 BIODIVERSITY RESOURCES, STATUS AND INITIATIVES

Nepal is located on the southern slope of the Himalayas between 26° 22' N to 30° 27' N latitude and 80° 40' E to 88° 12' E longitude. It is a mountainous country with a fragile and steep topography (about 200 m in the south to 8,848 m in the north). Nepal has land area of 147,181 sq. km, of which mountains, hills and plains occupy 35%, 42% and 23% area respectively. Nepal has a population of 25 million, 47% of which live in plains followed by 45% in hills and 8% in mountains. Nepal, the Himalayan Kingdom is popularly known for its natural and cultural heritage. The country is endowed with majestic mountains, lush valleys, luxurient forests, gushing rivers, crystal clear lakes, productive soil, adequate rain and diverse ethnic culture.

Seventy one percent of the Nepalese land has green coverage (29% forest land, 10.6% shrub land,

20% agricultural land, 12% rangeland). The Plains has the major agricultural land and the hill belt has the potential for horticultural development while the mountain zone has been considered suitable for livestock farming.

Nepal is hotspot of biodiversity and is placed 27th in the world in having biodiversity. The biodiversity profile project (1995) and WCMC (1992) ranked Nepal as having the tenth richest flowering plant diversity country in Asia. There are several books covering various aspects of Nepalese biodiversity (Choudhari, 1998; Jha et al., 2001; Joshi and Joshi, 2001; Lekhak and Lakhe, 2003). Nepal is endowed with at least 2.3% (465) of the lichen species, 2.4% (1822 spp.) fungi, 2.6% (687 spp.) algae, 5.1% (>1000 spp.) bryophytes, 3.4% (380 spp.) pteridophytes, 5.1% (28 spp.) gymnosperms, 2.7% (6200 spp.) angiosperms, 1.4% (168 spp.) platyhelminthes, 0.2% (144 spp.) spiders, 2.6% (640+2253 spp.) butterflies and moths, 0.7% (5052 spp.) other insects, 1.0% (182 spp.) fishes, 1.0% (43 spp.) amphibians, 1.6% (100 spp.) reptiles, 9.3% (863 spp.) birds and 4.5% (202 spp.) mammals in 0.09% of terrestrial surface of earth (0.03% of total earth surface) (Jha, 2005).

The biological resources are important sources of timber, medicine, fodder, ornamentals, fuel and industrial raw materials. Agriculture and forestry sectors contribute 39% of the national GDP (CBS, 2004). Livestock contributes up to 31% of agriculture GDP. There is high prospect for industries based on medicinal and aromatic plants. Herbal tea is produced and marketed. Herbs Production and Processing Company Ltd. (HPPCL) has encouraged farmers to cultivate aromatic plants, viz. Citronella, palmarosa, lemongrass, ginger, mentha, french basil, chamomile and holy brasil. Sustainable harvesting of *Asparagus racemosus* is under trial in some parts of Terai. Presently about one hundred species of medicinal plants are collected from wild and exported. Yarsa Gumba (*Cordyceps sinense*) is one of the most expensive medicinal fungi that grow on caterpillar (Jha, 2005). Tourism is considered potential foreign exchange earning sector for Nepal. Also, it has given employment to a large number of local people. Nepal has several national parks and wildlife reserves and has been identified as one of the best destinations for eco-tourism.

It is estimated that there are more than 6000 rivers and rivulets in Nepal. At present, water resources have been used to provide irrigation facility to about 1.1million ha of land. They are also used to generate 606 MW of hydropower. They are used to provide access to potable drinking water to around 71 % of the population. The available amount of water resources in the country is sufficient to provide year-round irrigation to the entire irrigable land, to generate huge amount of hydropower and meet the demands of other uses of water such as domestic, industrial, recreational, navigational, aqua-culture among others.

Nepal has yet to properly harness its biodiversity resources. Few eco-development plans such as alternative energy, plantations, conservation, economic development activities have been launched in buffer zones and some community forests. Research work is going on in different academic and research institutions, government departments and universities in Nepal to explore the floral, faunal and diversity of these bioresources. These activities include characterisation, conservation and utilization of these resources. For instance, bio-prospectings to explore ethno-medicinal plants are being carried out and include screening of plants extracts for antifungal, antibacterial and antiviral activity, isolation and elucidation of structure of active compounds and their quantitative study for population dynamics. Nepal Academy of Science and Technology is also doing research on molecular characterization of high value medicinal plants such as *Swertia chirayita* (*chirayito*), *Taxus bacatta*, tea cultivars, domestication and nutritional studies of wild edible mushrooms found in Khumbu region at the base of Mt. Everest. Increase in productivity of agricultural and livestock products, production of safe and environmentally friendly agricultural products and disease control of plants and livestock through biotechnological tools had been achieved. A status report on biotechnology activities in Nepal have been published (NAST, 2008c). Nepal Flora Project was implemented jointly by Nepal Academy of Science and Technology, Central Department of Botany of Tribhuvan University, Department of Plant Resources, GoN, Tokyo University and National Botanical Garden, Edinburgh. The report of this project will soon be published (Personal communication).

2.0 ENERGY RESOURCES: STRENGTHS, STATUS AND INITIATIVES

The energy resources of Nepal are traditional and renewable forms of energy. Biomass resources dominate the energy scenario. In other words, Nepal's energy supply is overwhelmingly dependent on bio-mass resources - firewood, agriculture residues and animal wastes. The most recent available energy consumption data show total energy consumption of 367.25 million GJ (WECS, 2006). The traditional sources (wood, agro residue, and animal dung) met more than 86% of the total energy demand. Petroleum products supplied 9.54%, coal (imported) 2%, electricity 2% and renewable energy sources (micro hydro, solar PV, biogas) 0.61%. Forest i.e. fuel wood supplies 77% of the total energy requirement, which is mainly consumed in rural areas. Nepal belongs to the least energy-consuming countries in the world. Notwithstanding, the energy demand is increasing by 7.5% annually and the deficiency in energy supply is growing continuously. There is no proven significant deposit of fossil fuels available. Nepal is net importer of petroleum products. Nepal became member of IAEA only in 2008 and so, going for nuclear energy, if at all, is decades away. Hydrogen fuel has not even entered the discussion in the scientific community.

At present hydropower, solar, biomass and wind energy are considered appropriate alternates of traditional energy sources in the Nepalese context. Renewable energy sources are clean, sustainable and environmentally friendly. The country is blessed with abundance of such renewable energy resources. The most important renewable energy technologies are related to pico-hydropower and micro-hydropower, biomass energy (biogas, briquettes, gasifier, improved cooking stove etc), solar photovoltaic energy (solar home system, solar PV water pumping, solar battery charging etc), solar thermal energy (solar water heater, solar dryer, solar cooker, etc) and wind energy (wind electricity generator, wind mill etc). Renewable energy technologies such as hydropower, biogas, solar water heating and solar PV is already at the stage of commercial applications whereas other technologies are at developmental stage.

2.1 Traditional Biomass

2.1.1 Biogas

More than 60% of the population live in rural area and agriculture is their main occupation. Cattle and buffalo dung, poultry litter and human waste are used for the production of biogas. Biogas program is appraised as one of most successful energy programs and stands as second largest energy program. Official biogas program was introduced in 1975, which marked the beginning of commercial of biogas program and this was aimed at controlling deforestation and preventing burning of cattle dung which otherwise is used as fertilizer (NBPG, 2007). Biogas is popular both in the private and public sector. In 1992, Biogas Support Program (BSP) was launched with a view to designing, developing and disseminating biogas technology as a commercially viable and market oriented industry. The Biogas Sector Partnership/Nepal (BSP/N) under the Alternative Energy Promotion Centre (AEPC) and with the cooperation from various donors like SNV and KfW is promoting installation of biogas plant in various parts of the country. There are now over 60 biogas plant manufacturers and over 16 biogas appliance manufacturing companies. These companies have given direct employment to over 2,900 people and generated total employment to 11,000 people. The current construction capacity of the biogas program is about 30,000 plants per year. By 2007, 186,000 biogas plants have been installed. One hundred and eighteen micro finance institutes are mobilized on biogas lending. The estimated total technical potential of biogas plant is about 1.3 million plants of which 600,000 plants are thought to be economically viable. The plants are now installed at different locations in the country. BSP/N has obtained ISO: 9000-2000 certificate for its quality system (NBPG, 2007). Biogas is mainly used for cooking (80%) and lighting (20%). Now the biogas technology has been accepted in the CDM market. Government has a plan to install 25,000 biogas plants in this fiscal year 2009.

2.1.2 Biomass briquette and other biomass technology

Bio-briquette is a composite fuel of biomass and coal at different proportions. Agriculture residues and wastes are conventionally used for biomass production. The bio-briquette is more energy efficient and environmentally friendly than the traditional fuel wood. In earlier time, the traditional way of bio-briquette making, animal dung mixed with straw, jute and other biomass materials were used to produce *guitha,* the traditional low pressure bio-briquette. Now a days agro waste biomass such as rice husk, rice straw, wheat husk, wheat straw, corn cobs, corn stalks, sugarcane leaves, bagasse, coffee waste like coffee husk, coconut husk, peanuts husk, cotton plantation waste, tea plantation waste like prunnings, leaves, waste from lentil production etc and forest wild plants and waste like *Banmara* (*Adenphorum eupatorium*), Tite Pati (Mugwort), Dhurmre Phool (Lantana Kamara), pine needle and cones, sawdust, wood shavings, wood chips, waste from herbs (*Bhojo, Jatamassi*, Taxus) processing, twigs, leaves, shrubs, bushes and all kinds of other forest wastes are being used for briquette making.

Nepal Bio-Extruder Industry established in 1982 in Kathmandu was the first biomass briquetting plant in Nepal with the annual production capacity of 900 metric tons. It was using the pyrolyzing technology based on rice husk. Following this, another factory with annual production capacity of 1200 tons of briquettes was established in Butwal about 400 km west of Kathmandu in 1984. A third factory followed, with Nepalese made machinery in Dharan (1984) about 500 km east of Kathmandu with annual production of 600 tons. There are now 24 briquette industries in operation (Personal communication, 2009). The screw extrusion type technology was introduced in the country only in 1986. Short life of screw due to wearing has been a haunting problem of screw extrusion technology. An improvement needed in bio-briquetting technology sector is the enhancement of the combustion characteristics of briquettes. Scientists and technologists are trying to use municipal waste like plastic and paper in making bio-briquette as this will reduce the bulk of municipality waste and environmental pollution (Bajracharya et al., 2008). In the related research work the waste plastics such as polyethylene and polystyrene is blended with low grade coal to produce ecofuel briquette. Such briquette has better combustion characteristics (lower ignition temperature than coal) and higher calorific values, since plastics are oil derived products. Some of the institutions involved in research and development (R&D) of briquetting technology are Nepal Academy of Science and Technology (NAST), formerly Royal Nepal Academy of Science and Technology (RONAST), Solid Waste Management and Research Mobilization Center (SWMRMC), Nepal Environmental and Scientific Services (NESS), Center for Energy and Environment (CEE), Center for Rural Technology (CRT) and International Council for Mountain Development (ICIMOD).

There is a huge potential for other biomass technologies like improved cooking stoves (ICS) and gasifiers. In 1981, Research Centre for Applied Science and Technology (RECAST) with the support of Community Forestry Development Project (CFDP) and assistance of FAO carried out research works for the designing of improved cook stoves suitable for Nepal. The ICS have potential to save fuel wood used for household cooking. About 11 million tons of fuel wood is burnt annually for cooking alone. The national ICS program (2002 – 2006) funded by Energy Sector Assistance Program (ESAP) of DANIDA and executed by AEPC has distributed about 200,000 ICS till June 2006 in 711 Village Development Committees (VDC) and 13 municipalities of 35 mid-hill districts. By this program altogether, 2,24,473 units have been installed till September 2007 (AEPC). There are more than 90 local NGOs involved in the program and about 50,000 ICS are being distributed every year. The government has a plan to install 2,500 improved iron cooking stoves and 50,000 improved earthen stoves in the current fiscal year. Institutional Gasifier Stoves (IGS) were introduced as they are more efficient and multifuel stoves, which can use a wide range of waste biomass such as corn cobs, pine cones, wood chips, and different types of briquettes among others.

2.2 Hydro-energy

Until now only a small fraction of the available water resources has been utilized for the benefit of the people and the country. The government has prepared National Water Plan (NWP) and integrated water resources management is one of its principal themes. This includes hydropower, environment, irrigation, tourism, mitigation of water induced disaster, sanitation among others. Action plan and program for this have been developed and are under execution (WECS).

2.2.1 Hydropower

Electricity is used for lighting and operating devices such as refrigerator, fan, air-conditioner, fax machine, other household equipment and heavy industrial machineries.

It is estimated that there flow more than 6,000 rivers and rivulets in Nepal, whose total length is approximately 45,000 km. The country has technically and economically feasible potential of generating 43,000 MW of electricity from the total potential of 83,000 MW. The present total installed hydropower capacity is 619 MW. The first hydropower plant was constructed in 1968, known as Pharping Hydropower Plant (500 KW). The second hydropower station was at Sundarijal with the installed capacity of 640 KW. The Trishuli Hydro Plant with 21 MW and Sunkoshi Hydro Plant with 40 MW were completed in early 70's. The storage plant of Kulekhani I with a capacity of 60 MW was commissioned in 1982 and its tail race scheme Kulekhani II with 32 MW was added in 1987. The Marshyandi project with 69 MW installed capacity was completed in 1989. Due to low water level in storage reservoir, destruction of transmission line from heavy flood last rainy season and currently the total power generation is 326 MW only (276 MW hydropower, 20 MW thermal power and 30 MW by import). The Present need (peak load) of electricity is 808 MW which is estimated to increase annually by 60-80 MW. Thus the country is now in acute shortage of electricity and the government has declared the state of emergency on electricity.

To meet the increasing demand of electricity and generate it for marketing purpose the government recently announced a plan to generate 10,000 MW of electricity in next ten years. The government's plan on the development of hydropower sector is as following:

- Completion of project and operation from 2009: Middle Marshyangdi (70 MW);
- Projects under construction: Chamelia (30 MW) , Kulekhani third (14 MW) ;
- Launching of construction of hydropower projects from this fiscal year of 2008/09: Upper Tamakoshi (456 MW), Upper Trishuli (60 MW), Rahughat (30 MW), Naumure (245 MW);
- Preparatory work to be launched in 2009: Upper Trishuli 3 B (40 MW), Tamor-Mewa (110 MW), Upper Seti (127 MW), Dudhkoshi (300 MW), Tamakoshi second and third (500 MW), West Seti (750 MW).

There exists a huge potential of micro and pico-hydropwer and as of 2003 there are a total of 2065 micro-hydro schemes, with an installed capacity of over 14,000 kW, installed in various parts of the country. The application of micro-hydro energy has been for basically two purposes, as locomotive i.e. substitute of human and animal power for transportation and thermal application i.e. substitute of fuel wood. Due to high cost of the electricity generated from micro-hydro power plants people still prefer to use wood for fuel, however where wood is not easily available there is no choice other than using the electricity. Of recent two largest Micro Hydro Demonstration Projects supported by REDP have been completed, viz. Labdi Khola MHDS (56 kW) in Gwalichaur VDC, and Bhim Khola MHDS (42 kW) in Bhimghithe VDC of Baglung districts. Similarly, Samundra Sagar peltric-set scheme (3 kW) in Bhimghithe VDC, Dalami peltric-set (3kW) in Rangkhani VDC and Bhim Danda peltricset (2kW) in Dhullubaskot VDC have been completed in Baglung district and benefiting 30, 27 and 22 households respectively. Likewise, Lekhpani Khola peltric-set (2kW) has been installed to benefit 19 households in Kharpachowk VDC of Kavre district. A pico-hydro of 0.2 kW capacity has been installed as an R & D initiative in Jalbire, Baramchi of Sindhupalchowk district. This has benefited 5 households. There are now total of 237 installations with total capacity of 3646.2 kW. Relevant technical data on projects completed during 1998-2006 and details about the status of micro-hydropower projects can be obtained from AEPC.

2.2.2 Water-mills

Nepal is a mountainous country with rugged terrain and several perennial streams, rivulets, and rivers. The natural energy of the gushing water has been traditionally used in the rural mountainous area for operating water mill which is used for milling cereals, grains and pulses. It is estimated that there are 25,000 to 30, 000 water mills in operation. By replacing traditional wooden runner with hydraulically better shaped metallic runner having cup shaped blades the efficiency of 903 water mills has been improved (Sharma et al., 2003). Since 2003 Improved Water Mills (IWM) Support Program has been initiated as part of government of Nepal and Netherlands Development Organization Program Support to develop and disseminate IWM in mid and high hills. Till June 2005 1,092 IWM have been installed (CRT/N, 2005).

2.2.3 Natural Force Ferry

There are large number of rivers and rivulets in Nepal, many of which divide villages. Not all rivers have bridge for crossing. In this case, there has to be some river crossing facility or mechanism. One such mechanism is offered by Natural Force Ferry (NFF). NFF is a special designed boat which utilizes river current for floating from one bank to the other. Nepal Academy of Science and Technology has designed such boat in cooperation with Japanese. This has been successfully used at some locations like Ghatbesi and Jyamirghat of Dhading district. Since the body of boat made up of fibre glass reinforced plastic is expensive research is underway to use locally available wood and test the performance of the boat (Ale et al., 2008).

2.3 Solar energy

The solar energy resource in Nepal is abundant, evenly distributed over the country and over the seasons. Seventy eighty percent of the land area of Nepal lies in high potential solar insolation area. The average insolation in Nepal is around 4.5 kWh/m^2. Nepal enjoys 6.8 sunshine hours per day and more than 300 sunny days in a year (Adhikari, 1998). A study recorded 1240 to 1290 W/m^2 of maximum radiation. The potential of solar energy is estimated to be 26 million MW.

The first use of solar PV in Nepal is believed to be in 1963 in Bhadrapur Airport for navigational purpose (Shrestha, 1998). Recorded use of solar PV power for domestic electrification started in 1991/92. Use of PV power for rural electrification gained momentum after the successful launching of Pulimarang Village Electrification Project in late 1993. The government announced renewable energy subsidy policy in October 2000. The subsidy will be provided to Solar Home System (SHS) of 10, 20, and 30-Watt peak or more. By now, SHS has reached 72 districts out of 75. In terms of power, the total capacity of all SHS has reached 2 MWp. In terms of system capacity, the most popular SHS having more than 1,000 installations are 36 Wp (30.7%), 40 Wp (25.1%), 45 Wp (8.2%), 20 Wp (6.0%), and 65 Wp (4.1%). By now there are 100,000 units of SHS installed in the country.

Apart from SHS, solar PV systems have been used for various other purposes such as telecommunications, pumping of water, operation of office equipment, lighting for educational institutions, operating vaccine refrigerator at health posts, etc. In order to tap the potential of PV technology in micro irrigation systems and drinking water supply schemes, the government has made a subsidy policy, providing subsidy up to 75 percent of the cost of solar pump of capacity up to 500 Wp. Institutions ranging from government, semi-government, NGOs, communities, research institutes to private business organisations had installed such systems exceeding total installed capacity of 1 MWp mark. Taken all systems together, total installed capacity has reached to 1.99 MWp. Individual households have emerged as the biggest sector that has used PV power mainly for domestic lighting. Another biggest sector is the government and semi-government offices which are using PV power for operating various equipment including telecommunication equipment, computers, etc. In solar photovoltaic sector the estimated market potential is huge and already over 2500 kWp is being used in various public and private sectors.

Solar PV industry is a nascent industry in Nepal. By mid-July 2002 altogether 16 registered companies were certified by AEPC/ESAP to carry on PV business. The solar PV industry has provided direct employment to 1,019 persons at the end of 2002. The solar PV sector is a growing sector in Nepal and suffers from certain constraints such as lack of awareness among rural people, high cost, and limited and lengthy lending mechanism. Despite these constraints, various factors such as growing involvement of the government, civil society, private sector and donors, recognition at policy level of the importance of solar PV for rural electrification, integrated approach taken by AEPC/ESAP with activities ranging from policy formulation, capacity building, quality control and monitoring and evaluation have contributed to the growth of this sector. Reports are available on status of solar PV sector (AEPC, 2003).

The government has the plan to install solar home system in 65,000 households in this fiscal year. The efficiency of solar cell depends on the irradiance which can be increased by using solar energy concentrator or mirror which concentrate the sun rays on solar panel. Research work in Nepal has shown the increase in power by 27 to 39 percent depending upon the number of mirror used (Poudel and Shrestha, 2008). Besides, White LED (light emitting diode) has shown inherent compatibility to solar battery charging system and posed as viable solution for rural lighting (Bista and Chetri, 2008).

Solar energy Support Program (SSP) has been created by Energy Support Assistance Program (ESAP) and Alternative Energy Promotion Centre (AEPC) which aim to develop commercially sustainable solar PV market system in Nepal. The SSP has program components of ISPS (Institutional Solar Photovoltaic System) for photovoltaic system of institutions, electrification of off-grid households and PVPS (Photovoltaic Pumping System) for village water supply and water pumping for irrigation. Detailed program framework and guidelines in this sector has been developed (AEPC, 2002).

Solar water heater, dryer and cooker are the three main types of solar thermal devices. Solar water heaters are widely used in big cities like Kathmandu but it is beyond reach of common people. Till 2005, there were around 61,000 solar heaters installed. Solar cookers were introduced by Research Centre for Applied Science and Technology (RECAST) in 1977 as parboiling cookers. Centre for Rural Technology/Nepal (CRT/N) took further initiative to promote solar cooker since early 1990's with the government subsidy channelled through Alternative Energy Promotion Centre (AEPC). This solar cooker is not popular in the rural areas because of high cost.

Solar drying of agricultural and forestry products is in use since 1970. Cabinet rack and tunnel type dryer has been used for drying of spices, herbs, fish, fruits and vegetables and non-timber forest products. The government has given emphasis upon its application in its 9th five-year- plan (1997-2002) by giving heavy subsidy on solar dryers and other incentives such as low interest rate loans on from banks. AEPC is the implementing body of this program. During the tenth five year plan (2002-2007) the government planned to commission 2700 solar dryers in 20 districts. In this fiscal year the government has planned to install 450 solar dryers and cookers. It is worth to mention that, there is R & D need as regards designing of efficient and feasible solar dryer for various types of agricultural commodities before successful commercialisation can be assured. Research has been undertaken to design appropriate low cost solar and hybrid solar crop dryers that use solar energy and biogas (Bajracharya, 2003; Shrestha and Patrabans, 2003).

2.4 Wind energy

Although appreciable potential of wind energy exists at various locations in Nepal it is one of the least harnessed energy sources. However, there is a lot of potentials in its application. For instance, 200 MW of electrical power (annual energy production of 500 GW) can be generated from wind along the 12 km valley between Kagbeni and Chusang in Manang district. Indeed when the development of wind energy started in 1989 two wind turbines of total capacity of 10 kW were installed in Mustang which failed in due course due to technical failure. Pilot projects with installation of 200-watt turbine for demonstration and transmission have been carried out at various places like Kavre, Tansen Palpa, Makwanpur, Chisapani (Karnali) and Udaypur by organizations such as Alternative Energy Promotion Centre and Intermediate Technology Development Group. Research

Centre for Applied Science and Technology has also installed 400-watt wind turbine at Nagarkot for demonstration purpose. The government has provision in the current fiscal year budget for the wind electricity generation program of 20 MW in Kathmandu Valley and surrounding localities under Public Private Partnership (PPP).

2.5 Geothermal energy

Hot springs are one of the hidden, clean, indigenous and renewable energy sources of Nepal. There are over such 40 hot springs located at places such as Kodari, Jomsom, Singha, Bhurung, Sadhu Khola, Jumla, Mayagdi, Rior, Surai Khola, Darchula, Bajhang, Chilime, Bhulbhule Khar, Syabru Besi, Kharpani and most are situated between 80.6 ° E to 85.3 ° E longitude and 27 ° N to 30 ° N latitude. It is recorded that more than 33 hot springs have surface temperature ranging from 21 °C to 73 °C and flow at the rate of 0.2 to 8.0 litre per second. Since the surface temperatures recorded are far below the temperature suitable for electricity generation more study need to be done to study their potential for energy generation. The hot springs are serving as attraction for tourists and local people as curing bath.

2.6 Agro and biodiversity products as energy source

2.6.1 Biofuel

Bio-ethanol and bio-diesel have been considered as new alternative sources of energy. In fact, constant steep rise in the price of petroleum products in the recent past forced the country to think over these alternatives. Nepal has good potential to produce bioethanol from sugarcane. There are 13 sugar factories in Nepal out of which 9 are operational having crushing capacity of over 19,000 MT per day. The molasses production in the year 2008 amounted to 43,756 MT. It could be computed that from so much of molasses 8,786.3 KL of ethanol could be produced. If 10% of ethanol would be used for blending with petrol the quantity of bioethanol required would be10,191 KL (NAST, 2008 a). Among the operating sugar factories, only one sugar factory i.e. Shree Ram Sugar Mill has installed ethanol producing plant with the production capacity of 3,000 KL of ethanol per day. The government had earlier formulated a policy of blending 10% ethanol with gasoline, which however has not been implemented. So bio-ethanol production and use has not come into practice.

Nepal has at least 7000 species of flowering plants of which at least 5891 species have been scientifically identified. Among these over 100 species are good sources of biofuel as these contain more than 30% fixed oil. Although there are lot of oil bearing plants available in the country, Jatropha has been much talked about as a potential source of biofuel. Jatropha curcas (L) belongs to the plant family Euphorbiaceae. The seed of jatropha contains 50% by weight of viscous oil which can be used for the manufacture of candles and soap, for cooking and lighting purpose or as a diesel substitute. With the funding support from the University College Northampton, Oxford Brookes university, U.K and the British Council in Nepal, RECAST was the first organization to undertake research to explore the possibility of industrial utilization of jatropha oil. They have demonstrated successfully the process of extracting bio-diesel and utilizing it for operating a diesel engine.

Nepal has vast area i.e. 2014,225 ha of potential land for jatropha cultivation. It is about 14 percent of total area which can produce 2371,000 KL of jatropha oil (NAST, 2008 b). It is estimated that the country has full capacity to produce diesel grade oil from jatropha enough to replace the diesel. In the past decades initiatives have been taken to produce jastropha oil but mostly at R & D level. Projects were run including jatropha plantation and oil production. The crude oil was used for lighting purpose in the villages. Occasionally people seemed to operate the motor vehicle with this oil. Somewhat large scale jatropha cultivation and oil extraction is being operated in Okhaldhunga district where the oil is being used by UMN (United Mission to Nepal) hospital. Recently, the government worked out a jatropha plantation action plan. The plan includes contract farming on long term basis and will utilize waste and barren land for jatropha plantation. The government has urged all the stakeholders like government departments, oil corporation, petroleum dealers, automobile dealers, farmers, community forestry groups for cooperation towards this end.

2.6.2 Plant hydrocarbon oil

This can be posed as another natural fuel resource having potential to be the substitute of kerosene and motor fuel. Particularly mid-hills lodge large number of plant species P*inus roxburgghii* and higher altitude hills are occupied by *Pinus wallichiana.* Both the trees ooze out copious resinous material during summer called oleoresin, which on distillation can yield about 16-20% w/w of transparent oil called turpentine oil (Singh and Mahoto, 2003).

Rural Energy System

The Rural Energy Development Programme (REDP) was initiated in 1996 with support of the United Nations Development Programme (UNDP). REDP has promoted rural energy systems in the form of micro-hydro, solar power, bio-gas and improved cooking stoves in the remote villages which do not have access to the national gridline. The support is significant in developing national and local capacity for decentralized energy planning and policy reform in the favour rural energy development in Nepal. That program aimed to deliver from 2007-2009 1.7 MW of community-managed micro-/mini hydropower, 2,200 toilet-connected biogas plants, 9,000 improved cook stoves, and 550 solar home systems to 20,000 households in 40 districts of Nepal through a holistic community development model and integrated rural energy planning.

3.0 SOCIO-ECONOMIC PROSPECTS

The government made attempt to develop CDM project in biogas in 2001. After a long course the final negotiation with the World Bank for the rate of carbon trading took place on March 3, 2006. Emission Reduction Purchase Agreement has been signed between Nepal and World Bank. Under this agreement Nepal has already earned 36 million US dollar in 2007 and the projects are likely to bring Nepal revenue up to US $ 677,500 for next several years.

The installation of biogas has benefited the people economically and environmentally. The residue slurry from biogas plant can be used as fertilizer in the field to grow vegetables. It has replaced 280 thousand tons of chemical fertilizer (NBPG, 2007). The biogas program is estimated to displace the use of 328 thousand tons of fuel wood and 5.2 million litres of kerosene.

The installations of solar home system (SHS) for lighting, television, telephone, computer among others had helped rural people to generate income and save money spent on kerosene used for lighting purpose. The solar converted electricity has been found useful in setting up and operation of small enterprises such as *thanka* painting, tailoring, weaving, small manufacturing industries such as building block and brick and manufacturing dairy products and paper (Kayastha, 2003).

Jatropha plantation has helped rural people earn money for living.

Beehive briquetting is a potential technology that can be propagated as an income generating activities. This technology will help to generate rural employment apart from providing environmentally clean fuel from sustainable residues replacing firewood.

4.0 ASSOCIATED ECOLOGICAL AND HEALTH IMPLICATIONS

Nepal is not isolated from the global climate change. The average temperature trend for the last three decades is higher than the global average. The average rise in temperature varied from 0.04°C in a location in the plains to 0.10°C in Kathmandu at an altitude of 1100 m (Burlakoti et al., 2008). The number of rainy days is decreasing. The high intensity of precipitation is resulting in increased landslide and floods. Haphazard wood cutting has led to soil erosion, flood and other disastrous ecological implications such as desertification, depletion of fodder plant for cattle and goat, decreased crop yield, lack of drinking water and water for household, industrial use and irrigation. Increased melting of glacier and strong probability of glacier lake outburst implicated by temperature rise has been the topic of regular discussion. The biodiversity of many national parks is decreasing due to forest degradation, poaching and collection of non timber forest products and grass, erosion and landslide, fire, invasion and illegal seizure of land for building huts etc. This is other part of the story that the people living near to forest are suffering heavy loss of crops and livestocks, seldom human life by falling prey to the wild creatures. This has put a challenge of balancing sustainability of human and biodiversity. The relation between temperature rise and outbreak of disease has been studied. Due to the increase in temperature the incidence of water borne disease Typhoid has been found to increase (Regmi et al., 2008).

Efforts are constantly underway to reduce human made ecological disaster but still not under full control. Due to heavy dependence on firewood for cooking and other purposes there is some emission of Green House gases. Use of fuel wood for coking is also associated with vulnerability of women to injury and violence when collecting wood in the forest, burns and fire accident. Due to increased use of biogas and improved cooking stoves the situation is improving. Less health hazard and conservation of forest has been observed. The women have now saved time (estimated 0.5 to 2 hours per day) spent on collecting wood after installation of biogas plant. Use of fuel wood and kerosene has reduced. The fuel wood consumption has been reduced up to 43%. Carbon emission has been reduced by 778260 g-c per annum (Khanal and Bajracharya, 2008). Diminished use of forest has improved carbon dioxide sequestration.

The health and environment-related benefits from biogas are improved indoor air quality due to non-emission of smoke generated from burning of fire wood for cooking and better sanitation due to the connection of toilet (people will be bound to build toilet for this purpose rather than go for free disposal) with biogas plant. Also, one biogas plant saves about 2.3 tons of fuel wood per year and thus saves forest from being depleted. Biogas helps reduce greenhouse gas emissions by displacing the consumption of fuel wood, agricultural residues and kerosene (Table 1).

TABLE 1: Net GHG savings per system (tCO2e/biogas system/year)

System size	Terai	Hills	Mountains
4 m^3	2.56	2.68	2.77
6 m^3	5.86	3.92	4.00
8 m^3	7.56	4.59	4.70
10 m^3	5.98	3.67	3.37

5.0 CHALLENGING ISSUES

There are a number of policy and institutional, technical and financial issues to be addressed as to properly develop the energy and biodiversity sector in Nepal. The challenges for exploiting the available energy and biodiversity resources to a maximum extent in a sustainable manner may be summarized as following:

1. Weaknesses in planning, unfavourable terms and conditions of donor and multilateral credit institutions, foul politics, wrong government decisions, national conflict, no construction of projects by license holders, corruption, unfavorable environment impact study rules of government, failure of PPP, high production cost, human factors, social hurdles in project implementation, leakage etc. responsible for underutilization of hydropower in the past. These weaknesses need to be overcome.

2. Installation cost of many energy projects such as micro-hydro power, biogas, solar water heater etc is high. Many of the energy programs are subsidy-based, and may fail in the absence of continuous financial support.

3. Material of construction of energy system is of inferior quality and the device and system

performance is not very satisfactory.

4. More cooperation among R & D institutions and industries is needed.
5. The operation of biogas plants during cold season and at high altitude is difficult due to low gas production. Hence, R & D on the development of low temperature resistant methanogenic bacteria for biogas production is needed.
6. Cleaner production system must be introduced and applied widely.
7. Effective biodiversity resources utilization calls for effective government initiatives, adequate qualified scientific manpower, enough physical facilities and funding.

REFERENCES

Adhikari, R.B. (1998). Income Generation Through Photovoltaics: Practices and Prospects in Nepal, Paper presented at Regional workshop on "Income Generation through Photovoltaics", Pondicherry, India, Centre for Scientific Research, Auroville.

AEPC. Alternative Energy Promotion Centre, Ministry of Environment, Science and Technology, Government of Nepal. http// www.aepc.gov.np

AEPC (2002). Guidelines for Institutional Solar PV Systems. Alternative Energy Promotion Centre:http// www.aepc.gov.np

AEPC (2003). Status of solar photovoltaic sector in Nepal. Alternative Energy Promotion Centre:http/ / www.aepc.gov.np

Ale, B., Kaini, B., Dahal, J., Shrestha, J.N. and Pradhan, S. (2008). Performance analysis of natural force ferry boat at different river conditions. Abstracts of Fifth National Conference on Science and Technology, November 10-12, Kathmandu, pp 239.

Bajracharya, S. (2003). Design and testing of low cost solar crop dryers. In: Proceedings of International Conference on Renewable Energy Technology for Rural Development, 12-14 October 2003, Kathmandu, Nepal, pp 190-196.

Bajracharya, S., Nakarmi, S.K., Maharjan, I.M., Singh, R.M., Shrstha, A. and Shrestha, K.B. (2008).. Briquetting machine for fuel production from municipality solid waste. Abstracts of Fifth National Conference on Science and Technology, November 10-12, Kathmandu, pp 248.

Burlakoti, C., Chatkuli, B. and Lakhey, P.B. (2008). Change in climate at three sites of central Nepal. Abstracts of Fifth National Conference on Science and Technology, November 10-12, Kathmandu, pp 283.

CBS (2004). Nepal Living Standard Survey 2003-04. The Central Bureau of Statistics, Kathmandu.

Choudhari, R.P. (1998). Biodiversity in Nepal: Status and Conservation. Saharanpur Press, Kathmandu.

CRT/N (2005). "National Energy Situation Report Nepal-Focus on Renewable Energy Poverty Reduction". Centre for Rural Technology/Nepal, Kathmandu.

Jha, P.K. (2005). Nepalese Journal of Development and Rural Studies Vol. 2, No. 2, 65-76.

Jha, P.K., Baral, S.R., Lekhak, H.D., Lacoul, P. and Baniya, C.B. (2001). Environment and Agriculture. biodiversity, Agriculture and Pollution in South Asia. Ecological Society, Kathmandu.

Joshi, K.K. and Joshi, S. (2001). Genetic Heritage of Medicinal and Aromatic Plants of Nepal Himalayas. Nepal Buddhist Academy, Kathmandu.

Kayastha, Y. (2003). The role of solar home system in the promotion of income generating activities: A case study of three selected village development committees of Kavre district in Central Nepal. Proceedings of International Conference on Renewable energy Technology for Rural Development, 12-14 Oct, 2003, Kathmandu, Nepal.

Khanal, R.C. and Bajracharya, L. (2008). Assessment of improved cooking stove in fuel wood consumption and reducing carbon emission: A case study from TMJ area, Nepal. Abstracts of Fifth National Conference on Science and Technology, November 10-12, Kathmandu, pp 277.

Lekhak, H.D. and Lakhe, B. (2003). Natural Resource conservation and Sustainable Development in Nepal. Kshitiz Publication, Kathmandu.

NBPG (2007). "Biogas Sector in Nepal". Nepal Biogas Promotion Group, Kathmandu.

NAST (2008 a). "Report on Preliminary assessment of the ethanol producing raw materials in Nepal". Nepal Academy of Science and Technology, Kathmandu.

NAST (2008 b). "Final report on assessing biofuel development for Nepal". Nepal Academy of Science and Technology, Kathmandu.

NAST (2008 c). "National State of Art Report on Biotechnology for Nepal". Nepal Academy of Science and Technology, Kathmandu.

Personal communication (2009). Ramesh Man Singh, Technical Officer, Nepal Academy of Science and Technology, Kathmandu, Nepal.

Poudel, A and Shrestha, J. (2008). Solar power concentrator using mirrors. Abstracts of Fifth National Conference on Science and Technology, November 10-12, Kathmandu, pp 250.

Regmi, B.R., Pradhan, B., Sharma, G.B.. Pandit, a. and Lama, P. (2008). Impact of climate change on human health. Abstracts of Fifth National Conference on Science and Technology, November 10-12, Kathmandu, pp 286.

Sharma, U., Munankami, R. and Karki, D. (2003). Improved water mills (Ghatta): An energy centre for rural communities. In: Proceedings of International Conference on Renewablw energy Technology for Rural Development, 12-14 October 2003, Kathmandu, Nepal, pp 289-292.

Shrestha, J.N., (1998), Status of PV Technology in Nepal, Institute of Engineering, Kathmandu. www.pvpower.com

Shrestha, K.R. and Patrabans, S. (2003). Integration of biogas into solar dryer for food processing. In: Proceedings of International Conference on Renewablw energy Technology for Rural Development, 12-14 October 2003, Kathmandu, Nepal, pp 420-426.

Singh, R.B. and Mahoto, P.K. (2003). Substitution of imported kerosene oil by indigenous and renewable plant bio-hydrocarbon in rural areas of mid-hills in Nepal. In: Proceedings of International Conference on Renewable energy Technology for Rural Development, 12-14 October 2003, Kathmandu, Nepal, pp 64.

WECS. "National Water Plan - Nepal". Water and Energy Commission Secretariat: http//www.wecs.gov.np

WECS (2006). "Annual Report, 2006". Water and Energy Commission Secretariat, Nepal.

WCMC (1992). Global Biodiversity: Status of the Earth's Living Resources. Chapman and Hall, London.

APPENDIX

I. Key actors in biodiversity conservation sector

National Bodies:

Ministry of Forestry and Soil Conservation/Government of Nepal
National Nature conservation Trust
Community forestry groups

International Bodies:

World wildlife Fund (WWF)
International Union for Nature Conservation (IUCN)
International Centre for Integrated Mountain Development (ICIMOD)
Ev-K2-CNR, Italy

II. Key actors in energy sector

Policy and Planning:

- Department of Electricity Development (DoED), Government of Nepal
- Ministry of Water Resources, Government of Nepal
- Ministry of Environment, Science and Technology/ Government of Nepal
- Water and Energy Commission Secretariat (WECS), Government of Nepal

Program Implementation and Support:

- Alternative Energy Promotion Centre (AEPC)/Ministry of Environment, Science and Technology/ Government of Nepal
- Biogas Support Program (BSP)
- Centre for Rural Technology/Nepal (CRT/N)

- Energy Sector Assistance Program (ESAP)
- Nepal Electricity Authority (NEA)
- Nepal Solar Energy Society (NSES),
- Rural Energy Development Program (REDP)/UNDP
- Solar energy Support Program (SSP)
- Netherlands Development Organization (SNV)
- German Development Bank (kfw)

R & D:

- Nepal Academy of Science and Technology (NAST)
- Centre for Energy Studies (CES)
- Research Centre for Applied Science and Technology (RECAST)/Tribhuvan University

III Some important government energy policy and plan

- Perspective Energy Plan (1991-2017)
- Hydropower Development Policy (1992)
- Renewable Energy Perspective Pan of Nepal (2000-2020)
- Water Resources Act (1992)
- Electricity Act (1992)
- Forest Act (1993)
- Environment Protection Act (1996)
- Environment Protection Rules (1997)
- Renewable Energy Subsidy Guidelines (2000)
- National Water Plan

18

Sustainable Utilization of Biodiversity Resources for Wealth Creation and Development

B.O. Solomon*, R.S.M. Gidado, A.A. Akinrelere, N.C. Ezebuiro and O.A. Adetunji
National Biotechnology Development Agency (NABDA),
No.16. Dunukofia Street, Former C.A.C Building, Area 11, FCT,
P.M.B. 5118, Wuse, Abuja, Nigeria
*E-mail: bosconsult@yahoo.com

Nature has been very generous to the developing countries in the distribution of biodiversity on the Earth. Tropical regions, which harbour most of the developing countries, contain more than half of the world's biodiversity; they harbour at least 50-70% of the total biodiversity of the world. Collectively, they have the unquestionable importance for human survival, industry, commerce and trade. Nigeria is endowed with enormous biodiversity resources, many of which have been successfully domesticated. Bioresources is an important and inimitable biological resource. The strength of most of the developing countries lies in the richness of their bioresources. It is therefore crucial that these bioresources be sustainably managed and utilized without ignoring the conservation of the same for future generations. Sustainability of these resources is central to our planning because our economy is dependent on agriculture (horticulture, animal husbandry, fisheries, and forestry), medicinal and bio-industrial products, and others. Plants and animals are important renewable resources and sources of food and nutrients for rural communities. They also provide significant and sometimes the only sources of income for the rural communities. These biological resources make valuable contributions to the rural economy. The poor amongst us must be empowered so as to improve their standard of living and enable them contribute their quota to the development of the country. The National Biotechnology Development Agency (NABDA) is mandated to develop enterprises by rational utilization of our rich bioresources through biotechnology application in Nigeria. These enterprises, which demand low capital inputs, require production skills that can easily be acquired and they are based on exploitation of the bioresources, and will fill available markets domestically and internationally with little adaptation. This paper presents the effort of the Agency at setting up various Bioresources Development Centres (BioDeC) across the ecological zones and the diffusive potential toward empowerment.

KEY WORDS: *Biotechnology, biodiversity, bioresources, renewable resources, utilization, sustainability, enterprise.*

1.0. INTRODUCTION

Nature has been very generous to the developing countries in the distribution of biodiversity on the Earth. Tropical regions, which harbour most of the developing countries, contain more than half of the world's biodiversity; they harbour at least 50-70% of the total biodiversity of the world. Collectively, they have the unquestionable importance for human survival, industry, commerce and trade. Nigeria is endowed with enormous biodiversity resources, many of which have been successfully domesticated. Bioresources is an important and inimitable biological resource. The strength of most of the developing countries lies in the richness of their bioresources. It is therefore crucial that these bioresources be sustainably managed and utilized without ignoring the conservation of the same for future generations. Sustainability of these resources is central to our planning because our economy is dependent on agriculture (horticulture, animal husbandry, fisheries, and forestry), medicinal and bio-industrial products, and others. Plants and animals are important renewable resources and sources of food and nutrients for rural communities. They also provide significant and sometimes the only sources of income for the rural communities. These biological resources make valuable contributions to the rural economy. The poor amongst us must be empowered so as to improve their standard of living and enable them contribute their quota to the development of the country.

1.1. Biotechnology

Biotechnology is one of the five key fields of technology that are driving economic growth and globalization in this 21st century. Others are Data management and software, Creative and interactive applications, Microelectronics and Telecommunications.

Biotechnology is broadly defined as "any technique that uses living organisms or substances from the organisms to make or modify a product, improve plants or animals, or develop microorganisms for specific uses" (Persley, 2000).

Our ancient practiced the art by which they produced our *garri*, *fufu*, *ogiri*, *ogi*, *lafu*, *burukutu* etc that we consumed. Their simple skill removed poison (cyanide from *garri*), produced new flavour (*ogiri* used as condiment in preparing soup), improved texture, taste and nutritive value (because of increase in biomass). However, their art, in the definition of biotechnology is not considered as biotechnology because they lacked "knowledge" of the principles behind the production of their products. The beauty of biotechnology is that it coordinates the skill and knowledge of many disciplines of science, viz: biochemistry, biology, microbiology, genetics, chemistry and chemical engineering. And with the recent developments in molecular biology, the sky remains the limit for the explorations and exploitations of biotechnology. It empowers scientists, through genetic engineering to break and reconstruct genes which express new proteins that can be targeted to specific uses.

There is no doubt that Biotechnology is a powerful technology currently driving the 21st century world bio-economy which Nigeria can safely engage in as one of the tools to revolutionize her agriculture, healthcare, industrial processing in a sustainable environmental condition.

1.2. Biodiversity

Biodiversity is the sum total of plants, animals and microorganisms existing as an interacting system in a given habitat. It is therefore the totality of genes, species and ecosystem in a region.

At the global level, it is estimated that there are around 17,980,000 species, but a working figure of about 12,250,000 species. These species exist on land, Free State with other organisms. It has been that only about 1,604,000 species of plants, animals and microorganisms have been described so far (WCMC, 1992) and that fewer than 1% of total biodiversity of the world have been studied for their ecological relationships and role in the ecosystems (Singh, 2004).

1.3. Bioresources (Biodiversity Resources/Biological Resources)

Biological resources commonly referred to as bioresources is the scientific terminology for the natural biological wealth that influences human life and well-being. Bioresources has therefore been defined as the living entities which include genetic resources, organisms (both macro and micro) or parts thereof, populations, or any other biotic component of the ecosystems with actual or potential uses or values to mankind. They are important components for progress and economic activities of a nation. Management of a nation's bioresources must be done in a systematic, sustainable manner so as not to endanger the resources of the future generation. Biotechnology is the tool deployed in the execution of the common purpose, which is to carry out well planned programmes to enhance the socio-economic status of the people.

The people in a community are provided with experts who enlighten them and teach them simple skills, which empower them to appreciate the enormous value of the biological resources around them and then exploit such resources to improve their economic and social status. The government regulates the use of the natural resources so that the natural biodiversity is conserved to serve as a sustainable source of raw material for continued development.

In planning the developmental projects, it is necessary to identify priority areas of activity since the societal problems are many, particularly, in the developing countries. Also, since the primary objective of bioresources development is to fight poverty, hunger and disease, it becomes imperative that the priority area of activity should include health, food, agriculture and environmental developments.

Health management must be carried out cheaply and sustainably too. There is need, therefore to develop drugs from the natural resources available in a locality. Therefore, phytomedicine and drugs developed from microorganisms take precedence over imported drugs. Phytomedicine is enjoying much prominence now because it offers many benefits viz cheapness, easy availability, less toxicity than most synthetic drugs, well tolerated and more in tune with the culture of the people. In addition, it offers opportunity for economic advancement, both at the individual and national levels. But before phytomedicines can be used, it must be properly standardized through the joint action of experts from different disciplines, in order to ensure safety.

Agriculture should be developed through biotechnological techniques to ensure improved plant breeding and better crop and food yield. The emergence of genetic engineering has revolutionalised plant-breeding methods because plant genes can now be manipulated to increase the quantity, as well as produce plants with desirable properties. Genes can also be manipulated to produce plants that are more resistant to pests than their progeny. With genetic engineering, it is possible to generate foods with improved nutritional quality.

Opportunities abound for the development of food industries as biotechnology offers alternative ways of increasing the quantity and quality of raw materials to generate better and more acceptable products. The bulk of our staple foods contain mostly starch. There is need to carry out research to develop protein enriched foods for better nourishment of our people. Nigeria is yet a virgin land in terms of industrial development because most of the raw materials and equipment needed for the industries are imported, at an enormous cost to the nation. Our challenge does not stop with only developing our industries but to protect our environment from degradation. Therefore, we must manage our biological diversity properly to ensure sustainable source of raw materials for industrial development. Pollution from industrial effluents and domestic wastes must be controlled. The seas also harbour abundant bioresources, which need to be explored, exploited and developed for human comfort. We are endowed with enormous natural and human resources. An important ingredient in bioresources development is purposeful and stable polity, which creates the enabling environment and empowers the people to gird their joins and take their destiny in their hands.

1.4. Bioresources Development

Bioresources development is the elaboration of mechanisms for exploitation of the nation's rich bioresources such as plants, animals and microbes. It includes the exploitation of natural products for drugs, cosmetics and other industrial uses.

2.0 HISTORICAL EVOLUTION OF NATIONAL BIOTECHNOLOGY DEVELOPMENT AGENCY (NABDA)

The development of biotechnology in Nigeria at government level dates back to late 1980s when Nigeria became a founding member of International Centre for Genetic Engineering and Biotechnology (ICGEB) with a view to capturing the broad aspects of the fast growing cutting-edge technology. This was followed by the development of local capabilities and Infrastructural facilities in Universities & Research Institutes which include:

- Establishment of National Center for Genetic Resources and Biotechnology (NACGRAB) in 1986 with the mandate to use biotechnology tools for the conservation of nation's genetic resources.
- Biotechnology Advanced Laboratory (BAL) at Sheda Science and Technology Complex

(SHETSCO) in 1993 with the mandate to provide a Center of Excellence for research and training in biotechnology & genetic engineering

In an attempt to latch on the fast moving biotechnology wagon, the Federal Government of Nigeria put in place a National Biotechnology Policy (NBP) in April; 2001. In a move towards effective implementation of this policy, the Government took a further step to establish the National Biotechnology Development Agency (NABDA) in November, 2001. The agency has been in existence for seven years overseeing the biotechnological activities in the Country. From the late 1980s, few national academic institutions got involved in the Plant tissue culture; these are University of Nigeria, Nsukka (UNN), National Center for Genetic Research and Biotechnology (NACGRAB) and National Cereal Research Institute (NCRI).

Mission Statement

Promote biotechnology activities that positively respond to national aspiration on food security, job/wealth creation, affordable healthcare delivery and sustainable environment.

Vision

To make biotechnology an engine for the industrialisation of Nigeria.

Mandates

NABDA has the mandate to coordinate, promote and regulate the development of biotechnology in the country. Specifically, the Agency is to, among other things:

1) Develop an indigenous critical mass of human resources and infrastructure for biotechnology in Nigeria.
2) Develop sustainable exploitation of bioresources for food and agriculture.
3) Develop mechanism for adequate funding of biotechnology activities through national and international funding agencies.
4) Promote indigenous competence in the development and application of biotechnology-based products and services.
5) Develop framework for ethical and profitable uses of biotechnology-based products and services.
6) Develop viable and commercial biotechnology and technologies through strategic investments in biotechnology R & D to support innovation and economic development.
7) Promote national and international collaboration between government/agencies and all other stakeholders and interest groups on matters relating to the development of biotechnology.
8) Develop mechanisms and activities to support the emergence of biotechnology enterprises

for the commercialisation of biotechnology products.

9) Develop appropriate legislations, compatible with international regulations, to promote biosafety, social and ethical use of biotechnology and to protect intellectual property, industrial property and farmers' rights.

3.0. THE ROLE OF NABDA IN BIORESOURCES DEVELOPMENT

Biotechnology is globally recognized as a rapidly emerging and far-reaching technology and is aptly being described as the "technology of hope" for its promise on food, health, industry and environmental sustainability. Although biotechnology has become well established in the developed countries, supporting their strong economies, modern biotechnology remains untapped in Africa, particularly in the Sub-Saharan Africa.

NABDA believes Nigeria can safely engage in this technology as one of the tools to revolutionize her agriculture, healthcare, industrial processing in a sustainable way. The Agency is involved in various projects targeted at wealth creation and National Development. The focus of this paper is however on NABDA's activities around biodiversity resources utilisation.

The vision of bioresources development is to transform the bioresource endowment of any human environment into a viable instrument of economic growth, of improved health and general well being of the great majority of the people. It is also to design programmes aimed at exploiting the enormous potentials inherent in bioresource to combat ignorance, poverty, hunger and disease and, at the same time, remain conscious of the proper management and conservation of the biodiversity of the ecosystem. It aims at defining a mechanism for expanding worldwide knowledge of health issues and the use of nature to create wealth. It offers an opportunity to help to generate truly global sense of well-being and to create an awareness of the need for healthier living. Some important features necessary for the realization of this dream include the following:

- Planning people- oriented programmes which focus on self-reliance.
- Proper management of biodiversity so that it can be used sustainably for the benefit of the people.
- Good information network for proper coordination of the activities involved in the management and exploitation of the treasures of biodiversity. It is a veritable instrument for the expansion of knowledge, creation of understanding and motivation of the mind to aspire to greater ideals.
- Establishment of technical infrastructures to enable scientists put into practice what they know.
- Proper data for guiding government in policy matters.
- Mapping out good marketing strategy.

3.1. Establishment of Bioresources Development Centres (BioDeC)

Establishment of Bioresources Development Centres in the six Geopolitical zones corresponding to the six ecological zones the country is in the front burner of NABDA's programmes. One of the objectives of this programme is for extension of technologies which are pro-poor, pro-women and pro-nature to the grass root level as a model for rural development. Farmers of target community are trained to acquire skills in their areas of natural trade of interest. One of such Centres that have been fully established is located at Odi, Bayelsa State of South-South zone of the country. The main occupation of the people has been farming and fishing, which has been disrupted by the activities of oil companies due to incessant oil spills that have polluted both land and water causing serious environmental degradation. Some of the skills being taught to the local people in the area include: modern fish farming (including fingerlings production), using ponding or concrete tank systems; grass-cutter rearing; snailry; mushroom production with application of tissue culture techniques. These skill acquisition training programmes are designed to encourage women and youth entrepreneurs to take up bio-enterprises that empower them for wealth creation and poverty alleviation. Other programmes of NABDA targeted towards achieving the Millennium Development Goal aimed at environmental sustainability and improved sanitation at grass root level include: production and use of biogas as alternative woodfuel as energy source for cooking; bio-fertilizers and bio-pesticides (as against chemical-based products that pollute the environment); Ecological Sanitation (EcoSan), which provides farmers with the option and opportunity for safe use of sanitized human waste for improved agricultural yields.

3.2. Specific programmes, projects and collaborative linkages of NABDA

Agricultural Biotechnology & Bioresources Development

Facilitating agricultural biotechnology and bioresources research, development and innovation in partnership with public and private laboratories. The following specific R & D programmes are being pursued:

a) Development of Improved Crops and Economic Trees Using Biotechnology Tools. NABDA is focused on facilitating the development of improved crops such as Cotton, Cowpea, Corn, Rubber, Palm-tree, Cassava, Cocoa and Rice using various genetic manipulation techniques. This is to obtain improved varieties of crops with higher yields and resistance to pests, diseases and environmental stresses.

   The agency has been collaborating with the USAID, National Root Crops Research Institute (NRCRI). Institute for Agricultural Research (IAR) & International Institute for Tropical Agriculture (IITA) on Cassava and Cowpea projects to develop transgenic cassava that is resistant to the African Cassava Mosaic virus and transgenic cowpea with resistance to the devastating caterpillar called *Maruca vitrata*. This will result in higher productivity per hectare and hence improve the economy of the farmers by reducing the need for pesticide application and thus protect the environment. In addition, improved Banana, Plantain, Pineapple and other Citrus fruits are being developed with model plantations to be established in several states of the Country. Kogi state was the first beneficiary in this respect with ten hectares developed in each Senatorial District. A significant development has been the response of the local farmers who are already copying our own farming system. Recently, the Agency has entered into a collaborative project to develop a Pineapple plantation in Delta State. This is to take advantage of the international demand for this commodity.

The agency is also facilitating the development of new economic trees such as the Baobab, Gum Arabic & Acacia, and Shear Butter for their commercialization.

b) Development of Improved Breeds of Animals for Livestock Industry including Domestication of Wild varieties. The livestock industry is also gaining our attention as we intend to improve on the various types of livestock e.g. Cattle, Goat & Sheep with capacity to grow fast and be meatier and with ability to produce more milk for our ailing diary industry through genetic improvement of local breeds. The Agency has made a giant stride in the domestication of the Grass Cutter, otherwise known as the Cane rat and has succeeded in training and establishing private Grass Cutter farmers in Odi, Bayelsa State.

c) Bioresources Development: This programme focuses on elaborating mechanisms for sustainable exploitation of the nation's rich bioresources.

The model Bioresources Development Centre at Odi, Bayelsa State commissioned in 2004 has the following Bioentrepreneurship projects:

a) A 95-family capacity Grasscutter house (a family is made of 1 male and 4 females);
b) Tissue Culture laboratory for the production of plantlets of economic crops;
c) Mushroom house for the development of local and exotic mushroom;
d) Aquaculture unit with earthen ponds, water re-circulatory system/flow through concrete fish tanks for 10,000 – fish capacity and 500,000 – fingerling capacity/year hatchery; and
e) Snailery for breeding indigenous snails.

The main objectives of the centre are to ensure:

i. Promotion of sustainable utilization of biological resources
ii. Conservation of plants, animals and microbial genetic resources
iii. Documentation of indigenous knowledge of bioresources
iv. Inventory of bioresources leading to development of database for our bioresources as well as information management and dissemination
v. Research and Development of our bioresources so as to discover and develop novel products

as well as improve the production of indigenous plants and animals

vi. Training and continuous education of experts in the field of bioresources

vii. Promotion of and collaboration in bioresources development with private and public sectors, researchers, industry and academia

viii. Development of bioresources entrepreneurship especially for natural medicine, agriculture and industry

ix. Capacity building for the non-formal education sector

The Bioresources Development Centres (BioDeC) will be established (replicated) in all geo-ecological zones of this country. Meanwhile two new centres were established in 2008 in Idofin-Isanlu, Kogi State for the North Central Zone; Katsina, Katsina State representing the North-western Zone and barring the availability of funds, those for the South-western, South-eastern and North-eastern Zones are in the offing.

Nigeria provides an interesting ecological biodiversity ranging from the Mangrove/Swamp in the South to the Near desert condition in the extreme North, Besides, unique temperate conditions exist on the high Plateau in Bauchi and Jos as well as the Mambila in Taraba State where special crops such as apple, Irish potato, vine grape, among others can be developed. Consequently, the Agency is developing a Tissue Culture Facility at the University of Jos to address the Temperate Ecological Zone, Tissue Culture Facilities at the Usman Danfodio University, Sokoto and Universities of Maiduguri to provide Planting material to ward-off the downward movement of the Sahara Desert, otherwise referred to as Desert Encroachment. Tissue Culture facilities are also being developed at the UNN, NACGRAB, and the FCT to provide disease-free planting materials to address the issue of Food Security in the respective areas.

4.0. BIOMEDICINE

Biomedicine deals with derivation of medically active principles from natural (biological) resources such as plants, animals and microorganisms which populate our ecosystem.

4.1. Microorganisms

Microorganisms are a prolific source of structurally diverse bioactive metabolites and have yielded some of the most important products of the pharmaceutical industry. These include:

i. Antibacterial agents such as penicillin, amino glycosides, tetracycline and polypeptides.

ii. Immunosuppressive agents, e.g. cyclosporins and rapamycin (from streptomyces)

iii. Cholesterol lowering agents such as mevastatin from penicillium spp

iv. Antihelmintics and antiparasitic drugs, e.g. ivermectins from streptomyces

4.2. Medicinal Plants Development

From the earliest times, plants have been highly treasured for their pain-relieving and healing abilities and today, man still relies on the curative properties of plants in about 75% of the world's medicines. Many useful drugs used by orthodox medical practitioners are obtained from plants. NABDA's initiative in the area of medicinal plant development includes the following:

i. Medicinal plants identification, analysis/characterisation and potentiation of active components.

ii. Commercial herbal farming in six geo-ecological zones (presently 1,000 hectares in Abakaliki, Ebonyi State and 1,300 hectares in Langtang South LGA in Plateau State

iii. Tissue culture laboratories in six geo-ecological zones for the mass propagation of endangered medicinal plants in the country using tissue culture techniques.

iv. NABDA's efforts aimed towards the protection and preservation of indigenous knowledge/ herbal medical practice in the country.

v. Shea butter (used traditionally in the treatment of arthritis, severe cough, bone-setting, etc.

4.3. Essential Oils from Plants

In Nigeria, there are several types of seeds or plants from which essential oils are extracted. These are ranging from seeds produced annually as crops to trees from which they are harvested yearly. The knowledge for the use of such oils unlike in other cultures is usually passed down orally from one generation to another. Though those oils with medicinal properties are often guarded secrets because of their commercial advantages, most often it is the elderly, especially women who are the custodians of such knowledge and herbal medical practitioners often referred to as "Native Doctors", (Zaruwa et al., 2008).

Among the most popular essential oils used in Northern Nigeria are oils extracted from the seeds of *Khaya senegalensis* (referred to the local language as "Mai or Man Madachi", which means bitter oil), *Ricinus communis* or castor seed oil (Man gelo), *Butyospermum parkii* or Shea butter oil (Man Kadanya), *Azadirachta indica* or neem seed oil (Man Dogonyaro), *Balanites eagyptica* and *Sesamum indicum* or sesame oil. (Zaruwa et al., 2008).

NABDA's has made concerted efforts to develop protocols for the mass production of these important plants which are not only for the better health care of the people but also are of economic importance.

4.4. Sickle Cell Disease

a) Ciklavit

Sickle Cell Disease (SCD) is a hereditary blood disorder caused by a single amino acid substitution

(Glutamic acid for Valine) due to a mutation at the sixth position of the Beta chain of hemoglobin A and the resultant hemoglobin is calied hemoglobin S.

In Nigeria, the initial institutional research and development on sickle cell disease includes:

i. *Fagara fagara (Orinata* in Yoruba) – Obafemi Awolowo University/ Prof. Sofowora
ii. Base research for Ciklavit- University of Port Harcourt / Prof. G.I. Ekeke.
iii. Base Research for Niprisan- Nigerian Institute for Pharmaceutical Research and Development (NIPRD)

Current management practices of SCD are mainly supportive, conservative and ameliorative and therefore do not address sickling. NABDA, in support of the efforts of NIPRD is collaborating with NIEMETH, the makers of Ciklavit through the sustainable exploitation of Bioresources which form the raw materials for Ciklavit production using Tissue Culture and the enhancement of active components using biotechnology tools. This drug's efficacy has been confirmed by clinical trials in Nigeria, Ghana and Gambia. Ciklavit is composed of nutritional food extracts with essential amino acids, vitamins and minerals.

b) The Drug Nicosan[tm]

NICOSAN is a phyto-pharmaceutical extract of a biological herbal mixture for the treatment and management of Sickle Cell Disease. NICOSAN is obtained from the mixture of four plant materials: *Piper guineese*, *Pterocarpus osun, Eugenia caryophyllum* and *Sorghum bicolour*.

These are all in a standard ratio which produces a hygroscopic, reddish-brown powder with a pungent odour.

The following activities of the drug were observed by investigators in-vitro:

— Reverses sickled red blood cells.

— Inhibits cell sickling.

— Shifts oxygen equilibrium curves of SS cells towards the left.

— Prolongs delay time of DEOXY-HbS prior to polymerization.

— Increases solubility of HbS

Certain novel biotechnology tools are now available for the treatment of sickle cell disease. These tools include Gene therapy, Stem cell technology/cord blood banking, Bone marrow transplant and Pre-implantation diagnosis/genetic manipulation.

NABDA is presently involved in the domestication of these technologies in Nigeria and also in the mass propagation of the raw materials (*Piper guineese*, *Pterocarpus osun, Eugenia caryophyllum* and *Sorghum bicolor*) using tissue culture.

4.5. Mushroom

Mushrooms are rich in protein, and are excellent ingredient in human diet. Mushrooms contain all the essential amino acids. Mushrooms are high in unsaturated fatty acids, low in cholesterol, high in fibre, and rich in vitamins. They are superb health foods (Chang and Mshigeni, 2004).

NABDA, convinced of the importance of mushrooms from economic, nutritional and medical perspectives, initiated the domestication of edible and medicinal mushrooms at the Bioresources Development Centre, Odi, Bayelsa State. Various trainings have been conducted on the cultivation of oyster mushroom (*pleurotus* species). Most species of pleurotus are adaptable to the tropical region.

It is understood that human beings have constantly searched for new substances that can improve the body's functions and make us fitter and healthier. Edible mushrooms can be taken regularly as part of the human diet and treated as health foods. The extractable products from medicinal mushrooms, designed to supplement the human diet but not used as regular food, are for the enhancement of health and fitness (Chang and Mshigeni, 2004).

Out of the 10,000 – 12,000 species of mushrooms in the world, around 1,800 has been estimated to have medical potency. The most prominent among them is *Ganoderma lucidum.*

At BioDeC, Odi *Ganoderma lucidum* has been produced at research level. Products from Ganoderma have attracted great attention globally. A Ganoderma database has been established at an International Research Centre. China, Korea, Japan and Taiwan are the main producers and suppliers.

Ganoderma has been used for the treatment of hepatopathy, nephritis, hypertension, hyperlipemia, arthritis, insomnia, bronchitis, asthma, gastric ulcers, diabetes etc.

With the now well-established fact that Ganoderma lucidum displays potent anti-HIV activities (Kim and Kim, 1999), more interests have been generated.

NABDA has assisted a private enterprise to establish an oyster mushroom facility (BAROS Farms) in Ondo State, Nigeria. This serves as a source of income and generates employment.

5.0 PLANT AND ANIMAL GENETIC RESOURCES EXPLORATION AND CONSERVATION

At the agency's seventh centre of excellence located in Ibadan, Plant and Animal Genetic Resource Exploration and Conservation are of high priority because of their economic importance to the National Development.

Establishment of shea butter tree gene bank (*Viterllaria paradoxa* (Gaertner f.j) Hepper) Shea butter tree populations are wild, nut production fluctuates from year to year (World Bank, 1989). Because of its enormous usefulness in the production of non-timber forest products, it is fast gaining attention of agroforestry researchers and breeders as a tree with very rich potential that are worth scientific investigation for the purpose of its domestication and subsequent improvement as an agricultural tree crop.

The seed of *V. paradoxa*, with a fat content of about 50% is used locally by many as a culinary fat or oil, for soap making, as an ointment, cosmetic and its root bark has medicinal applications.

Two hundred (200) accessions were collected from Guinea savanna zone of Nigeria during exploration and collection exercise in collaboration with Ladoke Akintola University of Technology (LAUTECH), Ogbomoso and Centre for Environmental, Renewable Natural Resources Management Research and Development (CENRAD), Ibadan. These 200 accessions have been successfully established on the centre's field genebank.

Herbarium: Over 200 different plant species including 12 different species of orchids were collected from Mambila Plateau in Taraba State and other forest zones of the Southern Nigeria. During the last quarter of 2007, another 180 herbarium specimens were collected from six Northern States and FCT which include: Kano, Kwara, Niger, Jigawa, Kogi and Kaduna.

Seed Genebank: Collection expedition of grain and legumes was carried out during the year and about 400 accessions of different crop species were collected from the Northern Nigeria and 100 maize inbred lines from Mexico. The unit was able to establish data base documentation using access programme for all germplasm accessions both in the short-term and long-term Genebank. These have actually helped in the information retrieval of seed from the seed store. Also data base documentation for regeneration has been established for development is more reliable and will assist in the data processing analysis, which will allow the unit to function efficiently. The unit was able to make a positive impact in the collation of passport data and identification of respective accession number/pedigree. Some valuable genetic materials in form of seeds were distributed to different university research students in Nigerian Universities (Solomon et al., 2008).

Constant animal genetic erosion especially in the areas of Aquaculture, Poultry, Ruminants and pseudo–ruminant gave birth to AGR Unit of the Centre. The recent exploit in Aquaculture most importantly catfish production demonstrated clearly that if these species are not protected, we may lose their pure lines as breeders are trying to manipulate each other with different crosses to get the best in terms of flesh, early maturing and good feed converters.

Presently, the AGR unit comprises of the fishery, the poultry (local fowls), goats (West African Draft) and the Snailery; emphasis is in feeds management and reproduction. Our target is to provide these animals to farmers at the cheapest and affordable price and as well conserve the genetic resources of our local bioresources. Success has been recorded in the use of browse plants to feed goats during the dry season.

Cryopreservation of catfish sperm has also been achieved for 3 months in collaboration with the Department of Animal Science, Obafemi Awolowo University, Ile-Ife (Solomon et al., 2008).

6.0. CONCLUSION

The level of capabilities in the science and biotechnology components of biodiversity and bioresources development has been used as a measure of the capability of a country to identify, conserve, utilize and commercialize her bioresources in a sustainable manner. Even though Nigeria is bioresources rich and biotechnology poor, it is encouraging that the efforts of agencies like NABDA, Nigeria Natural Medicine Development Agency (NNMDA), Nigerian Institute for Pharmaceutical Research and Development (NIPRD) and other Government Agencies has taken landmark steps in the sustainable utilization of these biodiversity resources.

These Agencies do not work in isolation but rather pull resources together to ensure the future of Nigerians unborn are safeguarded not only through the positive manner in which these bioresources are utilized, but through the creation of novel products via research and development. Advances in biotechnology have led to the establishment of genebanks for conservation of genetic resources. This is chiefly positive because due to overexploitation of biodiversity in the wild, many resources can no longer be found "in the field", and a great deal of these can now be accessed from Gene banks.

NABDA, in fulfilling her mandate on the development of the nation's enormous bioresources, is prepared to partner and establish close linkages among research scientists, extension workers, industry, the youths, the farming community, and consumers. These centres of excellence will serve as training hubs for our youths and also provide a platform for scientists to conduct research with the definitive aim of improving the standard of living of the people.

REFERENCES

Kim, B.K., I. H. Kim and Y.S. Kim (Eds.). Seoul, Recent advances in Ganoderma lucidum research. Pharmaceutical Society of Korea, pp. 61-92.

Onyia, O.C., Solomon, B.O. and Okpokwasili, G.O., 2008. *Sustainable Bioresources Management and Biotechnology Intervention.*

Persley, G.J.; Lantin, M.M. (eds.): Agricultural Biotechnology and the Poor. Consultative Group on International Agricultural Research, Washington, D.C, pp. 3-21.

S-T. Chang and K.E. Mshigeni, 2004. Mushrooms and Human Health. UNEP and UNOPS, Foreword Singh, B.K.2004, *Biodiversity Conservation and Management.* Mangaldeep Publishers, Jaipur p. 586.

Solomon, B.O., Gidado, R.S.M. and Adetunji, O.A., 2008. *The Role of Biotechnology for National Food Security* (A Plenary paper presented at the 32nd annual conference of the Genetic Society of Nigeria(GSN) at Usman Danfodio University, Sokoto-Nigeria).

WCMC (1992). Global Biodiversity: Status of Earth's Living Resources. World Conservation Monitoring Centre, WCMC London.

Zaruwa, M. Z., Manosroi, A., Gidado, R.S.M. and Manosroi, J. 2008, *Some Essential Oils from Plants and their Medicinal uses and Aromatherapy in the Northern Part of Nigeria.*

19

Acidification of Natural Resources: Concern for the Future of West Asia

Masoud Vahabi Moghaddam
University of Guilan, Rasht, IRAN
E-mail: mvmoghaddam@gmail.com

Anthropogenic emissions of the major atmospheric pollutants, namely sulphur dioxide (SO_2), nitrogen oxides (NO_x) and ammonia (NH_3) are increasing rapidly as industrialization proceeds and the use of fossil fuels increases in new geographical areas. These pollutants and their transformation products have long atmospheric lifetimes and can be carried by weather systems distances of up to a few thousand kilometres from their point of emission. This causes acidification or eutrophication far from the primary source of pollution, thus making it a regional problem and an international transboundary issue. Considerable advances in the understanding of pollutant emissions, transportation, deposition and consequent effects on ecosystems have allowed the development of *'critical loads'* concept, which has become an integral part of international negotiations, aimed at controlling emission levels. Economic progress in West Asia is associated with an increase in the oil and gas production, industrialization, and a higher demand for energy and food production. On the basis of previous experiences, such a progress will definitely lead to a considerable increase in pollutants emissions. Natural resources like freshwater, soil and forests will be obviously affected by these pollutants. A rational strategy for limiting the adverse effects of atmospheric pollutants should include consideration of *critical loads* and *critical levels*. So far, only preliminary estimates of *critical loads* are available for some parts of the region and there is a real need for regional comprehensive monitoring programs. Application of certain nuclear techniques could provide a measure of long-term variation in deposition rate of atmospheric aerosols as particles and/or droplets that is operationally easier to carry out than the direct measurement by artificial collectors over extended period of time. These methods averages over many thousands of precipitation events and are particularly valuable in quantifying aerosol and wet deposition processes at sites where conventional methods are not applicable.

KEY WORDS: *Acidification, Natural resources, Critical loads, Environmental monitoring*

1.0 INTODUCTION

The world is an acidic place, naturally. Most of the atmosphere has more acidic gases (*e.g.*, CO_2,

SO_2, NO, HCl, HNO_3) and particles (*e.g.* H_2SO_4), than basic gases (*e.g.* NH_3) and particles (*e.g.*, $CaCO_3$). So, it is not surprising that wet deposition in terrestrial and marine regions remote from human influence is acidic (Galloway, 1982). Most natural soils are acidic due to both organic matter production and decomposition, as are the headwaters that drain from acid soils. In the last 100 years, the acidity of many regions of the world has increased as a consequence of energy and food production (Galloway, 2001). In the last 30 years, the phenomena and impacts of 'acid rain' have been extensively investigated in Europe, North America and Asia. This paper will give a brief historical overview of acid deposition, acidification, eutrophication and their main impacts on the environment and the important role of international cooperation in establishing abatement strategies. The paper will conclude with a look towards the future of the West Asia in this regard.

2.0 POLLUTANT EMISSIONS AND CONSEQUENCES

Since the beginning of this century, accelerated rates of anthropogenic emissions of pollutants have led to a substantial increase in acidity of aerosol particles, cloud droplets, and precipitation in and around industrialized regions, particularly in Europe and North America. Enhanced acidification of soil and water as well as eutrophication (over-enrichment through the deposition of nitrogen compounds) is environmental threat, which has had significant adverse effects on many elements of both terrestrial and aquatic ecosystems. The impact on human health and material corrosion (predominantly stone decay which has been affecting Europe's heritage buildings and monuments) are other areas of concern.

The major pollutants involved are sulphur dioxide (SO_2), nitrogen oxides (NO_X) and ammonia (NH_3). However, the full range would include sulphur species (SO_X), oxidized nitrogen species (NO_Y), volatile organic compounds (VOCs) and ozone as well. *'Acid deposition'* is the term used for the removal of the atmospheric pollutants of an acidic, or potentially acidic nature whether they are deposited in rain or snow (wet deposition) as gases and aerosol particles (dry deposition) or in cloud droplet interception (occult deposition). The phenomenon of acid deposition (commonly identified term is *acid rain*) appears to have been discovered first by an English chemist, Robert Angus Smith, in the middle of the nineteenth century in Manchester (Cowling, 1982).

Rainfall acidity measurements have revealed that precipitation is more acidic in industrialized regions of the world. Figure 1 shows the global pattern of rainfall acidity with a range of pH values from 4.0 in central Europe and eastern North America to 6.3 in Australia and India.

Diatom analyses of sediment cores obtained from numerous lakes provided clear and quite dramatic evidence of marked acidification of surface waters in the last several decades. The diatoms found in sediments can be used as a biological pH meter, indicating how the acidity of a lake has changed over time. Furthermore, the sediment studies also reveal that this modern-day acidification coincides in time with rising sediment levels of substances like heavy metals and soot particles which indicate pollution deposition deriving from the combustion of fossil fuels and from industrial emissions. Renberg and Hellberg (1982), for example, analyzed diatom assemblages in Lake Gardsjon sediment cores and found that the pH of the lake had decreased from about 7 to about 6 in the period of 12500 BC to 1950s due to natural long-term acidification. However, a more marked decrease in pH values from 6 to around 4.5 occurred in the period of 1950s to late 1970s (Figure 2). Similar investigations in the UK (Battarbee *et al.,* 1985), Norway (Berge *et al.,* 1991), Sweden (Renberg *et al.,* 1991) and North America (Charles, 1991) have revealed temporal trends in lake water pH in the post-1850 period in response to industrialization and increasing use of fossil fuels.

It is well established that the major impact of enhanced acidification, due almost entirely to anthropogenic activity, has been on freshwater ecosystems in the acidic bedrock regions of the northern hemisphere (AWRG, 1986; Brodin and Kuylenstienna, 1992; Havas and Rosseland, 1995). On the other hand, the impact of acidification on terrestrial ecosystems is more controversial. The potential of acidic deposition to change the soil environment of trees is obvious, but, one characteristic feature of the health condition of forest areas is the complex nature of pathological phenomena and the positive feedback of the negative results of action of many simultaneously occurring factors, both natural (droughts, winds, frosts, fires, insects, fungi etc.) and paranatural, the latter including chiefly emissions of environment-polluting substances that cause acidification and eutrophication of sites, toxic deposition of heavy metals and high concentration of tropospheric ozone (Tamm, 1989; Roberts *et al.,* 1989; Aamid *et al.,* 1992; Matzner and Murach, 1995). Where the concentrations of air pollutants are very high, the connection is clear. The Europe-wide survey of leaf and needle losses in 1995 and 1996 revealed that 25% of the sample trees could be classified as damaged with a defoliation of more than 25 per cent. The highest degree of damage has been found in Central Europe, namely in Poland, Czech Republic, Slovakia and eastern Germany, where local emissions of pollutants are substantially high (Cerny and Paces, 1995; Elvinsen, 1996; UN ECE, 1997). Figure 3 shows the pattern of defoliation intensity of trees in Europe in 1996. It is quite likely that even low concentrations of air pollutants or prolonged acidification of the soil will tend to stress the trees and make them susceptible to damage from drought, frost, insect attack, etc. Where, however, acidic emissions have been reduced and where mitigation measures has been employed, there is evidence that ecosystems can recover (Aamlid *et al.,* 1992; Rasmussen *et al.,* 1995; Visser, 1995; Elingson, 1996).

In late 1960s, a Swedish soil scientist named Svante Oden hypothesized the broad ecological consequences of long-range transport of acidifying air pollutants based on measurements through Scandinavian network of surface water chemistry (Oden, 1968). The international perspective of his findings and hypotheses had an important role in calling the world's attention to the phenomena and stimulation of further national and international research on identifying emission sources, pollutant deposition processes and patterns, acidification of ecosystems and source-effect relationships (OECD, 1977; Overrein *et al.,* 1980; Cowling, 1982; Gorham, 1989; Cowling and Nilsson, 1995). Sulphur dioxide and nitrogen oxides are derived mainly from the combustion of fossil fuels (coal and oil products) in power stations, industrial boilers and motor vehicles. The major sources for ammonia, by contrast, are animal wastes and the application of fertilizers. These pollutants and their transformation products have long atmospheric lifetimes and can be carried by weather systems

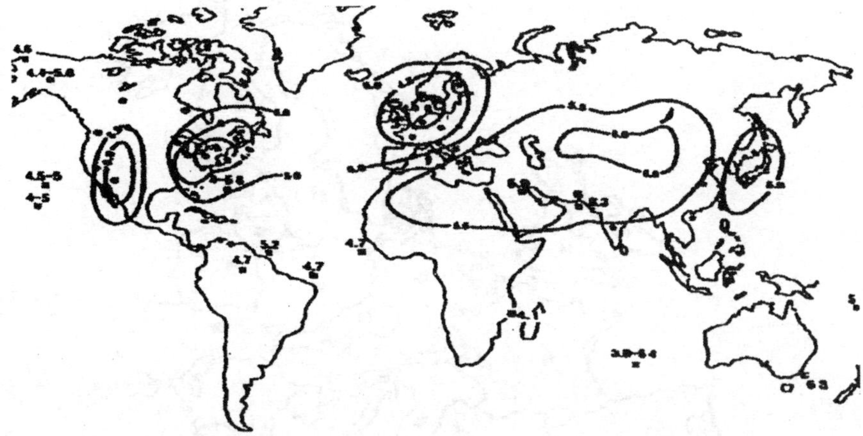

FIGURE 1: Global pattern of rainfall acidity (WMO)

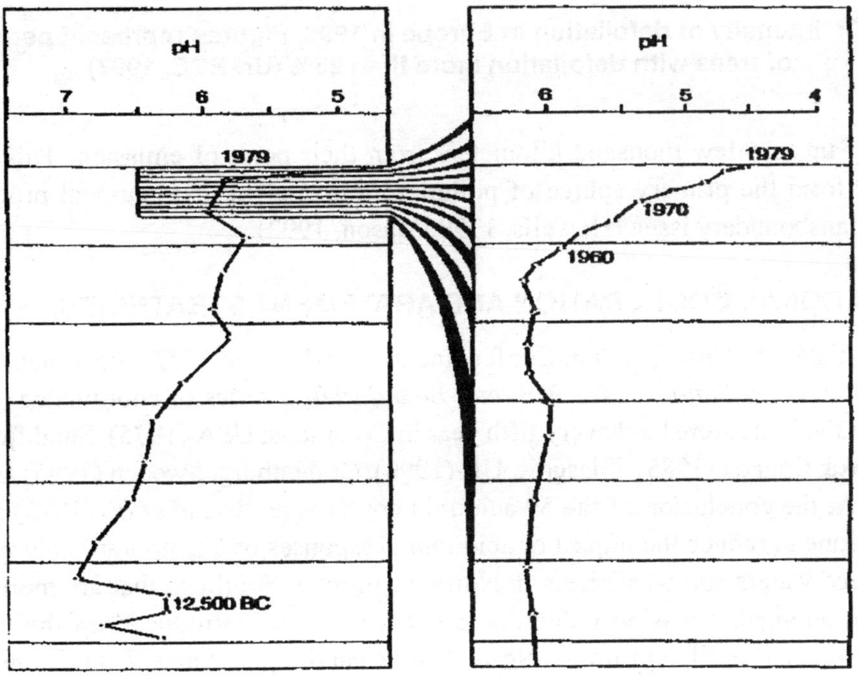

FIGURE 2: The pH of Lake Gardsjon, SW Sweden, over a time-scale from 12500 BC to 1979 AD (Elsom, 1987)

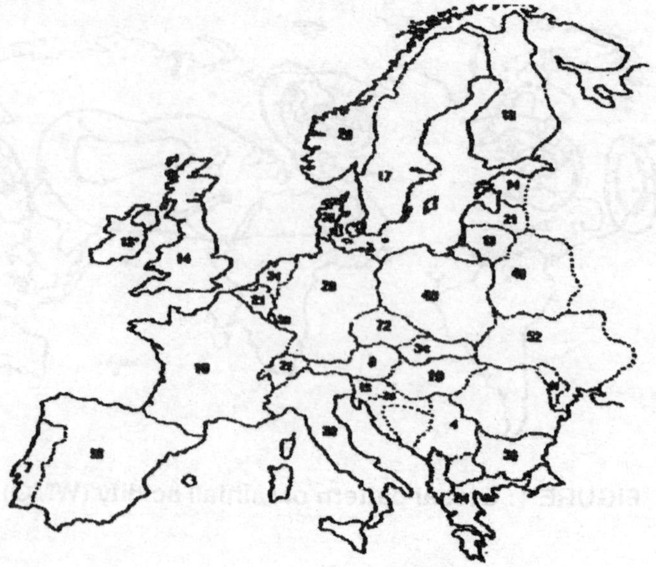

FIGURE 3: Intensity of defoliation in Europe in 1996. Figures represent percentage of trees with defoliation more than 25% (UN ECE, 1997)

to distances of up to a few thousand kilometers from their point of emission. This causes acid deposition far from the primary source of pollution, thus making it a regional problem and an international transboundary issue (Howells, 1990; Mason, 1992).

3.0 INTERNATIONAL COOPERATION AND ABATEMENT STRATEGIES

Following the historical United Nation Conference in Stockholm in 1972, concerned with *Human Environment* and *Air Pollution across National Boundaries*, a series of coordinated international conferences on the issue were held every fifth year in Columbus, USA (1975), Sand fjord, Norway (1980), Muskoka, Canada (1985), Glasgow, UK (1990), Gothenburg, Sweden (1995) and Tsukuba, Japan (2000). At the conclusion of the 5th acid rain conference, Rodhe *et al.* (1995) summarized what is being done to reduce the impact of acid rain. Responses of a temporary nature include the liming of surface waters and watersheds in Northern Europe. Solutions that are more permanent have also been adopted, emission reduction targets have been introduced on the national and international levels, primarily in Europe, North America and Japan. Since that meeting, China has also implemented reduction targets, which is one of the reasons why SO_2 emissions from China are planned to be significantly less in the future than that they were projected to be in 1995 (Streets and Waldhoff, 2000). Although the emissions of sulphur dioxide have fallen generally both in Europe and North America and reduced fallout have been recorded even in the so-called Black Triangle (the area of Central Europe where Germany, Poland and the Czech Republic converge), in many areas the depositions are still five or ten times greater than nature can withstand. There are as yet no signs of any decline in the emissions of nitrogen oxides and ammonia; the same can be said of the concentrations of low-level ozone, which is formed from nitrogen oxides and VOCs (Rodhe *et al.*, 1995; Cerny and Paces, 1995; Satake *et al.*, 2001). Furthermore, in some areas of the developing world, such as east and Southeast Asia, southern Africa and South America, recent industrialization has created major increases in pollutant emissions and acidity (Zhao and Sun, 1986; Foell and Green, 1990; Galloway, 1995). Figure 4 represents the continental emissions of SO_2, NO_X and NH_3 projected for the year 2020. Ammonia has been also considered here because it may cause acidification of soils and surface waters even though in the gaseous phase it is alkaline and, thus neutralizes airborne acids. In addition, ammonia and nitrogen oxides may act as fertilizers as well and cause eutrophication in aquatic ecosystems. At the 6th acid rain conference in Tsukuba, Japan, it was concluded that acidic deposition problems in East Asia would become increasingly serious in the foreseeable future along with socio-economic developments in the region. The need for increasing international cooperation was addressed and various monitoring network systems currently on-going and also those being planned, including the East Asian Monitoring Network (EANET), were introduced and comments were made on them (Satake *et al.*, 2001).

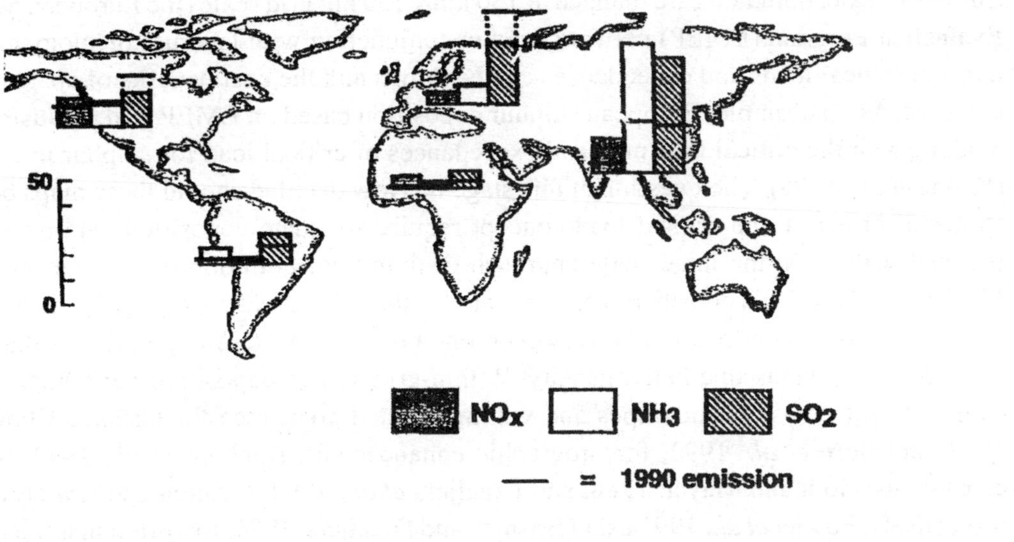

FIGURE 4: Map of projected 2020 emissions by continent of SO_2, NO_X and NH_3 in Tg S yr^{-1}, Tg S yr^{-1}, and Tg N yr^{-1}, respectively; bar represents 1990 emissions (Galloway, 1995)

During the last few decades, considerable advances in the understanding of acidic emissions; transportation, deposition and consequent effects on ecosystems have allowed the development of the concept of *'critical loads'*. These are quantitative thresholds, which help scientists and policy-makers in their assessments in identifying areas most at risk from continued deposition, while evaluating the impact of various pollution-reduction strategies. These thresholds are defined individually for particular pollutants and their receptors (i.e. water, soil, vegetation or materials), which once exceeded, will cause chemical changes leading to long-term harmful effects on ecosystem structure and functioning (Brodin and Kuylenstierna, 1992; Sverdrup *et al.*, 1992; Henriksen *et al.*, 1992; CLAG, 1994). The definition of critical load adopted by UN ECE (1988) is: *'a quantitative estimate of exposure to one or more pollutants below which significant harmful effects on sensitive elements of the environment do not occur according to present knowledge'*. The term *'exposure'* refers to a dose of pollutant deposition to a defined area over a specified time period and is expressed as deposition per unit area; *'pollutants'* refers mainly to atmospheric emissions of sulphur and nitrogen compounds; *'harmful effects'* refers to biotic aspects of ecosystems considering both direct or indirect responses; and finally, *'sensitive elements'* may be chosen by individual countries and entire ecosystems or single species may be selected (Sverdrup *et al.*, 1990; Brodin and Kuylenstierna, 1992). *'critical levels'* (the thresholds for gaseous concentration exposures) have been also defined in a similar way.

Critical loads/levels have become an integral part of international negotiations aimed at controlling emission levels. Maps of critical loads for acidifying substances and the areas where the loads are exceeded play a vital part in deciding where and by how much, reductions needs to be made. At a European scale, pollutants are mapped at 150 km x 150 km grid scale (the European Monitoring and Evaluation Program (EMEP) network) and in conjunction with land use, geology and vegetation, maps of critical loads and exceedances can be drawn and the cost benefits of abatement measures explored. As an example, European sulphur deposition based on EMEP grid is illustrated in Figure 5, along with the critical load map and exceedances of critical load for sulphur in Europe in 1990 (Rodhe *et al.*, 1995). The question at this stage is: How detailed should these maps be? Abatement strategies based on the critical load concept require adequate deposition data on both local and regional scales. On the large-scale approach dealing with pollution deposition over Europe, the EMEP model has been considered as adequate for the quantification of transboundary exchange of pollutants (Lovblad *et al.*, 1992). However, one problem of these large grids is that they mask a great degree of landscape heterogeneity. Within-grid square deposition variability could be high over different types of landscapes and within forested areas (see for instance Choularton *et al.*, 1988, and Dore *et al.*, 1992, for orographic enhancement; Bobbink *et al.*, 1992, for heath land ecosystems; Godt and Mayer, 1988, and Draaijers *et al.*, 1992, for high and low elevation forests, respectively; Fowler *et al.*, 1991a, and Erisman and Draaijers, 1995, for different terrestrial surfaces). This is especially the case for coniferous forests where deposition may be much larger than to open land. For forests growing near the ocean, or on mountain slopes, which are frequently immersed in fog or clouds, interception deposition of droplets can remove significant quantities of water and pollutants from the atmosphere (Fowler, 1984; Lovett and Reiners, 1986; Fowler *et al.*, 1989; Collett *et al.*, 1991; Fowler *et al.*, 1992). In upland areas, enhancement in deposition through this mechanism may influence the pollutant inputs to catchments (RGAR, 1997). Furthermore, forests are often grown on soils, which are more sensitive to acidification and are thus the determining receptor areas for critical loads (Hultberg & Grennfelt, 1992).

4.0 CONCERN FOR WESTERN ASIA

Economic progress in West Asian countries is associated with industrialization, a higher demand for energy and food production and for the majority of them, an increase in oil and gas production. On the basis of previous experiences, such a progress will definitely lead to a considerable increase in emissions of acidifying as well as eutrophying pollutants. For a sustainable development, it is important that the environmental impacts of these emissions are estimated and taken into account in the planning process. The impact study is especially very crucial for highly sensitive ecosystems such as the southern Caspian watershed. If less polluting techniques are applied in the region, environmental problems, like those experienced in Europe and North America may be lessened. A rational strategy for limiting the adverse effects of atmospheric pollutants should include consideration of *critical loads* and *critical levels*. So far, only preliminary estimates of critical loads are available for some parts of the region and there is a real need for regional research programs in this regard. Experience from policy actions, especially in Europe, could also be very useful for West Asia. In general, we need more accurate emission inventory data, long-term atmospheric deposition measurements, and integrated monitoring of terrestrial and aquatic ecosystems.

The rates of both dry deposition of particles and occult deposition are largely dependent on surface properties and in the case of deposition to forests are rather uncertain (Erisman *et al.,* 1995). Neither dry deposition nor cloud droplet interception is easily quantified and many studies have been devoted to an assessment of their contribution to total deposition to forests. However, there is a lack of large-scale, long-term measurements in this regard and there is a large gap between the results from field experiments, wind tunnel studies, and model estimates (Ruijgrok *et al.,* 1995).

In the process of establishing critical loads, soil acidification in forests as well as the influence of upland forests on pollutant inputs to catchments needed to be taken into account. Thus, it is important to be able to quantify the atmospheric input to forests in a reasonable way. The quantification is necessary on a local level to make a linkage between modelled deposition estimates and soil loads. Moreover, if the intent is to help protect sensitive areas then there is no substitute for good on-site data (Hicks, 1995).

Wet deposition is known to be rather evenly distributed over large areas and its measurement is more straightforward than the measurement of dry and occult deposition by using precipitation collectors. The exception is in upland areas where orographic enhancement of rainfall occurs and an assessment of the extent of seeder-feeder effect has to be performed.

Several methods exist to estimate the dry and occult deposition to forests on small spatial scales. However, direct measurement of aerosols and cloud droplet deposition is difficult except by micrometeorological methods that suffer mainly from their limitation to certain terrain situations.

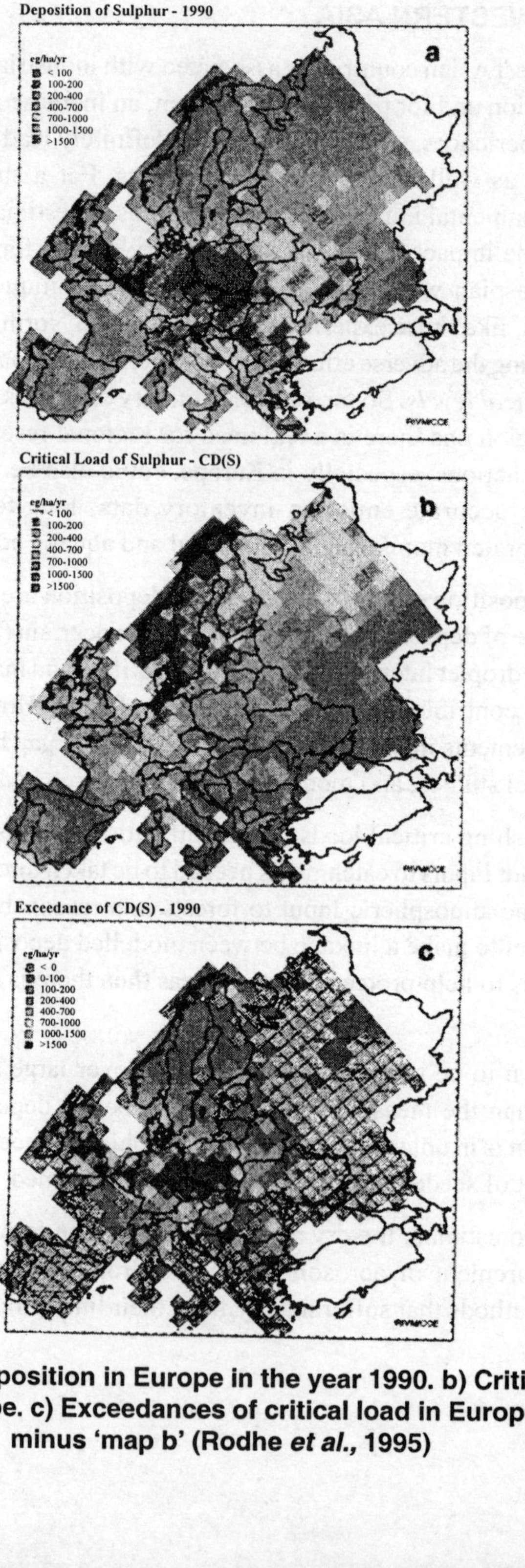

FIGURE 5: a) Sulphur deposition in Europe in the year 1990. b) Critical load of sulphur deposition in Europe. c) Exceedances of critical load in Europe, i.e. 'map a' minus 'map b' (Rodhe *et al.*, 1995)

5.0 The Radioactive Tracer Technique

Natural and artificial radionuclides present in the atmosphere (radon and its daughters, cosmogenic origin beryllium-7, fission products) have long been widely used as atmospheric tracers. These radionuclides are associated with nonradioactive aerosols and hence can serve to trace the fluxes of aerosols to various surfaces. Lead-210 and caesium-137 have been shown to be particularly useful because they are associated mainly with submicron-sized aerosols, which contain the bulk of the pollutant sulphur and nitrogen (Fowler *et al.*, 1995; Branford *et al.*, 1998; Vahabi Moghaddam *et al.*, 2000). The ^{210}Pb isotope (half-life: 22 y) is the decay product of ^{222}Rn that belongs to the ^{238}U decay series and readily diffuses from soil into the atmosphere. When scavenged from the atmosphere along with carrier aerosols, ^{210}Pb and the stratospheric origin ^{137}Cs (half-life: 30 y) are retained by the organic-rich surface horizons and clay layers of soil that acts as an efficient integrating collector.

We can use measurements of the soil inventory of atmospherically derived naturally occurring radioisotope ^{210}Pb and the anthropogenic one, ^{137}Cs, to quantify the effects of aerodynamic roughness and orography on the deposition of atmospheric aerosols as particles and/or droplets. This independent technique may enable improved estimates of the effects of land use on long-term inputs of pollutants in precipitation, cloud droplets or as aerosol. It provides a measure of long-term variation in deposition rate of atmospheric aerosols as particles and/or droplets that is operationally easier to carry out than the direct measurement by artificial collectors over extended period of time. The method averages over many thousands of precipitation events and is particularly valuable in quantifying aerosol and wet deposition processes at sites where conventional methods are not applicable.

There is, indeed, an obligation for environmental scientists to promote communication and cooperation with policy-making and decision-making sectors of regional states hopping that they can have better, if not the best, policy options to prevent further environmental deterioration. As pointed out by Satake *et al.* (2001), it is our common recognition as well as resolution that there is the need to guard the environment. "A peaceful and beautiful green planet for the generations in the coming centuries" is important.

REFERENCES

Aamlid, D. *et al.*, 1992. Forest decline in Europe. In: D.W. Johnson and S.E. Lindberg (eds.), Atmospheric Deposition and Forest Nutrient Cycling. Springer-Verlag, New York: 543-570.

Allard, B. *et al.*, 1994. Acidifying substances and tropospheric ozone: Effects and critical loads. Report 4352, Swedish Environmental Protection Agency.

Amann, M., 1990. Energy use, emissions, and abatement costs. In: J. Alcamo et al. (eds), The RAINS model of acidification. Kluwer Academic, Dordrecht: 61-113.

Asman, W.A.H., 1992. Ammonia emission in Europe: updated emission and emission variations. National Institute of Public Health and Environmental Protection, Bilthoven, The Netherlands.

AWRG, 1986. Acidity in United Kingdom fresh waters. UK Acid Water Review Group report, Department of the Environment.

Battarbee, R.W. *et al.*, 1985. Galloway: A palaeoecological test of competing hypotheses. Nature, 314: 350-352.

Beilke, S., 1982. Acid deposition: The present situation in Europe. In: S. Beilke and A.J. Elshout (eds.), Acid deposition. CEC Environmental Research Program, Berlin: 5-32.

Branford, D. *et al.*, 1998. Spatial variation of wet deposition rates in an extended region of complex topography deduced from measurements of ^{210}Pb soil inventories, *J. Environ. Radioactivity*, 41, No. 2, pp. 111-125.

Brodin, Y. and Kuylenstierna, J. 1992. Acidification and critical loads in Nordic countries: A background. Ambio, 21, No. 5: 332-338.

Cerny, J. and Paces, T. (eds), 1995. Acidification in the Black Triabgle Region. Czech Geological Survey, Prague.

Charles, D.F., 1991. Effects of acidic deposition on North American lakes: Palaeolimnological evidence from diatoms and chryophtes. Phil. Trans. R. Soc. London, B327: 403-412.

Choularton, T.W. *et al.*, 1988. The influence of altitude on wet deposition: Comparison between field measurement at Great Dun Fell and the predictions of a seeder feeder model. Atmos. Environ., 22, No. 7: 1363-1371.

Cowling, E.B., 1982. A historical resume of progress in scientific and public understanding of acid precipitation and its consequences. Environ. Sci.Tech., 16: 110A-123A.

Cowling, E. and Nilsson, J., 1995. Acidification research: Lessons from history and visions of environmental future.Water, Air and Soil Pollution, 85: 279-292.

Cox, R.A. and Penkett, S.A., 1983. Formation of atmospheric acidity. In: S. Beilke & A.J Elshout (eds), Acid Deposition. Proceedings of the CEC workshop, Berlin: 58-83.

Elvingson, P., 1996. Evidence of continued decline. Acid News, 5: 6-7.

EMEP, 1997. Transboundary air pollution in Europe. MSC-W Report 1/1997. Norwegian Meteorological Institute, Oslo.

Erisman, J.W. *et al.*, 1994. A generalised description of the deposition acidifying pollutants on a small scale in Europe.In: P.M. Borrel et al. (eds), The proceedings of EUROTRC symposium '94. Academic, The Hague: 588-596.

Erisman, J.W. *et al.*, 1995. Particle deposition to forests, In: G J Heij and J W Erisman (eds), *Acid Rain Research:Do we have enough answers?* Elsevier Science, Amsterdam, pp. 115-126.

Foell, W.K. and Green, C.W., 1990. Acid rain in Asia: An economic, energy and emission overview. Second workshop on acid rain in Asia, Asian Institute of Technology, Bankok, 19-22 November, 1990.

Fowler, D., 1995. Atmospheric inputs of nitrogen. In: M. Hornung et al. (eds.), Mapping and Modelling of Critical Loads for Nitrogen- a workshop report. Institute of Terrestrial Ecology, Edinburgh: 128-141.

Fowler, D. *et al.*, 1995. Quantifying fine-scale variability in pollutant deposition in complex terrain using ^{210}Pb inventories in soil, *Water, Air and Soil Pollution*, 105, pp 459- 470.

Galloway, J.N. *et al.*, 1995. The composition of precipitation in remote areas of the world, *j. Geophys. Res.*, 87: 8771.

Galloway, J.N., 1995. Acid deposition: Perspectives in time and space. *Water, Air, and Soil Pollution*, 85: 15-24.

Galloway, J.N., 2001. Acidification of the World: Natural and Anthropogenic. *Water, Air,and Soil Pollution*, 130: 17-24.

Gorham, E., 1989. Scientific understanding of ecosystem acidification: A historical review. Ambio, 18, No. 3: 150-154.

Graedel, T.E. *et al.*, 1995. Global emissions inventories of acid related compounds. Water, Air, and Soil Pollution, 85:25-36.

Havas, M. and Rosseland, B.O., 1995. Response of zooplankton, benthos and fish to acidification: An overview. *Water, Air and Soil Pollution*, 85: 51-62.

Henrikson, A. et al., 1992. Critical loads of acidity: Nordic surface waters. Ambio, 21, No. 5: 356-363.

Hicks, B.B., 1995. On the determination of total deposition to remote areas", in: G J Heij and J W Erisman (eds), *Acid Rain Research: Do we have enough answers?* Elsevier Science, Amsterdam, pp 163-173.

Hov, O., 1987. Transport and acid deposition. In: O. Hov et al. (eds), Evaluation of atmospheric processes leading to acid deposition in Europe. CEC Report: EUR 11441, Lillestrom, Norway: 69-85.

Hov, O. *et al.*, 1987. Evaluation of atmospheric processes leading to acid deposition in Europe. Rep. 10, EUR 11441, CEC, Brussels.

Howells, G., 1990. Acid rain and acid waters. Ellis Horwood, Chichester, England.

Hultberg, H. and Grennfelt, P., 1991. Sulphur and sea salt deposition as reflected by throughfall and runoff chemistry in forested catchments. IVL, B-1009, Swedish Environmental Research Institute, Goteborg, Sweden.

Kamari, J., 1986. Linkage between atmospheric inputs and soil and water acidification. In: J. Alcamo and J. Bartnicki (eds), Atmospheric computations for assessment of acidification in Europe, RR-86-5, International Institute for Applied Systems Analysis, Laxenberg, Austria.

Krahl-Urban, B. *et al.*, 1990. Forest decline. Assessment Group for Biology, Ecology and Energy,Julich Nuclear Research Centre, German Ministry of Research and Technology.

Lee, D.S. *et al.*, 1997. Estimations of global NO_x emissions and their uncertainties. Atmos. Environ., 31: 1735-1749.

Lovblad, G. *et al.*, 1992. Deposition of sulphur and nitrogen in the Nordic countries: Present and future. Ambio, 21, No.5: 339-347.

Mannion, A.M. and Bowlby, S.R., 1992. Environmental Issues in the 1990's. John Wiley; Chichester, UK.

Mason, B.J., 1992. Acid rain. Claredon, Oxford.

Matzner, E. and Murach, D., 1995. Soil changes induced by air pollutant deposition and their implication for forests in Central Europe. *Water, Air and Soil Pollution*, 85: 63-76.

MSC-W, 1997. Transboundary Air Pollution in Europe. MSC-W Report 1/1997. Norwegian Meteorological Institute,

Blindern, Norway. Oden, S., 1968. The acidification of air and precipitation and its consequences in the natural environment. Ecological Committee Bulletin 1, Swedish Natural Science Research Council, Stockholm. Translations Consultants, Ltd.Arlington, USA.

OECD (Organizattion for Economic Co-operation and Development), 1979. The OECD program on long range transport of air pollutants: Measurements and findings. OECD, Paris.

OTE (Office of Technology Assessment), 1984. Acid rain and transported air pollutants: Implications for public policy. Washington, DC.

Overrein, L.N. et al., 1980. Acid precipitation: Effects on forest and fish. Final report of the SNSF project 1972-1980.

Oslo, Norway. Prinz, B., 1989. Air pollution and forest ecosystem research. In: A.H.M. Bresser and P. Mathy(eds.), Monitoring Air Pollution and Forest Ecosystem Research. Report No. 21, CEC, Brussels, Belgium: 7-28.

Rasmussen, L. *et al.*, 1995. Experimental studies and modelling of enhanced acidification and recovery. *Water, Air,and Soil Pollution*, 85: 77-88.

Renberg, I. and Hellberg, T., 1982. The pH history of lakes in SW Sweden, as calculated from the sub-fossil diatom flora of the sediments. Ambio, 11: 30-33.

Renberg, I. *et al.*, 1991. Recent acidification and biological changes in Lilla Oresjon, southwest Sweden, and the relation to atmospheric pollution and land-use history. Phil. Trans. R. Soc. London, B327: 391-396.

RGAR , 1997. Acid deposition in the United Kingdom, 1992-1994. Fourth report of the Review Group on Acid Rain, Department on Environment.

Roberts, T.M. *et al.*, 1989. Causes of type-1 spruce decline in Europe. Forestry, 62, No. 3: 179-222.

Rodhe, H. *et al.*, 1995a. Acid Reign '95? Summary Statement from the 5th International Conference on Acidic Deposition Science and Policy. Gothenburg, Sweden, 26-30 June 1995: Kluwer Academic.

Rodhe, H. *et al.*, 1995b. Global scale transport of acidifying pollutants. *Water, Air and Soil Pollution*, 85: 37-50.

Ruijgrok, w. *et al.*, 1995. Dry deposition of particles: Implications and recommendation for mapping of deposition over Europe, *Tellus*, 47B, pp 587-601.

Satake, K. *et al.*, 2001. Acid Rain 2000 – Conference Summary Statement – Looking Back to the past and Thinking of the Future. *Water, Air, and Soil Pollution*, 130: 1-16.

Smith, F.B., 1991. An overview of the acid rain problem. Meteorological Magazine, 120, No. 1426: 77-91.

Spiro P.P. et al., 1992. Global inventory of sulphur emissions with $1^{o}x1^{o}$ resolution. J. Geophys. Res., 97:6023-6036.

Sverdrup, H. *et al.*, 1990. Mapping critical loads. Guidance to criteria, methods and examples for mapping critical loads and areas where they have been exceeded. Report 1990:14, Nord 1990: 98, Nordic Council of Ministers, Copenhagen.

Sverdrup, H. *et al.*, 1992. Critical loads for forest soils in the Nordic countries. Ambio, 21, No. 5: 348-355.

Tamm, C.O., 1989. Comparative and experimental approaches to the study of acid deposition ffects on soil as substrate for forest growth. Ambio, 18, No. 3: 184-191.

UN ECE, 1997. Forest condition in Europe. Results of the 1996 crown condition survey, 1997 Technical Report. Brussels, Belgium.

United Nations Economic Commission for Europe, 1988. Conclusions and draft recommendations of The workshop on critical levels for forests, crops and materials and on critical loads for sulphur and nitrogen, EB.AIR/R.30, UN ECE, Geneva.

Vahabi-Moghaddam, M. *et al.*, 2000. Study of aerosol deposition at a wind exposed forest edge using ^{210}Pb soil inventories", *6th International Conference on Acidic Deposition*, Tsukuba, Japan, 10-16 December 2000.

Visser, H., 1995. Forest condition in Europe and North America: What have we learnt over the past ten years? In: G.J. Heij and J.W. Erisman (eds), Acid Rain Research: Do we have enough answers? Elsevier Science, Amsterdam: 465-472.

Wellburn, A., 1988. Air Pollution and Acid Rain. Longman, Essex, UK.

Westling, O. *et al.*, 1992. Air pollution in southern and central Sweden: Deposition and Effects. IVL Report B-107 Swedish Environmental Research Institute, Aneboda, Sweden.

Whelpdale, D.M., 1987. Atmospheric deposition of acidic pollutants. In: Int. Symp. on acidification and water path Bolkesjo, Norway, 4-8 May,1987.

Zhao, D. and Sun, B., 1986. Air pollution and acid rain in China. Ambio, 15, No. 1: 2-5.

20

Exploring the Therapeutic Potentials of African Medicinal Plants

Charles Wambebe
International Biomedical Research in Africa, Abuja, Nigeria
E-mail: wambebe@yahoo.com

Prior to the advent of colonialists, Africans depended entirely on African traditional medicines for the management of all their prevailing diseases. Currently, about two thirds of Western medicines are derived from natural sources attesting to their therapeutic potentials. Although Africa has over 50,000 plant species and a rich cultural heritage of indigenous medical knowledge (IMK), less than 2% of African medicinal plants have been subjected to scientific/clinical validation. According to WHO, the criteria for evaluating traditional medicines, like those for conventional medicines, are quality, safety and efficacy. Research and development of standardized phytomedicines based on Africa's IMK involve some strategic issues. The application of WHO Good Laboratory Practice (GLP) principles leading to generation of solid scientific data which can be published after patenting the new innovation is crucial. Ethical review systems need to be strengthened in health institutions. The mandatory evaluation of study protocols and subsequent monitoring of clinical trials by national drug regulatory authorities are generally weak. The issue of IMK ownership and equitable and fair benefit sharing must be addressed. The involvement of the private sector in research and development efforts by biomedical researchers is almost non-existent. Similarly, investment by national government in health research is poor and inconsistent. The national governments need to formulate national Traditional Medicine and Biodiversity policies and enact appropriate laws. A robust biodiversity policy, which among others, covers the cultivation and conservation of medicinal and aromatic plants as well as appropriate instruments for combating bio-piracy needs to be developed by various African countries. Furthermore, clinical evaluation of any standardized phytomedicine must be done in accordance with WHO Good Clinical Practice principles. The example of wealth creation through the development of a standardized phytomedicine (NIPRISAN) by the National Institute for Pharmaceutical Research and Development for the management of sickle cell anaemia will be discussed.

KEY WORDS: *African traditional medicine, Scientific/clinical validation, Production.*

1.0 INTRODUCTION

African Traditional Medicine has been defined by Sofowora (1982) as "the aggregate blend of explicable and inexplicable knowledge and practice used in diagnosing, preventing or eliminating a physical, mental or social disease and may be wholly dependent upon past experiences and observations passed down from generation to generation in writing or by spoken word while bearing in mind the original concept of nature, which includes the material world, the sociological environment, whether living or dead and the metaphysical forces of nature". According to the Bible "Their (plants) fruits will be used for food and their leaves for medicine" (Ezekiel 47 verse 12b). In African Traditional Medicine (ATM), food plants are widely used as medicines. It has been estimated that about 67% of the new chemical entities introduced worldwide from 1981 to 2002 were derived from natural sources (Mills *et al.*, 2005). The oldest herbal text, *Prescriptions for 52 diseases* (Wu Shi Er Bing Fang) was written in 168 BC (Harper, 1997). According to some scholars, Traditional Chinese Medicine (TCM) has been practiced for more than 10,000 years (Taylor, 1998). It is widely believed that the original home of man is Africa. Thus, African Traditional Medicine, though not documented should be, if not older than TCM, but as old. In spite of this historical fact, research in Traditional Medicine is not attracting attention it rightfully deserves because of the following reasons:

Chemical structures for novel pharmaceutical substances are patentable and consequently potentially highly profitable. On the other hand, chemical structures from natural sources are considered common knowledge and cannot be patented (Moss, 2007). However, it is possible to obtain "use" patent for a novel application of a natural compound but such patents are far less profitable than "structure" patents and suffer from complex intellectual property rights issues (Sampath, 2005). The pharmaceutical companies recover some of the research and development expenditures partly from structure patents-current cost for developing one single medicine is about $1billion (Kaitin *et al.*, 2006). The second challenge confronting research in Traditional Medicine (TM) is that funding for medical research comes from pharmaceutical companies that need to invest their resources in patentable compounds so as to remain in business (Hilbrich and Sleight, 2006). Few companies involved in African medicinal plants are only interested in isolating active compounds which can be used for derivatives for patenting leading to their use outside ATM context (holistic), become very expensive and laden with side effects. The third issue is that the big pharmaceutical companies are looking for a "silver bullet" miracle medicine for effective management of chronic disorders and rapid cure for acute but recurrent infections/diseases. Generally, traditional medicines are effective in synergistic combinations thereby eliminating or reducing side effects of their single components while significantly enhancing their curative potencies *vis-à-vis* single components of the plants. This paper will review challenges confronting the research and development of standardized phytomedicines from African Medicinal plants. African case studies will be highlighted while some strategic issues will be treated which are crucial to wealth creation of biodiversity resources *vis-à-vis* development of medicines.

2.0 DRUG DEVELOPMENT CHAIN

A simplified drug development chain encompasses discovery phase, pre-clinical/non-clinical phase, clinical phase, regulatory phase, marketing phase, post-marketing phase and policy practice (Figure 1). Traditional medicine belongs to the discovery phase. If a proper ethno medical survey is conducted accompanied by clinical observational study following WHO guidelines, it is possible to save substantial funds and drastically reduce the time needed to obtain credible data for proceeding to non-clinical phase or stopping at that stage. Essentially, the rest processes are exactly the same as for research and development of conventional medicines. This is not surprising since according to the WHO, the criteria for evaluating plant-derived and classical structure-based medicines are the same viz; quality, safety and efficacy. However, there are some unique challenges associated with the research and development of plant derived medicines which will be highlighted below.

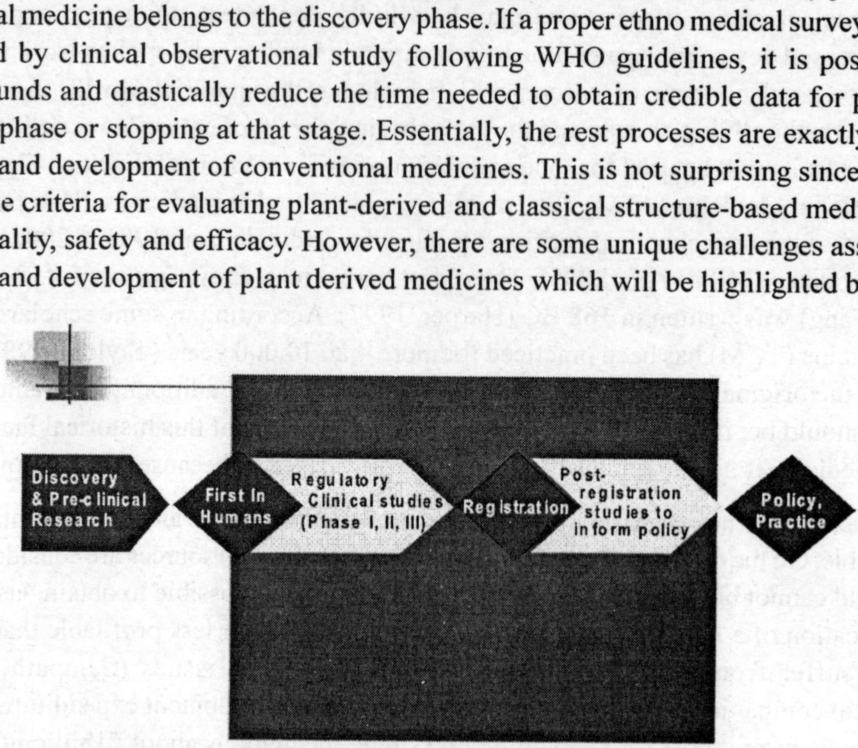

FIGURE 1: Simplified Drug Development Chain

The value chain for research and development of plant derived medicinal products is shown in figure 2. This figure should be used taking cognizance of the WHO classification scheme for African Traditional Medicines:

- Family recipes
- Popular community recipes
- Phytomedicines
- Imported traditional medicines

In order to add maximum value to the potential therapeutic agent, the target should reach phytomedicine. This may require collaboration within Africa and beyond using a multi-disciplinary approach.

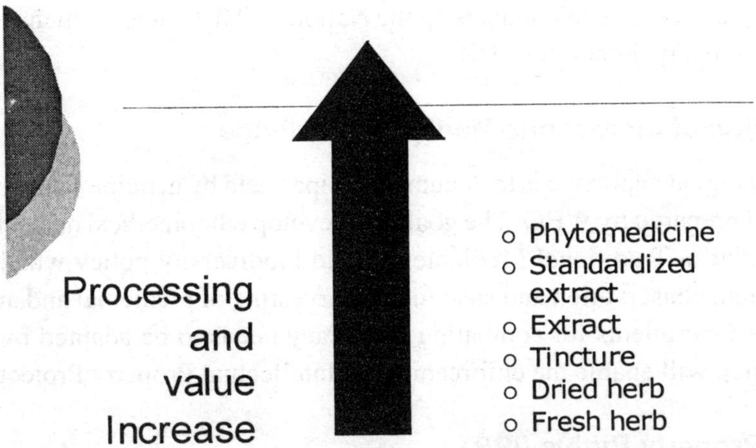

FIGURE 2: Value Added to Plant-Derived Medicines

3.0 CHALLENGES

3.1 Official Recognition of Traditional Health Practitioners

For the purpose of exploring the therapeutic potentials of African Traditional Medicine, issues related to the Traditional Health Practitioner (THP) should be adequately addressed. One key element which can facilitate the cooperation of THP is recognition. The national health authorities need to utilize the various documents developed by the WHO for institutionalization of traditional medicine and recognize the THPs as holders of indigenous medical knowledge (IMK). For example, the generic document on National Policy on Traditional Medicine can be appropriately modified and adapted.

3.2 Traditional Medicine Policy

Less than half of the countries in the WHO African countries have policies on Traditional Medicine. In order to promote the role of traditional medicine in public health care delivery systems, the WHO developed some generic documents for adaptation and adoption by Member States. One of such documents is on policy. The policy, among others, has provisions for the Traditional Health Practitioners Council, registration of THPs, professional curriculum for THPs. Subsequently, Traditional Medicine Law and applicable regulations can be developed. Furthermore, relevant government structures including directorate or department of Traditional Medicine should be established. These instruments plus declaration in a Government gazette *vis-à-vis* official recognition of Traditional Medicine will encourage the THPs to collaborate with the Government and biomedical researchers. In such collaboration arrangements, appropriate Memorandum of Understanding (MOU) which recognizes the THPs as partners in the product development process, owner of the IMK and consequently beneficiaries of the commercial benefits of the IMK. Similarly, the THPs will be trained in the proper use of code of conducts by the National THP Council which, among others, has the powers to sanction any defaulting THP.

3.3 Documentation of Indigenous Medical Knowledge

The immediate and urgent challenge is to document recipes held by genuine THPs using appropriate documentation tool prepared by WHO. The goal is to develop ethnomedical data which contains the best practices in African Traditional Medicine. A solid biodiversity policy which, among others, covers the cultivation, conservation and sustainable harvesting of medicinal and aromatic plants as well as appropriate instruments for combating bio-piracy needs to be adapted by various African countries. Such a step will enable the enforcement of Intellectual Property Protection policy.

3.4 Intellectual Property Rights (IPR)

One of the most crucial factors regarding the research and development of therapeutic potentials of African indigenous medical knowledge is intellectual property rights. This will require an appropriate Memorandum of Agreement (MOA) to be developed which must include material transfer mechanism, fair and equitable sharing of benefits, etc. The process of developing the MOA should be transparent and allow for development of mutual trust. Closely linked to IPR is the important issue of benefit sharing arrangement that is acceptable to all the stakeholders. There should be both monetary and non-monetary components in the IPR document. Furthermore, the mechanisms for implementing the benefit sharing arrangement should be included in the IPR Memorandum of Agreement (for example Trust Fund).

3.5 Ethical

The drug industry is among the most regulated industries in the world. In most countries in Sub-Saharan Africa the regulatory structures, resources and experts with requisite knowledge and skills on the mandatory ethical issues are very few. In some countries where such experts exist, they are usually not provided with the resources to perform their functions. Consequently, issues related to the acceptable administration of informed consent, ethics review committees (institutional, hospital and national) and regulatory approvals are usually not treated properly in accordance with international standards. Furthermore, mandatory regulatory monitoring of clinical trials is hindered by inadequate resources and limited expertise.

3.6 Legal

The legal issues related to research and development of plant-derived medicines include acceptable memorandum of understanding for research collaboration between the traditional health practitioners (THPs) and researchers/institutions, intellectual property rights, appropriate policies, laws and regulations. Unfortunately, these crucial issues have not been addressed adequately in most African countries.

3.7 Social

Generally, there are multiple THP Associations-they need to be properly organized into one Federation of THP Associations (eg Ghana, Zimbabwe) so as to enhance their negotiating capacity. There should be clearly defined agreement regarding the involvement of the local community where the traditional medicine recipe originates in the research and development process which may include social facilities and capacity building initiatives.

3.8 Resources

Research and development of plant-derived medicines although not as expensive as development of orthodox medicines (about $1 billion), is still a high capital intensive investment. In Sub-Saharan Africa, there are many competing demands on the limited available resources. Furthermore, skilled human resources with passion for research are few. The political will to regard research and development initiatives as the engines for development is generally weak.

3.9 Infrastructure

In order to undertake research and development of plant-derived medicines, the mandatory requirements for basic research equipment, regular supplies of good quality water and energy cannot be over emphasized. Standard physical facilities which will guarantee generation of quality research data are usually severely limited.

3.10 Scientific procedures

The criteria for evaluating traditional medicines, like those for conventional medicines, are quality (i.e. medicines should be well formulated and possess consistent quality from batch to batch as specified in the official pharmacopoeia chosen as a standard), safety (i.e. medicines should not present risks that are disproportionate to their benefits) and efficacy (i.e. medicines should be effective for the indications claimed and should be clinically significant and useful). In order to generate quality research data which can be reproduced anywhere in the world, the WHO Good Laboratory Practice (GLP) was developed. Most research scientists in Sub-Saharan Africa have not undergone appropriate training in GLP. Even the biomedical scientists, who have been trained, find its application in their routine research a real challenge due to limited resources and sometimes lack of political will from the management.

3.11 Clinical methodologies

Due to the complex mixtures inherent in traditional medicine preparations, performing clinical pharmacokinetic studies pose a challenge to the researcher. There are multiple factors which can influence the safety and efficacy of plant-derived medicines in clinical studies include. These include:

- Proper identification of plant species

- Plant habitat
- Age of plant
- Time of plant collection
- Harvesting conditions
- Process technologies
- Packaging and storage conditions
- Standardization of dosage, formulation and dosage regimen.

In addition to the issues indicated above, the mandatory discipline for prompt record keeping associated with conducting; managing and monitoring clinical trials pose additional challenges in Sub-Saharan Africa. Furthermore, training in the application of WHO Good Clinical Practice principles is mandatory.

3.12 Leadership

Political will, professional understanding and academic leadership are crucial to mobilizing the resources needed for research and development of plant-derived medicines.

4.0 WHO SUPPORT

The following summarizes the interventions of WHO regarding their support for promotion of Traditional Medicine in sub-Saharan Africa:

- Regional Strategy on Promoting Traditional Medicine in Health Systems in WHO African Region, 2000.
- Abuja Declaration by African Union on Research of African Traditional Medicine as a priority in 2001.
- African Union Declaration of Decade of African Traditional Medicine (2001-2010).
- Adoption of global Traditional Medicine Strategy, 2002-2005.
- Support for clinical research projects on their safety and efficacy, particularly with reference to diseases such as malaria (Ghana, Kenya, Nigeria) and HIV/AIDS (Burkina Faso, Zimbabwe, Nigeria).
- Advocating the rational use of TM by promoting their evidence-based use.
- Promoting the safety, efficacy and quality of traditional medicines through expanding the knowledge-base on traditional medicine and providing guidance on evaluation, regulation and quality assurance standards. For example, *Model Generic Guidelines on Clinical Evaluation of Traditional Medicines in the WHO African Region, WHO Good Laboratory Practice, etc.*

5.0 AFRICAN CASE STUDIES

5.1 HIV/AIDS

A standardized herbal medicine was developed by University of Ougadougou under the leadership of Prof. Jean-Baptiste Nikiema and clinically evaluated at Centre Medical Saint Camille, Ougadougou, Burkina Faso with WHO support. The study protocol was based on *Generic Model Guidelines for Clinical Evaluation of Traditional Medicines in the WHO African Region.* The Traditional Health Practitioners were selected based on anecdotal evidence, community recognition and interviews. The number of HIV/AIDS patients recruited for the study was 1,246. The pilot study was undertaken for one year and 1200 study participants completed the trial. The CD4 values were significantly elevated in 75% of the participants whose CD4 values were tested (i.e. 82). Similarly, viral load was estimated for 82 study participants and their values were reduced significantly and undetectable in a few patients. Due to financial constraints, the controlled, randomized comparative clinical evaluation of the herbal medicine has not yet been undertaken.

5.2 Malaria

Gamaniel et al (1998) studied the non-clinical efficacy and safety profiles of a standardized phytomedicine called NIPRD AM1 and was safe and efficacious against *P. berghei* in mice. The extract had an LD50 > 2000mg/kg per oral in rats and mice and did not show any significant toxic activity on the organs in the 28 day study. There was no abnormality in the liver enzymes AST, ALP and ALT in rats. There was no effect on both conjugated and total bilirubin which often results from jaundice or liver disease. The extract did not alter the levels of creatinine or urea which are sensitive indicators of kidney function. Furthermore, the extract had no effect on triglycerides, cholesterol and glucose (products of carbohydrate and fatty acid metabolism). These agents increase in liver disease, renal damage, hypertension and diabetes mellitus and are risk factors in coronary heart disease. Lack of effects of the product even at high doses indicates relative safety in laboratory animals.

In 2000, Gamaniel and his colleagues conducted a comparative randomized clinical trial using NIPRD/ AM1 against symptomatic but uncomplicated malaria in human volunteers at two district hospitals located in Abuja, Nigeria. The study had three arms with 30 study participants per arm; the test arm received 300 mg of standardized extract orally two times daily for five days, the second arm was given chloroquinne 600 mg base orally once daily for three days and the last arm used sulphadoxine/pyrimethamnie (Fansidar). The results indicate that NIPRD/AM1

was efficacious against uncomplicated malaria. The parasite clearance was better than chloroquine and there were no threats of serious side effects affecting the organs or tissues. The randomized, controlled, comparative clinical trial of the herbal medicine is presently in progress using three University Teaching Hospitals in Nigeria. The comparator medicine is Artemisinine combination product.

5.3 Sickle Cell Anaemia

About 70% of sickle cell anaemia subjects reside in Africa, estimated at over 12 million. The prevalence of sickle cell anemia (SCA) is estimated at over 2% (i.e. about 2.8 million) while those with sickle cell traits is about 25% of the general population in Nigeria (Lesi, 1999). Furthermore, infant mortality is about 8% and survival rate of SCA babies in rural areas by five years of age is about 20%. These statistics indicate that SCA is probably the most neglected serious public health disorder with serious mortality and morbidity rates in Africa. The general objective was to undertake pre-clinical and clinical assessments of a herbal extract *vis-à-vis* management of sickle cell anemia using Good Laboratory Practice and Good Clinical Practice (GCP) principles respectively. In Sub-Saharan Africa, there is no standard treatment for sickle cell anemia; only palliative management is generally available. In view of this situation, sickle cell anemia patients consult traditional health practitioners who give them herbal medicines. Ethnomedical evidence indicated that Rev. P.O. Ogunyale used herbal recipe that apparently reduced the frequency of SCA–related crises in patients. Subsequently, the National Institute for Pharmaceutical Research and Development (NIPRD) developed a Memorandum of Understanding (MOU) with Rev. P.O.Ogunyale (Traditional Health Practitioner-THP). In 1993, a clinical observational study was undertaken which indicated a promising result. The first crucial investigation was safety evaluation. The short term toxicity study was undertaken and the data indicated that NIPRISAN was safe in laboratory animals (Awodogan *et al.*, 1996). Subsequently, the pharmacological profile of NIPRISAN in animals was conducted (Gamaniel *et al.*, 1998). The phytochemical screening of the four plants used in the formulation of NIPRISAN was also undertaken. NIPRISAN is a standardized extract from four medicinal/food plants: *Piper guineenses* seeds, *Pterocarpus osun* stem, *Eugenia caryophylum* fruit *and Sorghum bicolour* leaves. *In vitro* data showed that NIPRISAN reversed sickled red blood cells (RBC) and protected RBC from being sickled when exposed to low oxygen tension (Figure 3). It was observed that NIPRISAN dose-dependently delayed polymer formation of haemoglobin S using UV technique from 6.6 to 38 minutes. The observed inhibition of sickling was time-related while cytochrome p50 was slightly reduced from 34.1 to 31.6 mmHg. Furthermore, NIPRISAN did not affect the Hill coefficient but induced 85% increased solubility of deoxy haemoglobin S. It is important that NIPRISAN did not cause lysis of the red blood cells; membrane associated denatured Hb or methaemoglobin formation.

The *in vivo* efficacy study was undertaken at Children Hospital of Philadelphia, USA. Histological examination of lungs of control Tg transgenic mice carrying human sickle haemoglobin showed entrapment of massive numbers of sickled cells in alveolar capillaries. NIPRISAN significantly cleared the lungs of sickled cells (Figure 4). Furthermore, figure 4 shows the profound effect of NIPRISAN (HEMOZIN) on the survival time of Tg mice under hypoxic conditions. In the absence of NIPRISAN all the Tg mice die within 15 minutes of exposure to the hypoxic conditions. However, NIPRISAN dose-dependently (0-500 mg/kg po) improves the survival time of the Tg mice under the same hypoxic conditions from 0 to 100% survival ($P<0.0001$) (Iyamu *et al.*, 2003). One possible cautious extrapolation of this significant effect of NIPRISAN is protection of SCA subjects from acute chest syndrome which is a major cause of sudden death among SCA subjects.

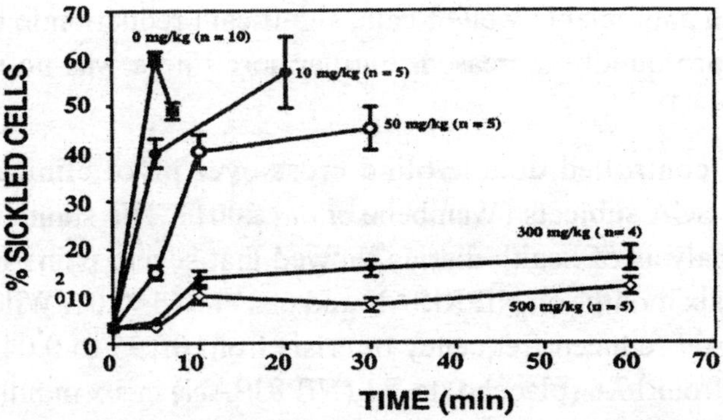

Fig 3. Degree of SS cell sickling in Tg sickle mice over time, upon hypoxic exposure. Five groups of Tg sickle mice were treated with 0 (n = 10), 10 (n = 5), 50 (n = 5), 300 (n = 4) or 500 (n = 5) mg/kg/d Nix-0699 for 7 d. Thereafter, the mice were exposed to hypoxia for 60 min. During hypoxic exposure, aliquots of blood were collected from the tail vein under venous oxygen pressure, and the cells were fixed in 2% glutaraldehyde solution without exposure to air. The percentage of sickled cells was determined by the computer-assisted image analysis system as described in Materials and Methods.

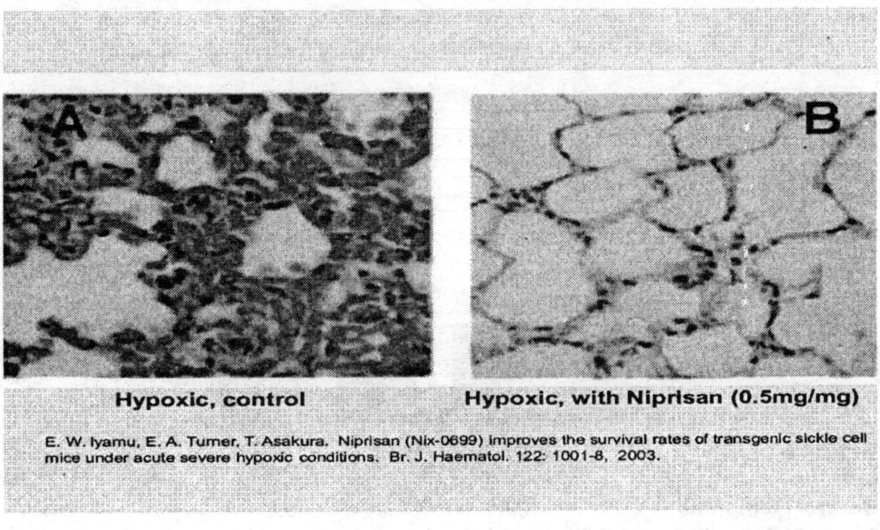

FIGURE 4: Effect of NIPRISAN on Trapping of Sickle Cells in Mouse Lung

SCA subjects aged between 2 and 45 years were recruited for phase II A clinical trial. The data indicated that all the subjects benefited from NIPRISAN with no serious adverse effect. About 80% of the subjects did not experience any crisis during the period of the study (12 months) while the remaining subjects had less severe and less frequent SCA-related crises. The subjects experienced improved appetite with appreciable weight gain, significant reduction in hospital admission while attendance at school profoundly increased. Furthermore, there was no evidence of kidney/liver damage.

Randomized placebo-controlled double-blind cross-over pivot clinical trial (Phase IIB) was undertaken using 100 SCA subjects (Wambebe *et al.*, 2001). The summary of the clinical data is shown in Figure 5. Analysis of health diaries showed that severe pains were reported 7.9 and 21 times respectively for six months on NIPRISAN and placebo ($p<0.01$, Wilcoxon Signed Rank test). Furthermore, NIPRISAN reduced frequency of crisis from 0.195 to 0.045 per patient per month. Absenteeism reduced from 19.6 (placebo) to 7.1 (NIPRISAN) in six months. Frequency of hospital admission reduced from 0.22 (placebo) to 0.15 (NIPRISAN) per month. In addition, no changes in nutritional and metabolic analytes-total protein, albumin, total calcium, fasting total cholestrol, fasting triglyceride, fasting blood glucose and uric acid. Similarly, AST and ALT levels remained normal indicating no acute liver toxicity. Alkaline phosphatase and GGT were unchanged indicating that there was no cholestatic and chronic toxicity. Interestingly, creatinine and BUN levels remained same as placebo group suggesting that there was no kidney toxicity. The summary of this study (Figure 5) as well as previous pilot clinical trials is that NIPRISAN is safe and efficacious for the management of SCA.

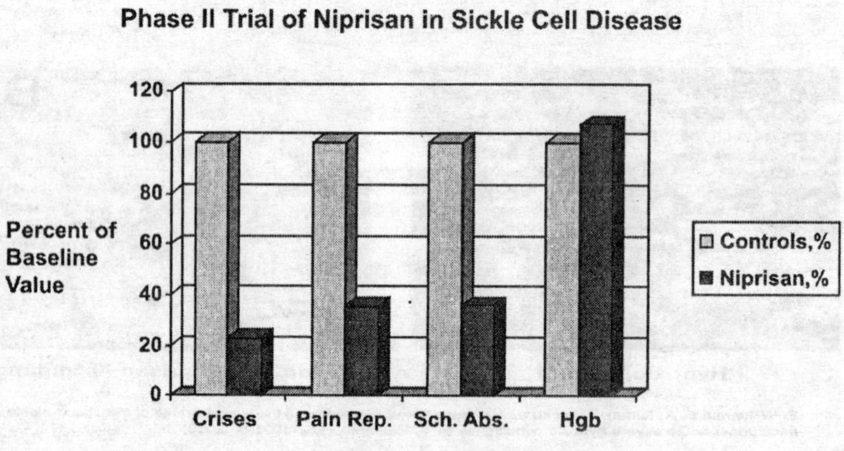

	Crises	Pain Rep.	Sch. Abs.	Hgb
Control / 6 months	**1.17**	**22.1**	**19.6**	**8.6**
Treated / 6 months	**0.27**	**7.9**	**7.1**	**9.2**

C. Wambebe, E. A. Bamgboye, B. O. Badru, et al. Double-blind, placebo-controlled, randomised, cross-over, clinical trial of NIPRISAN in patients with Sickle Cell Disorder. Curr. Ther. Res. 62: 26-34, 2001.

FIGURE 5: Summary of Phase II Clinical Data of NIPRISAN in Sickle Cell Subjects.

Patenting of the process technology and potential therapeutic use of NIPRISAN was undertaken through funding from UNDP in USA (Wambebe *et al.*, 1998), England, India and 42 other countries. In June 2002, the Federal of Ministry of Health granted an exclusive license for the manufacture and global marketing of NIPRISAN to XECHEM Inc, USA. Subsequently, XECHEM Inc established a factory at Abuja for the manufacture of NIPRISAN for the global market. Such an action has obvious positive impacts on the Nigerian economy, capacity building and foreign earnings and serves as a good example of a private-public partnership arrangement. XECHEM Nig has employed about 150 Nigerian scientists and professionals in the factory, administration, Quality Control and Marketing departments. NIPRISAN was granted orphan drug status by the United States Food and Drug Administration (2004) and the European Medicine Evaluation Agency (2005). On July 2006, the then President of Nigeria, Chief Olusegun Obasanjo commissioned XECHEM Manufacturing Plant at Abuja and launched NIPRISAN (renamed NICOSAN) for use in the health sector.

6.0 STRATEGIC ISSUES

6.1 Vision-Innovation based on indigenous medical knowledge and biodiversity.

6.2 Mission-selection of priority disease to be targeted for research and development must be technology driven.

6.3 Ethical Reviews at institutional level and regulatory clearance by the National Drug Regulatory Authority are mandatory.

6.4 **Multi-disciplinary approach**-Research and development of plant-derived medicines is complex and involves various experts (chemists, taxonomists, pharmacists, pharmacologists, clinicians, statisticians, etc). Collaboration at national and international levels is recommended due to local shortage of relevant skilled personnel.

6.5 **Private public partnership**- the success story with NIPRISAN/NICOSAN exemplifies the significance of involving the private sector.

6.6 **Intellectual property rights** issues including equitable sharing of benefits to all the stakeholders must be clearly agreed at an early stage in the research and development process.

6.7 **Resources**-physical facilities, skilled human personnel and financial resource constitute the most critical factors in research and development of plant-derived medicines in Africa.

7. CONCLUSION

Apparently, less than 10% of the plant species in Africa (over 50,000 plant species) have been subjected to some scientific validation *vis-a-vis* their phytochemical constituents. Furthermore, the assessment of the biological effects of isolated compounds/standardized purified extracts is essentially lacking. Careful consideration of the strategic issues mentioned above is crucial to more success stories from Africa on research and development of plant-derived medicines based on African

Traditional Medicine. Such sustained efforts will lead to acquisition of valuable skills in research and development of plant-derived medicines, preservation of valuable indigenous medical knowledge, improved quality of life through scientific/clinical exploration of African medicinal plants, job opportunities and wealth creation both on local and foreign currencies. New medicines developed in Africa by African scientists will also add to the global therapeutic profile which has inherent positive political impact.

ACKNOWLEDGEMENTS

My sincere gratitude to:

- UNDP, Governments of Nigeria and Japan for generous funding for R&D studies.
- WIPO for guidance on IPR issues
- Late Rev P.O. Ogunyale for entrusting me with the original recipe for NIPRISAN
- NIPRD Research Fellows, Abuja, approx 20
- Philadelphia: NIH Hemoglobin Reference Lab Toshio Asakura, MD, PhD, Hematology and Biochemistry
- O. Iyamu, MD, O. Abdulmalik, many others
- New Brunswick, N J, Ramesh Pandey, PhD, Medicinal Chemistry, CEO Xechem Inc
- P. Tripathi, R. Misra, many others approx 10
- Richmond,V.A., Donald Abraham, PhD + >15
- WHO for funding clinical studies of NIPRD AM1.
- Above all, the Almighty God for divine inspiration and establishing the works of my hands.

REFERENCES

Awodogan, A.O. et al., Acute and short term toxicity of NIPRISAN in rats: A biochemical study, *Journal of Pharmaceutical Research and Development*, Vol. 1 (1), pp. 39-45, 1996.

Gamaniel, K., Amos, A., Akah, P.A., Samuel, B.B., Kapu, S., Olusola, A., Abayomi, O., Okogun, J.I. and Wambebe, C. (1998). "Pharmacological Profile of NIPRD 94/002/1-0: A Novel Herbal Antisickling Agent," *Journal of Pharmaceutical Research and Development*, Vol. 3 (2), pp. 89-94.

Harper, D. (1997). Early Chinese Medical Literature: *The Mawangdui Medical Manuscript*, Kegan Paul International.

Hilbrich, L. and Sleight, P. (2006). Progress and problems for randomized clinical trials: From streptomycin to the era of mega trials. *European Heart Journal*, Vol. 27 (18), 2158-2164.

Iyamu, E.W, Turner, E. and Asakura, T. (2003). NIPRISAN (Nix-0699) improves the survival rates of transgenic sickle cell mice under acute severe hypoxic conditions. *British Journal of Hematology*, Vol. 122, pp. 1001-1008.

Lesi, F.E.A. Sickle Cell Disorder: *A Handbook for Patients, Parents, Counselors and Primary Health Care Practitioners*, Edited by A.F. Fleming, Churchill Livingstone Publishers, London, pp. 1-23, 1999.

Mills, E., Cooper, C. and Kanfer, R. Traditional African medicine in the treatment of HIV (2005). *The Lancent Infectious Diseases*, Vol. 5, pp. 465-467.

Moss, R. (2007). Patents over patients. *The New York Times*, April 1.

Kaitin, K.L. et. al., (2006). Cost to develop new biotech products is estimated to average $1.2 billion.

Tufts Center for the study of drug development impact report, Nov/Dec 8(6).

Sampath, P.G. (2005). *Regulating Bioprospecting*. United Nations University Press.

Sofowora, Abayomi (1982). *Medicinal Plants and Traditional Medicine in Africa.* John Wiley and Sons Ltd, New York.

Taylor, M. (1998). Chinese Patent Medicines: A Beginner's Guide. Global Eyes International Press.

Wambebe, C., Ogunyale, P.O., Gamaniel, K.S., Nasipuri, R.N., Okogun, J.I., Samuel, B., Olusola, A. and Orisadipe, A. (1998). *Piper guineense, Pterocarpus osun, Eugenia caryophyllata and Sorghum bicolor* extracts for treating sickle cell disease. United States Patent No. 5,800, 819 of September 1, 1998.

Wambebe, C, Khamofu, H, Momoh, J.A.F, Ekpenyong, M, Audu, B.S, Njoku, S.O. Bamgboye, E.A., Nasipuri, N.A., Kunle, O.O, Okogun, J.I, Enwerem, N.M, Gamaniel, K.S, Obodozie, O.O., Samuel, B, Fojule, G and Ogunyale, P.O. (2001). Double-blind, placebo-controlled, randomised cross-over clinical trial of NIPRISAN in patients with Sickle Cell Disorder. *Phytomedicine.* 2001 July 8(4):252-61.

Wambebe, C., Bamgboye, E.A, Badru, B.O, Khamufu, H., Momoh, J.A., Ekpenyong, M., Audu, B.S., Njoku, S.O., Nasipuri, N.R., Kunle, O.O., Okogun, J.I., Enwerem, N.M., Gamaniel, K.S., Obodozie, O.O., Samuel, B., Fojule, G. and Ogunyale, P.O. (2001). Efficacy of NIPRISAN in the prophylactic management of patients with sickle cell disease. *Cur. Ther. Res. Clin. And Exper.* Vol. 62, No. 1: 26-34

21

Biofuel and Biodiversity Prospects in Tanzania

Hulda Gideon
Tanzania Biodiversity Information Facility (TanBIF), Tanzania
E-mail: hgideon@costech.or.tz

Bio-energy is gathering more and more attention as a feasible way of reducing dependence on imported oil and gas in Tanzania. Biofuel is being hailed as one of the potential weapons in the battle against global warming. The government of Tanzania, like other governments in developing countries, has opened doors for investors to invest in biofuel production. Areas for investment have been identified and more than nine companies have already shown their interest to invest in this sector. This paper is exploring the issue of biofuel in depth, to discover what production and growth in the 'new industry' could really mean to the sustainability of biodiversity in Tanzania.

1.0 COUNTRY PROFILE

Tanzania is located in East Africa on the western Indian Ocean. To the north of the country are Uganda and Kenya; to the west, Burundi, Rwanda, and Congo; and to the south, Mozambique, Zambia, and Malawi. Tanzania is the biggest of the East African countries (i.e. Kenya, Uganda, Rwanda, Burundi and Tanzania). It has a surface area of 945,000 km^2 of which the Zanzibar Island covers 2, 000 km^2. 6.5% of this area is covered by inland waters. The country has both the highest altitude (Kilimanjaro Mountain) and the lowest point (floor of Lake Tanganyika) in Africa. Tanzania has a population (2007 est.) of 38,139,640 with the population growth rate of 1.8%. Agriculture is the dominant sector in Tanzanian economy, providing livelihood, income and employment to over 80% of the population. Other sectors which contribute to the national economy include tourism, mining and fisheries. Widespread poverty limits capabilities to adapt new technologies.

Tanzania has a tropical type of climate. In the highlands, temperatures range between 10ºCand 20ºC during cold and hot seasons, respectively. The rest of the country has temperatures never falling lower than 20ºC. The hottest period spreads between November and February (25ºC - 31ºC) while the coldest period occurs between May and August (15ºC - 20ºC). Two rainfall regimes exist over

Tanzania. One is unimodal (December-April) and the other is bimodal (October-December and March-May). The former is experienced in southern, south- west, central and western parts of the country, and the latter is found to the north and northern coast. In the bimodal regime the March-May rains are referred to as the long rains or Masika (in Kiswahili), whereas the October-December rains are generally known as short rains or Vuli (in Kiswahili).

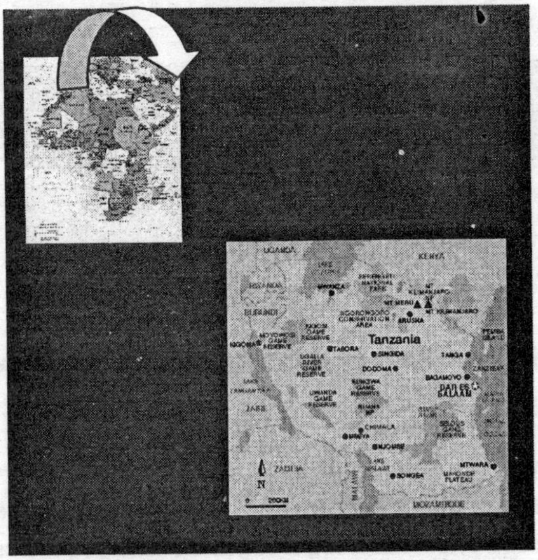

PLATE 1: Map of Tanzania

2.0 AN OVERVIEW OF ENERGY DEMAND AND SUPPLY IN TANZANIA

Tanzania has per capita electricity consumption of 46/KWh per annum, which is growing at the rate of 11 to 13 per cent (http://www.tanzania.go.tz/energy.html 24th February 2009). Petroleum, hydropower and coal are the major source of commercial energy in the country. Currently, electricity contributes about 1.2 per cent of total energy consumption (Mngodo, 2008), and only three quarters of the country (mainly urban areas) is connected to the national grid. The biomass energy resource, which comprises fuel-wood and charcoal from both natural forest and plantations, accounts for 90 per cent of total energy consumption.

Besides wood fuel, the economy of the country has considerable biomass resources in form of agriculture and forest residue and animal wastes which, all together, account for about 90% of the national energy requirements (Mngodo 2008). Commercial energy is basically provided by electricity while in rural households it is predominantly from biomass. The major sources of electricity are generation of hydro, thermo and natural gases power plants, with imported petroleum products contributing significantly. The energy products are consumed in all sectors of the economy, including transport, industry, household, mining and agriculture.

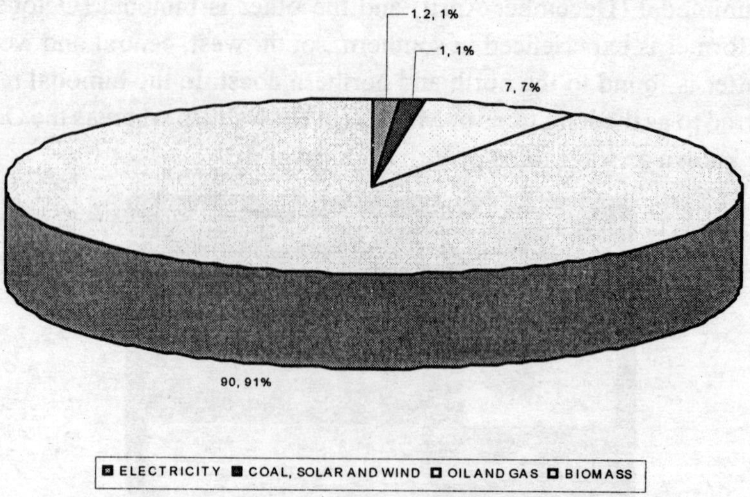

Figure 1: Energy Consumption in Tanzania

Commercial energy is dominated by unpredictable and increasing prices in the world market, with its heavy reliance on imported petroleum products. This has created insecurity and uncertainty in the sectors that entirely depend on these sources and has significant impact on the overall economy. In the past five years, the energy crisis has been a common feature particularly with the hydropower plants frequently facing acute shortage of water due to inadequate rainfall, which has led to very low power production, high cost of generating power using thermal plants that are fuelled by petroleum products.

Nevertheless, Tanzania has large reserves of indigenous energy resources including natural gas, coal, water and bio-energy to meet the demands of the power sector for many years. There is also undetermined potential of geothermal energy. The Government of the United Republic of Tanzania and the Tanzania Electric Supply Company Limited (TANESCO) have identified some key projects such as Mchuchuma deposits which can provide fuel for 400MW generation capacity for up to 35-40 years. There is also an existing small coal mine at Songwe-Kiwira which started production in 1988 with initial production planned at 150,000 tonnes and reaching a maximum of 300,000 tonnes at maturity of the project.

Natural gas reserve located in the Songo-Songo Island offshore on the Indian Ocean is another identified source for production of energy. The gas reserves are estimated to be 726 billion cubic feet. There is also a significant gas field at Mnazi bay near Mtwara. Ocelot Trans Canada Pipelines (OTC), a consortium formed by two Canadian companies are already involved in a project to build a gas pipeline to deliver gas to Dar es Salaam for use in power generation. Stiglers Gorge hydro power project located at narrow gorge on the Rufiji River is yet another project which has been identified to substitute the hydropower generation in Tanzania. Two power stations will be developed with total capacity of 2,300MW. The Government is also planning for grid interconnection with neighbouring countries namely Zambia, Kenya and Uganda.

In summary, there is a big potential of the energy sector in Tanzania. However, at the moment, demand for energy exceeds supply. The Government should continue spearheading the campaign of producing reliable and clean energy. Producing energy from hydropower plants is not trustworthy, as this depends on water resources in particular rainfall. With the current trend of climate change, water shortage will be a "ringing tone" to many of the Sub-Saharan countries. Coal and natural gas are non-renewable resources, and therefore their investment is a short-term one. Therefore the Government, investors, scientists and other partners all have the urgent need to explore how bioenergy can substitute the current energy sources in Tanzania in a sustainable manner.

3.0 BIOFUELS POTENTIAL IN TANZANIA

Bio-energy is a term used to describe energy derived from organic materials, living plants and plant components. This type of fuel appears to be carbon-neutral as the carbon they emit to the atmosphere when burnt is claimed to offset by the carbon that plants absorb from the atmosphere while growing. Biofuel is renewable as fresh supply can be grown as needed and capable to be cultivated in many environments. In addition, biofuel is an integral part of the emerging bio-economy where plant materials are used to produce specific chemical and bulk industrial chemicals. In future, biofuel may replace energy derived from fossil fuels. The full picture, however, is much more complex as different biofuels have widely differing environmental, social and economic aspects.

Like in some other countries in the world, Tanzania is embracing biofuels, and biofuels are already entering the country's investment planning. The driving force, among other things, is their potential to improve energy security and to contribute to climate change mitigation. Nine regions have been identified as potential areas for biofuel plants cultivation in Tanzania (Figure 2). These regions include, Arusha Kilimanjaro, Tanga, Coast, Morogoro, Kigoma and Kagera. Some of these regions such as Tanga, Kilimanjaro, Coast and Arusha are among the biodiversity hotspots in the country. Thirty seven investors have shown interest to invest in biofuel. These companies include thirteen foreign companies such SEKAB of Sweden, PROKON of Germany, WILMA from USA, Mitsubishi Corporation of Japan, FELISA (Tanzania-Belgium Partnership) and DILIGENT (a Dutch firm) etc. Ten local organizations such as TATEDO, Sun Biofuels and others have showed their interest to invest on biofuel cultivation.

The diplomatic community (as represented by Belgium, Brazil, Canada, Japan, Germany, India, Malaysia, Switzerland, UK, USA), international development agencies (CIDA, DfID, EU, GTZ, JICA, ILO, SIDA, UNDP, UNEP, USAID) and international development partners (financial institutions) (AfDB, IMF, World Bank) are key constituencies supporting biofuel initiative in Tanzania. For example, SIDA has provided funds which have enabled the National Biofuel Task Force (NBTF) to conduct initial meetings, while GTZ has commissioned the first ever comprehensive study on the prospects of biofuels for the transport sector in Tanzania.

Globally, there are two common strategies of producing agrofuels. One is to grow crops high in sugar (sugar cane, sugar beet, and sweet sorghum, or starch (corn/maize), and then use yeast fermentation to produce ethyl alcohol (ethanol). The second is to grow plants that contain high amounts of vegetable oil, such as oil palm, soybean, algae, jatropha, or pongamia pinnata. When these oils are heated, their viscosity is reduced, and they can be burned directly in a diesel engine, or they can be chemically processed to produce fuels such as biodiesel. Wood and its byproducts can also be converted into biofuels such as woodgas, methanol or ethanol fuel. It is also possible to make cellulosic ethanol from non-edible plant parts, but this can be difficult to accomplish economically.

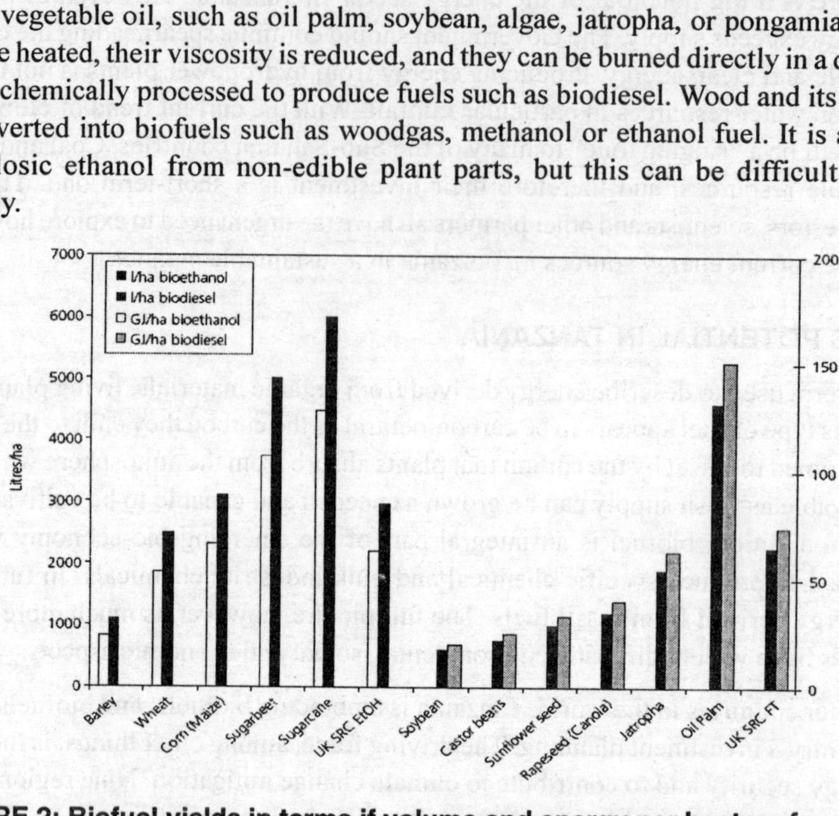

FIGURE 2: Biofuel yields in terms if volume and energy per hectare for selected ethanol and biodiesel feedstock (Adapted from IEA 2004)

Currently, the oil from *Jatropha curcas* seeds is used for making biodiesel fuel in Philippines and in Brazil, where it grows naturally and in plantations in the Southeast, and the North/Northeast Brazil. Likewise, jatropha oil is being promoted as an easily grown biofuel crop in hundreds of projects throughout India and other developing countries. In Africa, cultivation of Jatropha is being promoted and it is grown successfully in countries such as Mali, South Africa, and Zimbabwe etc.

Estimates of Jatropha seed yield vary widely, due to a lack of research data, the genetic diversity of the crop, the range of environments in which it is grown, and Jatropha's perennial life cycle. Seed yields under cultivation can range from 1,500 to 2,000 kilograms per hectare, corresponding to extractable oil yields of 540 to 680 litres per hectare (58 to 73 US gallons per acre).

Generally, investors in Tanzania are investing in Jatropha (Figure 3). Jatropha is grown in more than twelve regions of the Tanzania mainland. Other crops cultivated include sugar cane, palm oil, sisal,

PLATE 2: Jatropha Farm

cotton etc. The economic use of Jatropha oil as fuel (direct or as biodiesel) is due to the fact that it can grow in arid conditions (even deserts), on any kind of ground, and does not require irrigation or suffer in droughts. Jatropha is fast growing and it begins yielding oil in the second year and for the next forty to fifty years. Optimal yields are obtained from the sixth year, and spaced at 2 metre intervals; around 2500 plants can be cultivated per hectare. It absorbs large amounts of carbon dioxide from the atmosphere and therefore earns carbon credits. Therefore, unlike corn and sugar cane, Jatropha is very easy to cultivate even on poor lands of Tanzania and will provide great social and economic benefits for the country.

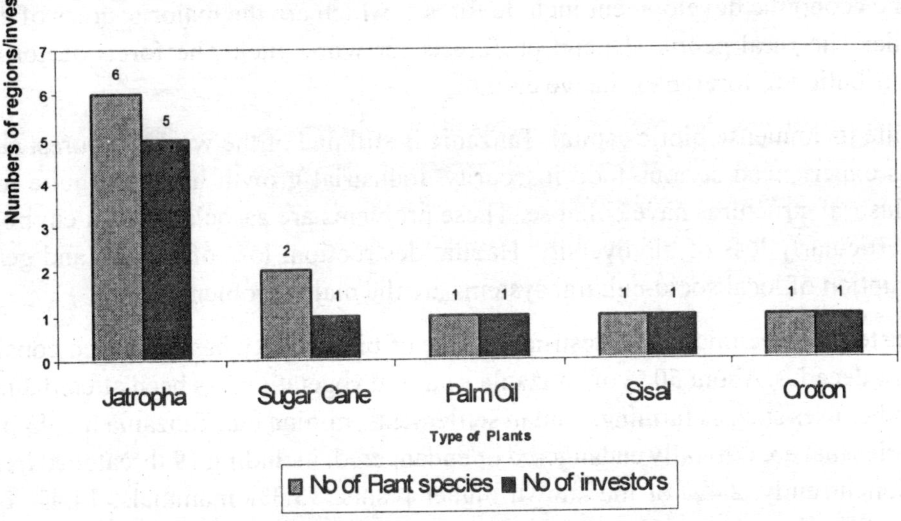

FIGURE 3: Biofuel Plants Areas and Investors in Tanzania

4.0 BIODIVERSITY PROSPECTS IN TANZANIA

Tanzania is endowed with a number of biological resources of which its population relies for survival. The country boasts a vast array of biodiversity, with much of it being endemic and is home to numerous species including 20 primates; 43 antelopes; 290 reptiles; 40 amphibians; and many fish. In addition, Africa's richest and most diverse flora is found in Tanzania. Overall, the country contains six biodiversity hotspots that have value as centres of high species diversity and endemism, namely the Eastern Arc Mountain Forests, the coastal forests, the great lakes for Cichlid fishes, the ecosystems of the alkaline Rift Valley Lakes, and the grassland savannahs for large mammals: in addition, wetlands cover about 10% of the country's total surface and are home to a multitude of aquatic flora and fauna.

The country's biodiversity is a source of major socio-economic and ecological goods and services. Tanzania's economy is heavily dependent on availability and use of biological resources. Agriculture is the main economic activity of the country. It contributes 42.5% of country GDP and employs more than 80% of the rural population. Coffee, cotton, tea, tobacco, cloves, sisal, cashew nuts, maize, livestock, sugar cane, paddy, wheat and pyrethrum are the main crops cultivated. Agricultural productivity and diversity of production systems depends, to the large extent, on availability of wide variety of biological resources and systems. Climatic change and ecological changes that influence agricultural activities are regulated by biological diversity.

The tourist industry, which forms a significant part of country's economy, depends on wildlife resources, particularly large wildlife animals' species such as elephant, rhino, giraffe, buffalo, lion, cheetah, leopard, etc, in national parks and game reserves. Tourism ranks second to agriculture in foreign exchange earnings and greatly contribute to the GDP. Other components of biodiversity that contribute to the economic development include forests, which are the major sources of energy in the country side. The rural people depend on forests for wood fuel. The forest sector makes a substantial contribution to foreign exchange earnings.

However, despite its immense biotic capital, Tanzania is still one of the world's poorest countries. Recently it has experienced serious food insecurity. Industrial growth has been quite marginal; some of its industrial structures have collapse. These problems are associated with environmental degradation, particularly loss of biodiversity. Habitat destruction, loss of species and genes, and associated disruption of local socio-cultural systems are the major problems.

Biodiversity is steadily declining. The destruction rate of biodiversity has increased considerably over the past two decades. About 50 % of Tanzania's natural vegetation has been altered due to land conversion to other uses such as farming, human settlements, mining etc. Tanzania has 43 plant and 123 animal species that are critically endangered or endangered, including 19 threatened freshwater fish species. Concurrently, 2.4% of the known higher plants, 13.3% mammals, 14.4% breeding birds and 5.1% of fish have been or are being lost and/or threatened. Even though the Eastern Arc Mountain Forest is a biological hotspot, about 77% of the original cover has been converted to other

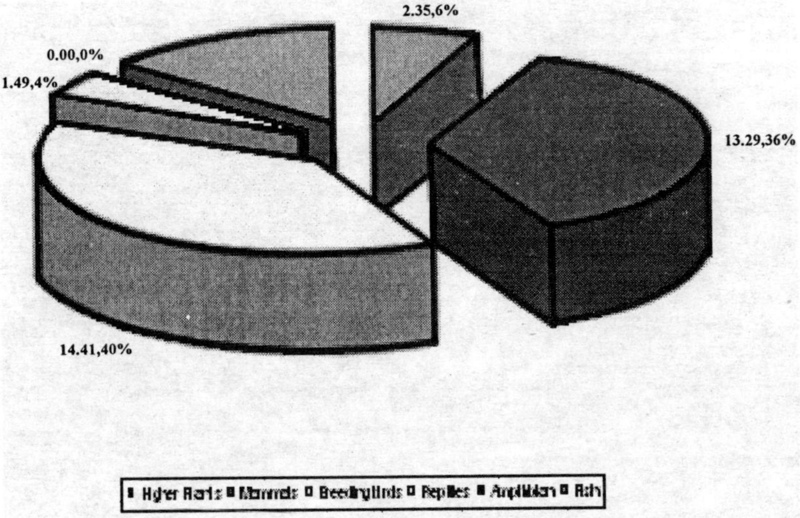

FIGURE 4: Biodiversity Status in Tanzania (Adapted from World Resources Institute, Earth Trends Portal 23rd February 2009)

land uses. In addition, the Miombo Woodlands which cover about 90% of forests in reserves are being degraded at a high rate. Due to the severity of Tanzania's habitat loss, the overall risk to the country's ecosystems and species is high.

The causes of biodiversity degradation are many, complex and interactive. Climate change, human population growth and inappropriate land tenure are major factors causing biodiversity degradation. Population growth has increased the demand for fuel wood and other forest products and land cultivation and grazing, leading to deforestation and encroachment into protected and steep sloped areas. Globalization and the global free market have led to the intensification of agriculture and introduction of new mono-cropping species, but, in general, the over exploitation of plants and animals has reduced the agro-ecosystem and natural forest diversity. Poaching and illegal hunting for bush meat and for sport is a perennial problem. Increasing negative impact of invasive species is another concern to the loss of biodiversity.

Ultimately, inadequate baseline data, and limited human and financial resources plague biodiversity. Thus, the decision makers from state, local government, private sector, and conservation organizations must become better informed on how to incorporate biodiversity information into their activities and how to assess the potential effects of their decisions on biodiversity. Without such biodiversity knowledge, appropriate decisions cannot be made to effectively utilize and protect the Tanzania biodiversity even when the decision maker desires to do so.

Needs for sound taxonomic information and expertise for the successful implementation of biodiversity policies and management programmes has been expressed widely at national and international fora, both at scientific and political levels. With the prevailing political focus on the

PLATE 3: Kihansi Toad, *Strongylopus kitumbeine*

establishment of an effective global network of protected areas for biodiversity conservation to achieve the internationally agreed Millennium Development Goals, especially the "2010 target" in reducing the rate of loss of biodiversity, efforts supporting efficient inventory and monitoring of biodiversity are particularly pertinent.

Even where knowledge on biodiversity exists (mostly in scientific community), to be effective, it must be accessible, available and known to potential users including decision makers. The Tanzania Biodiversity information Facility (TanBIF) established in August, 2007 to furthering technical and scientific efforts to promote the mobilization, sharing and utilization of biodiversity data and information in the country, in service to science, economy and public good. TanBIF through its gateway provides online service of the country data and information held in various database over the world. Currently more than 8,000 records of country biological resources are accessible via the TanBIF Portal.

In conclusion, Tanzania has potential in terms of biological resources. With its high level of biodiversity, bio-economy and tourism industry are the potential areas where government and investors should be focusing for promotion. However, to make sound decision on utilization of biodiversity and development of Tanzania in general, researches on biodiversity should be promoted. An information system such as TanBIF should be strengthened so that while scientists are investing on research, decision-makers can use the research output in taking sound decisions.

5.0 BIOFUEL AND BIODIVERSITY

The climate is changing; the planet is warming; the ice caps are melting - surely we need to find an alternative energy source to dirty fossil fuels. Many believe the answer lies in the development of renewable and sustainable biofuels derived from crops. In response to this, much of the world's rainforests have been replaced, whether directly or indirectly, by such plants. One particular crop, oil-palm, has overtaken soybean as the most traded oilseed crop in the world. Oil-palm plantations cover over 13 million hectares (over 50,000 square miles) and studies claim that up to 16% of recent deforestation in Indonesia is down to these plantations.

The production of bio-energy is gathering more and more attention as a feasible way of reducing dependence on imported oil and gas and is even being hailed as one of the potential key weapons in the battle against global warming. However, recent study has shown that biofuel plantations, replacing forests, could take up to 600 years to offset the carbon lost through land conversion. Likewise, the study, soon to be published in Conservation Biology, estimated that it would take between 75 and 93 years for an oil-palm plantation to start having an effect on carbon emissions, depending on how the forest was cleared. This would also vary depending on the type of forestland it is replacing. Whereas it would only be expected to take around 10 years to offset the carbon balance when a plantation replaces degraded grasslands, the estimate rockets to 600 years when carbon-rich peatland is replaced.

What is more, the biodiversity of forestlands is also being threatened. Whereas forestlands house a rich diversity of species, few can survive the harsher habitat of the oil-palm plantations, let alone the human impact of agricultural intensification. Pollutants include fertilisers, insecticides, herbicides and palm oil mill effluent, all of which put pressure on existing wildlife. It is claimed that oil palm staggering about 80% species is lost to such forest conversion. Frighteningly, these estimates are likely to be conservative, if anything, due to the difficulty of species-counting in complex forestlands. There is also a time lag between habitat loss and extinction; wildlife that could perhaps survive for a short time may not be there for long.

A recent assessment of the overall status of birds in Europe by Bird Life International indicates that 43% of all bird species are now in trouble, and that the situation has worsened over the past ten years. Farmland birds in the Europe have fared particularly badly experiencing a decline of 32% between 1980 and 2002. Many common farmland birds such as Corn Bunting *Milaria calandra*, Yellow hammer *Emberizia citrinella*, Barn Swallow *Hirundo rustica* and Tree Sparrow *Passer montanus* have already gone through a sharp or even catastrophic decline in Europe and this is due largely to the intensification of agriculture.

The current rapid expansion of energy crops in Europe often utilizes set-aside land, which had previously been withdrawn from production under the Common Agricultural Policy. Farmers were paid subsidy to provide set-aside land, and these pockets of natural land are currently vital for the survival of birds and other wildlife. In many areas of intensive agriculture, set-asides comprise virtually the only habitat left for many grassland-dependent species, such as the Little Bustard *Tetrax tetrax* – in France this species is breeding almost exclusively on pockets of habitat provided by set-aside land. Farmers are already allowed to grow energy crops on set-aside land, and if further incentives are given to do this, Europe might lose hundreds of thousands of hectares of key wildlife habitat within a few years.

The great appeal of bioenergy is that it is theoretically a renewable source of energy. Crops can be converted to energy either by being processed into liquid fuel for the transport sector (biofuels) or by being burnt in power plants (biomass). With combustion, carbon dioxide (CO_2) is released. Because they are derived from biological material, the carbon released on combustion comes from CO_2 absorbed by plant material through the process of photosynthesis. Effectively, producing energy from biofuels or biomass could be seen as 'recycling' CO_2."

"However, this rosy picture is over-simplistic. Bioenergy production is never a neutral process when it comes to "greenhouse gases" (Sharpe, 2009). During production, processing and transportation of the crops there are many other inputs to consider. Fossil fuels are used to power tractors for working the land; fertilizers and pesticides are needed to grow the energy crops, one of the most potent greenhouse gases N_2O, is released from fields that are intensively fertilized, and a lot of energy is needed to convert the crop into fuel. In fact, life-cycle analyses of bioenergy production have shown that with poor management methods, production can actually result in a net increase in the emission of greenhouse gases."

In summary, no-one seriously questions that renewable energy offers global environmental benefits over the alternative of fossil-fired power stations, as a means of generating the power we need. These global benefits, however, often come at a price that has to be paid at the local level, in terms of the ecological and environmental damage associated with specific projects. If not managed well biofuel production will negatively affect Tanzanias' biodiversity. Expansion of biofuel projects to the biodiversity hotspots regions such as Arusha, Tanga, Kilimanjaro, Coastal would simply threaten life of majority of living organisms. Biofuel policy makers should make sure that while promoting biofuel, its production should not degrade habitat of both fauna and flora species. Thorough studies should be carried out to identify potential areas for biofuel, biofuel crops and impacts of biofuel on biodiversity. The government should not opt to set aside biodiversity hotspot areas for biofuel plantation. In addition, monitoring and evaluation of biofuel projects should be a continuous process.

6.0 CONCLUSION AND RECOMMENDATIONS

If managed sustainably, biofuels can help to cut greenhouse gas emissions. Furthermore, biofuel will help developing countries like Tanzania to improve their economies and livelihoods of their citizens. However, without adequate regulation and a well developed policy on the production and industry, bio-energy could have devastating impacts on biodiversity. It is therefore, paramount to conduct research and inventories on the biodiversity species in areas identified for biofuel. Knowledge gathered should be used to take sound decisions. Furthermore, biofuel projects should be monitored and evaluated to see their impacts to the national economy, greenhouse effect, livelihood and biodiversity.

REFERENCES

Kamanga (2009) The Agrofuel Industry In Tanzania: A Critical Enquiry Into Challenges and Opportunities (research report).

Marwa N. (2008) The economic Potentials of Liquid Biofuels Production: Challenges and Opportunities for Smallholders Farmers in Tanzania (A Discussion Presented for Strategic Production and Marketing of Biodiesel to improve Socio Economic and Environment Conditions in Rwanda, on 17th July 2008).

Mngodo, J. (2008): Tackling the nexus- Bioenergy and Food Security in Tanzania (Paper presented at the Second FAO Technical Consultation on BEFS, Rome 5-th February 2008).

The Royal Society (2008) Sustainable Biofuels: prospects and challenges.

Sharpe, S. (2009) Bio-energy: dream or nightmare? In Biophile Magazine (http://biophile.co.za/energy/bio-energy-dream-or-nightmare 27th February 2009)

Bird Life International (2009): Bio-energy: dream or nightmare http://www.birdlife.org (27th February 2009)

Carers in Green Energy (2009) Just how "green" is renewable energy? http://www.careersinrenewables.co.uk/news2.htm (27th February roo9):

Centre for Jatropha Promotion and Biodiesel (2009): http:www.jatropha.org

REFERENCES

Kamunga (2009) The Agrofuel Industry in Tanzania: A Critical Enquiry into challenges and Opportunities (Research report).

Marwa N. (2008) The economic Potentials of Liquid Biofuels Production: Challenges and Opportunities for Smallholder Farmers in Tanzania (A Discussion Presented for Strategic Production and Marketing of Biodiesel to improve Socio-Economic and Environmental Conditions in Rwanda, on 12th July 2008)

Mngodo, J (2009), Tackling the nexus-Bioenergy and Food Security in Tanzania (Paper presented at the Second FAO Technical Consultation on BEFS, Rome 24th February 2009)

The Royal Society (2008) Sustainable Biofuels: prospects and challenges.

Sharpe, S. (2009) Bio-energy: dream or nightmare? In Bioplus Magazine (http://bioplus.co.za/energy/bioenergy-dream-nightmare 27th February 2009)

Bird Life International (2009) Bio-energy: dream or nightmare (http://www.birdlife.org 27th February 2009)

Gators in Green Energy (2005). Just how "green" is renewable energy? http://www.conservationwatch.org/news2.htm (23rd February 2009)

Centre for Jatropha Promotion and Biodiesel (2007) http://www.jatrophaworld.org

Index

A

acidic emissions 298, 301
acidification 296
activated carbon industry 133
adult and child labour 69
Africa 79, 90, 167
    largest producers of biofuels in 179
African medicinal plants 310, 311
African traditional medicine 311, 312, 313
agricultural 109
    and forestry products 273
    and tree growing development 67
    biofuels 189
    biotechnology 288
    crop 28, 44
    lands 51, 52, 59, 62, 63
    production 63, 65
    productivity 209
        and diversity 330
    wage labour rates 35
    waste 67, 129, 148
agriculture 12
agro and biodiversity products as energy source 274
agrochemicals 123, 124
agro-ecosystems 28, 331
agro-fuels visa vie food production 171
agronomy 178
AHEC 243
alcohol producing workshops 163
Aleurites moluccana 158
Algae 222
Alternative Energy Promotion Centre (AEPC) 273
Americas 171

anaerobic digestion 6
Andhra Pradesh 215
Angelsen 27
animal dung 48
animal fats 119
animal husbandry 282, 283
animal population 67
animal waste 3, 148, 298
arable land 171
arthritis 293
Asia 79, 171
asthma 293
atmospheric lifetimes 296
Australia 297

B

Bagasse 97
bamboo 48
Bangladesh 145
below the poverty line (BPL) 79
BioDeC 293
biodiesel 107, 113, 119, 123, 145, 150, 221, 274, 275
  production of 119, 231
  economics of 121
  production from plant and fat oils 157
  technologies 229
biodigester operator 11
biodigesters 21
biodiverse country 26
biodiversity 27, 28, 87, 90, 179, 284
  degradation 331
  knowledge 331
  of the world 282
  policy 310
  resources 26, 27, 28, 29, 29, 30, 32, 33, 34, 36, 265
bio-economic model 52
bioenergy 81, 110, 123, 324, 334
  creating market demand for 115
  policy 114
  production and utilization 125
  technologies 81, 116
bioethanol 107, 116, 118, 124, 145, 150, 221, 274
  emerging technologies in 117
  feedstock 112
  fuel 151
  production and technology of 116
bio-fertilizers 288
biofuel 178, 274
  advantages of 154
  and biodiversity 332
  and gel fuel 181
  crops 184
  cultivation 179
  development 154, 165
  energy 167
  industry 186, 189
    development of the 187
  in Uganda 171
  manufacturing 164
  plantation 334
  programme 259
  producing crops 172
  production 162, 176
  projects, current and future 190
  requirements vs land requirements 190, 194
  standards 182
  traditional 166
  what are 154
biogas 107, 114, 121, 124, 145, 148, 235, 268, 276, 277
  and slurry and purification 248
  bottling of 248
  digesters 7, 10, 13, 16
  health benefits of 20
  plant 14, 20, 21, 23, 149
    currently in operation 13
    in Botswana 9
      distribution of 9
  production 6, 12
      methods of improving 122
  program 21
  resource activities 23
  system 3, 15, 22, 132
  technologies 4, 7, 22
  technology dessemination 12
  technology in developing countries 3
biological resources 266
biomass 46, 81, 221, 228, 235
  consumption 131
  energy 180 267
    resource base of Nigeria 110
  feedstock 6
  fuels 145
  power 243
  resources 109
  sources 163
  technology 268
  traditional 268

biomedicine 290
bio-pesticides 288
bioresources 282, 284
    development 285, 287, 288
    Development Centres (BioDeC) 288, 290
biotechnological activities 286
biotechnological techniques 285
biotechnology 179, 189, 283, 285, 286
birth control program 58
black African population 26
Black Triangle 300
blood disorder 291
Botswana 3, 5, 7, 9, 11, 13, 15, 17, 19, 21, 23
    energy balance 4
    energy sector 4
Botswana College of Agriculture (BCA) 12
brassware industry 131
Brazil 165, 254, 328
British Council in Nepal 274
bronchitis 293
Burkina Faso 316
Burundi 324

C

Camellia sasanqua 158
Carbon Capture and Storage (CCS) 211
carbon credits 88
carbon energy resource 215
Cassava plantation 158
casual wage labour 34
Central Europe 298
charcoal 45, 138
Chattisgarh 215
China 7, 145, 165, 216
Chinese fixed dome 9
Chinese model biogas plant 7
cholesterol 293
Clean Development Mechanism (CDM) 236
climate change 27, 79, 90, 173, 212, 331
    and sustainable development 88
    at global level 91
    intergovernmental panel on climate 234
    strategic knowledge for 214
clinical methodologies 315
CO_2 emission scenarios 209, 225
CO_2 stabilization scenarios 210, 214
Coal 81
Coconut plantation 158
Commercial ethanol production 151
Congo 324
conservation 179
    of forest 22, 277
conventional energy sources 126
cooking environment 139
cooking fuels 74
cooking oils 119
cooking stoves 277
coronary heart disease 317
cost-benefit analysis 3, 21
Council of Scientific and Industrial Research (CSIR) 227, 230
cow dung 11
    collection 12
crops in the Hoima districts 64
cross-cutting issues 253
cultivated lands 46
cultivation 47
    of Jatropha 184
Cumberland Hotel biogas plant 14
Czech Republic 298, 300

D

deforestation 46
degraded lands 35
Dendro power 132
    energy 132
    plant 130
    projects 129
developing countries 3
developing world 27
diabetes 293
diabetes mellitus 317
diesel 15, 19, 21
    fueled water heater 15

digester 11
  dome 7
  drum 7
  flow 7
digesters 10, 12
Diphawana Biogas Plant 9
diseases 311, 316
distillation process 118
diversify export 108
domestic wastes 285
domestic water heating 95
Downdraft Gasifier 148
drought 298
Drug Development Chain 312
drug industry 314
drug Nicosan 292
dry-land cultivation 63

E

East Africa 136, 324
  acidic deposition problems in 301
Early oil Production Scheme (EPS) 169
East Asian Monitoring Network (EANET) 301
ecological and health implications 276
ecological disaster 276
ecological zones 288
economic 27
  development 25
  growth 28
  social and environmental benefits 186
ecosystem goods 25
ecosystem services 35
ecosystems 296
education 27, 58, 209
  of children 32
  standard 57
effective generation/ utilization 231
efficient lighting 226
electricity 48, 195
  access programme, rural and urban-poor 259
  consumption 99
  demand 97
  from bagasse 99
  from coal 100
  from hydro 99
  from wind 99
  generation 23, 213, 236
  infrastructure 101
  supply 104
Electricity Regulatory Authority (ERA) 253, 259
electropollution 79
employment 68, 108
  opportunities 33, 127
energy 22, 32, 34
  and food production 297, 303
  conservation 81, 101, 132
  consumption 101
    India ranks 209
  demand growth 127
  efficiency 81, 91, 212
    in Indian industry 216
    measures 205
    programme 259
    surveys 203
  efficient buildings 225
    devices/systems 224
  generation 79, 85
  imports of 186
  infrastructure 87
  investment 108
  management programmes 95
  plants 108
  policy 98
  production of 19
  programme, wastes for 259
  renewable 23
  requirements 71
  savings 101
  security 159, 195, 218
  self-sufficiency 109
  services programme, modern 259
  storage batteries 223
  supply in India 236
Energy Affairs Department (EAD) 9
Energy and Resource Institute (TERI) 87

Energy Support Assistance Program (ESAP) 273
utilizing 153
environment 81, 179
protection 159
biodiversity and conservation 188
policy on 115
environmental advantages 20
environmental and social impact 19, 22
environmental conservation and management 146
environmental degradation 74, 75, 92, 288
environmental goods 26
and services 25
environmental impact 187
Environmental impact assessment (EIAs) 169
environmental pollution 84, 108, 123, 124
environmental problem 235, 303
environmental protection 125
environmental quality 80
environmental sustainability 95
ethanol and biodiesel feedstock 328
ethanol and biodiesel production 179
ethanol producing technology 163
ethanol production for fuel blending 194
ethanol production for Gel fuel 195
Ethiopia 136
Europe 171, 172, 297, 300, 302, 303, 333
European Monitoring and Evaluation Program (EMEP) 302
European Union (EU) 99, 254
Europe-wide survey of leaf 298
exploitation 80

F

farm employment 59
farm laborforce 58
farmer-land ratio generates unemployment 59
farmland ownership 57
feedstocks 178, 183
fertilizer, organic 19
fertilizers 20, 298
fire 276
firewood 45, 138
fish farming 288
Fish fat 159
fisheries 282, 283
floods 276
fodder 45, 61
production 62
requirements 52
food 26, 32, 45, 61, 293
and agriculture 286
and bio-fuel production 175
and wood production program 62
consumption 60
crop production 62, 64, 65
grains 52
industries 285
plants 311
preparation 50
production 45
security 27, 110, 179, 187, 193
shortages 74
sufficiency 60
foreign exchange 73
forest 90, 296
degradation 44, 73, 276
development 75
dwellers 87
ecological system 51
ecosystem 51
products, non-timber 276
forestry 45, 55, 109, 282, 283
system 61
fossil fuel 81, 96, 109, 126, 296, 298, 333, 334
fossil-fired power stations 334
fragile upland systems 45
freshwater 296
ecosystems 298
scarcity problems 90
frost 298
fuel 60
cell technology 231
consumption 52
production and consumption 51
wood 4, 52, 109, 127, 134

consumption 60, 277
fuels increases 296
fuelwood 29, 32, 44, 45, 51, 61, 62, 69, 70, 72, 74
collection 71
demand 48
expenses 139
farming 71
option 69
harvest 70
shortages 46

G

gas 48
production 113
and collection 7
gastric ulcers 293
genetic engineering 285
geothermal power 169, 274
Germany 165, 298, 300
Ghana 316
Gliricidia energy plantations 128
Gliricidia plants 132
global CO_2 emissions 210
global energy consumption 210
global environmental impact 21
global free market 331
global issues, three burining 153
global primary energy economy demand 87
global renewable energy supply 108
global warming 79, 235
potential 173
globalisation 27, 331
goal-seeking system 68
Good Clinical Practice (GCP) 316, 318
Good Laboratory Practice 318
Good Laboratory Practice (GLP) 310, 315
green house gas effect 79
green house gas emissions 87
Green House Gases (GHG) 209
emissions 211
Green India 213
greenhouse gas 22, 84, 173, 334
mitigation 92
groundwater 90
GTZ's approach 139

H

health 27, 209
and fitness 293
and sanitation standards 145
hazard 277
land ecosystems 302
hepatopathy 293
herbal medicines 318
Himalayas Ecosystem 213
HIV/AIDS 27, 316, 317
pandemic 26
Hoima district 47, 49, 59, 66, 68
rural energy in 47
home consumption 31
homogenous land units 53
horticulture 282, 283
hotel biogas plant 14
hotel wastes 3
household laborforce 58
human and animal population 52
human and labour rights 179
human economic activities 25
human encroachment 167
human fertility rate 57
human population growth 331
human poverty value 80
Human resource training 165
human resources 25
human social ecosystem 51
human waste 288
hunting, poaching and illegal 331
hydro power 104
hydro-energy 269
hydrogen 223
hydrogen generation and storage systems 231
hydropower 97, 99, 270
projects 126
hyperlipemia 293

hypertension 293, 317

I

IEC standards 243
IIT Roorkee 243
illegal seizure of land 276
incomes 50
India 20, 165, 216, 297
- energy scenario in 236
- increasing population and fast-growing economy 235
- nuclear power programme 215
- population and fast-growing economy of 234
- R&D challenges/opportunities for 218

Indian floating dome 9, 10
Indian nuclear fuel 215
Indian Ocean 324
Indian Renewable Energy Development Agency (IREDA) 216
Indigenous Medical Knowledge (IMK) 310, 314
industrial development 285
industrial waste 148
industrialization 296, 303
infants 57
insect attack 298
insomnia 293
Intellectual Property Rights (IPR) 314
intellectual resources 25
Inter-governmental Panel on Climate Change (IPCC) 15, 210
International Energy Organization 210
invasion 276

J

Japan 165
Jatropha 184, 329
- cultivation 275
- oil 190, 193, 329
- plantation 185, 276
- varieties 190, 193

job market 36
job-opportunities 61

K

Kang BH 1500 biogas plant 12
Kang brigade Biogas Plant 11
Kang clinic biogas plant 10
Kathmandu 269, 273, 276
Kathmandu Valley 274
Kenya 136, 137, 139, 140, 254, 316, 324
Kenya Improved Stove Association (KEISA) 138
kerosene 48, 277
- stove option 72
- substitution 44, 74

Khadi and Village Industries Commission (KVIC) 7
Kigorobya sub-county 47
kitchens 15
Kyoto Protocol 91, 133, 251, 252

L

labour 48, 69
- hiring activities: 54

Ladoke Akintola University of Technology (LAUTECH) 294
land clearing/abandoning activities 54
land cultivation 331
land degradation 27
land deterioration 74
land use 179, 188, 194
landfill biogas 18
landfill gas 16
landholdings 68
landscapes 302
landslide 276
- protection 62

landuse 62
- practices 74

Lanka Sustainable Energy Authority 132
Laos 145
Latin America 79, 90
livelihood 209
- strategies 27, 37

liver disease 317
livestock feed to fuel 46
livestock population 67

LPG 18, 19, 21
  gas for cooking 14

M

Malaria 317
Malawi 324
Mali 328
management practices 90
marine and freshwater resources 28
market 34, 35
Maurice Ile Durable Fund 104
Mauritius 98, 102
  sugar industry in 99
medicinal and aromatic plants 310
medicinal and bio-industrial products 283
medicinal plant 34
  vendors 34
  development 291
medicine 32
meteorology department 253
Mexico 294
Microalgal Bio-diesel 92
microorganisms 290
Middle-east 171
mitigation and reduction 28
mitigations 124
Mogwalale Biogas Plant 10
molasses 48
Mozambique 324
municipal and industrial activities 109
municipal and industrial wastes 235
municipal solid waste 23, 148
municipal waste water biogas project 17
municipal wastes and landfill gas 100
Mushroom 293
Myanmar 145, 150
  energy development in 147
  environment policy 147
  environmental quality management in 146

N

national biofuels strategy 178, 182
National Biotechnology Development Agency (NABDA) 282, 285, 286
National Biotechnology Policy (NBP) 286
national energy policy 178, 180
  new 181
National Environment Management Authority 253
National Environment Policy of Myanmar 147
National Forestry Authority (NFA) 48, 253
national programme on energy recovery 246
national water mission 213
National Water Plan (NWP) 269
Natural Force Ferry 271
natural gas 327
natural gas and biomass 80
natural gas flaring 87
natural growth rate 55
natural resource 37, 85, 90, 95, 108, 296
  exploitation of 80
Nepal 267, 271, 272, 273, 275
  energy and biodiversity sector in 277
  renewable energy sources of 274
  rural energy development in 275
Nepal Academy of Science and Technology (NAST) 269
Nepal Bio-Extruder Industry 269
Nepalese land 265
nephritis 293
network systems 301
new renewable energy (NRE) 237
NGOs 37
Nigeria 109, 114, 254, 282, 285, 286, 291, 292, 316
  aggregate annual crop production 110
  biomass energy in 111
  biomass resources of 108
  development of biotechnology in 285
  energy options 108
  Renewable Energy Options 110
Nigeria Natural Medicine Development Agency (NNMDA 295
Nigerian households 110
Nigerian Institute for Pharmaceutical Research and 295
Nigerian National Petroleum Corporation (NNPC) 115

noise pollution 123, 124
non-renewable energy 80
Non-conventional Renewable Energy (NCRE) 127
non-food crop biomass 109
Non-Governmental organizations (NGOs) 254
non-renewable sources of energy 81
North America 297, 298, 300, 303
Northampton 274
NSG (Nuclear Supplier Group) 215
nuclear energy 215
nuclear power programme 215
nutritional quality 285

O

oil and gas production 296
oil-palm plantations 333
oils from plants 291
Oval Biofuels 195
Oxford Brookes university 274
oxygen tension 318

P

Patel, Jashu Bhai J. 7
peanut cultivation 157
peasant households 58
Pelwatte Sugar Industries 129
Pharmaceutical Research and Development (NIPRD) 318
phytomedicines 284
Pitsane biogas plant 13
Pitsane household biogas 13
plant and animal genetic resources 293
plant hydrocarbon oil 275
plant tissue culture 286
plantation material 229
plant-breeding methods 285
planting standard 66
poaching 276
Poland 298, 300
political and cultural structures 27
pollutant emissions 296
pollutant emissions and consequences 297
pollution 79, 87, 145, 285
    reduction strategies 301
poor farmers 67
poor households 32, 37
poor lands of Tanzania 329
poor nutrition 27
poor people 45
population density 57
population growth 45, 57
    in Hoima 46
population growth rate 44
population pressure 59
poultry biogas digesters 17
poverty 27, 33
    alleviation 27, 28, 33, 36
    and well-being 27
    elimination 28
    levels 26
    line 71
    mitigation 36
    prevention 28
power demand 236
power generation 99, 147, 235
    in Uganda's lakes and rivers 166
    programme 258
power market and industry 98
powerlessness 27
Pressurized Heavy Water Reactors (PHWRs) 215
private-public partnership 22
processing 178
processing plants 190, 193
public health care 313
public health disorder 318
Public Private Partnership (PPP) 274
public transport systems 90
PVPS (Photovoltaic Pumping System) 273

R

Radioactive Tracer Technique 305
raw material resources 25
red blood cells 318
rehabilitation 75

renal damage 317
renewable biofuel 171
renewable energy 90, 95, 98, 108, 215, 216, 228, 234, 247
  resources 129
Renewable Energy (RE) (Pty) Ltd 15
  financing 234
  for economic development 128
  policy 254
  resources 127
  sources 92, 209, 267
  specific issues 252
  subsidy policy 272
  technologies 175, 265
Renewable Energy Resources and Technology in India 237
renewable fuel 119
renewable power 237
  grid interactive 237
  off-Grid 246
    distributed 247
rice husks 48
Richmark Poultry Plant 14, 15
rural dwellers 55
Rural Electrification Agency (REA) 253, 259
Rural Electrification Board (REB) 259
rural employment generation 87, 90
rural energy 44
Rural Energy Development Programme (REDP) 275
rural energy system 275
rural fuel 75
rural households 35, 45
Rural Industries Innovation Centre (RIIC) 9
rural laborforce 58
rural poor alleviates 32
rural poverty 74
rural unemployment 55, 57, 59
Rwanda 136, 324

S

Safety-Net 28
Seed Genebank 294
Sesame (Jatropha curcas L.) 158
Shackleton 27
shelter 32
Sickle Cell Anaemia 318
Sickle Cell Disease 291, 292
silvi-cultural 67
Slovakia 298
small hydro power 239
small-scale hydropower 128
small-scale trading 33
small-scale vendors 32
social 27
  cultural attitudes 58
  economic impact 187, 189
  economic prospects 275
  problems 75
soil 296
  depletion problem 66
  environment of trees 298
  erosion 22
  fertility 26
  protection 22
solar biomass dryers 131
solar energy 97, 100, 127, 218, 228, 271, 273
  potential of 272
Solar Home System (SHS) 272, 276
solar mission 212
solar power 81, 126, 134, 169, 245
solar PV systems 272
solar water heaters 97, 104
  dryer and cooker 273
solid waste, recycling of 103
South Africa 7, 9, 30, 33, 35, 328
  biodiversity resources to rural live 29
  poor live in rural areas 26
South Asia 90
Southern Africa 45
soya bean cultivation 157
Sri Lanka 127, 128, 134
  Broadcasting Corporation 132
  current economic development policy 126
  energy needs 129
  Sustainable Energy Authority 131

success stories and experience of 133
stove programmes in Kenya and Uganda 138
stove technology 45
Sub-Saharan Africa 44, 314
sugarcane industry 183
sugar industry 99
sugarcane plantation 159
sugarcane processing 97
Sunderlin 27
sustainable agriculture 214
sustainable development 81, 87
sustainable energy use pattern 108
sustainable habitat 213
sustainable strategy 108
Swaziland 254
Sweden 298

T

Tanzania 136, 254, 324, 328
biodiversity 330, 331, 334
Biodiversity Information Facility (TanBIF) 332
biofuels potential in 327
economy 324, 330
energy consumption in 326
energy demand and supply in 325
natural vegetation 330
poor lands of 329
Tanzania Electric Supply Company Limited (TANESCO) 326
technological development 86, 238
terrestrial ecosystems 298
Thailand 145, 165
thin-film solar cells 220
timber 46, 61
production 62
toilet waste 8
tourist industry 330
Traditional Chinese Medicine (TCM) 311
traditional health practitioners (THPs) 313, 314, 218
traditional medicine and biodiversity policies 310
traditional medicine in sub-Saharan Africa 316
traditional medicine law 313
traditional medicine policy 313
transmigration 59
transport 225
tree biomass 54
tree-farming 44, 74
practice 50
projects 45
tree planting 62, 65

U

Uganda 44, 136, 138, 139, 140, 169, 176, 201, 324
bio-fuel development in 175
electricity 166
transmission company 253
energy situation in 167
imports 73
lighting fuel use in 202
main source of electricity in 168
population 166
Quest for Bio-fuels in 170
Renewable Energy Policy (2007) 174
UMN (United Mission to Nepal) 275
underemployment 48
unemployment 48, 57
households 58
rate 55
United Nations Development Programme (UNDP) 275
United Nations Environment Programme (UNEP) 91
United Nations Framework Convention on Climate Change 88
University College Northampton 274
urban households 45
urbanization loops 57
USA 165, 216, 254
nuclear deal with 215

V

vegetable oils 119
vegetation 47
vegetative biomass 52, 54
vietnam 159

biofuel application research in 156
biofuel development in 159
biofuel production situation in 153
biofuel production technology of 162
consumption and demands of energy in 155
violence and crime 27
Vision up to 2025 162
vulnerability 27

W

wage 52
labour rates 34
waste 21
Abattoir 18
materials 108
water treatment plants 3
wasteland utilization 179
wastes, organic 19
waste-to-energy 246
water 26
catchment areas 45
pumping 10
scarcity 85
water borne diseases 22
Water-mills 271
West Asia 296, 297
Western Asia 303
WHO 312, 313, 315, 316
wild biodiversity enterprise 34
wild biodiversity resource 36-37
safety-net functions 32
local trade in 33
wild fruits 29
wild natural resources 31
wild vegetables 29
wind 228
energy 97, 220, 237, 273
power projects 238
women 36, 58, 277
and youth entrepreneurs 288
poor urban and peri-urban 36
rural 36
vulnerability of 276
wood 26
cutting 276
energy 50
energy demand versus supply 4, 5
energy resources 71
fuel 138
production 65
model 65
scarcities 74
wooden utensils 29
woodlot technology 74
woodstoves 44, 45, 48, 74
option 72
workers 36
World Bank 275
World Energy Outlook (WEO) 2009 210
world's biodiversity 283
world's food supply 92

Z

Zambia 178, 193, 324
agrofuel production in 190
benefits of the biofuels for 186
biofuels association of 183
biofuels development in 183
biofuels feedstocks in 183
biofuels industry in 179, 187
biofuels programme in 182
biofuels research in 187
biomass energy 180
bureau of standards 182
energy consumption 181
sugar industry in 194
wastelands for jatropha production 188
Zambia Agricultural Research Institute (ZARI) 184
Zimbabwe 316, 328
zymotic technology 162